Computational Methods in Elementary Numerical Analysis

J. Ll. Morris

University of Waterloo, Ontario, Canada

A Wiley–Interscience Publication

JOHN WILEY & SONS

Chichester · New York · Brisbane · Toronto · Singapore

Library of Congress Cataloging in Publication Data:
Morris, John Ll.
  Computational methods in elementary numerical analysis.
  'A Wiley–Interscience publication.'
  Includes bibliographical references and index.
  1. Numerical analysis. I. Title.
  QA297.M647   519.4   82-2778
                    AACR2

ISBN 0 471 10419 1 (cloth)
ISBN 0 471 10420 5 (paper)

British Library Cataloguing in Publication Data:

Morris, J. Ll.
  Computational methods in elementary numerical analysis.
  1. Numerical analysis—Data processing
  I. Title
  519.4   QA297

ISBN 0 471 10419 1 (cloth)
ISBN 0 471 10420 5 (paper)

*Typeset in Great Britain by J. W. Arrowsmith Ltd., Bristol
and printed at the Pitman Press Ltd., Bath, Avon.*

To the memory of
Tom Meirion Williams

Contents

Preface

This book has evolved out of lecture notes used by the author for elementary numerical analysis courses at the Universities of Dundee (Scotland) and Waterloo (Canada). Prerequisites for most of the book are sophomore courses in mathematics; in particular calculus and linear algebra should have been studied by the reader. The reader should also be able to program in a scientific programming language; Fortran, Algol, or Pascal are particularly appropriate.

The title of the text suggests elementary numerical analysis with an emphasis on computation. The interpretation of what is elementary is of course very much an individual one. The author feels that what is described should be comprehensible to all who have the patience to study the contents of the book; I have tried to keep to a minimum the need to digress to other more elaborate texts for clarification of the contents of any section. However, I have also tried to curtail the unnecessary discussions of proofs of theorems for their own sake—here the reader will need to explore other texts if he wishes fuller account of the proofs. The computational aspect of the text is important! It is my firm belief that in this subject learning is by doing. Hence there are many exercises which need the facilities of your local computer (even if it be your personal micro with a high-level language processor). However, I have not padded the text with my own programs or subroutines. There is, of course, a case for such texts—they often give valuable insight into how (and how not) to write good scientific routines for solving numerical problems. An excellent example of such a text is that by Forsythe, Malcolm, and Moler, the full reference to which may be found in the reference section at the back of this book. It has been my experience that students have required a fuller explanation of the methods used for solving problems with sufficient analysis to explain why (and why not) they work. It is my feeling that "software" books fail to give this important background. It is therefore my contention that any successful study of the subject should be based on texts both of the variety of the present one and those of the type exemplified by Forsythe, Malcolm, and Moler. In addition to appropriate texts for studying numerical analysis, there are now available excellent software libraries which most computing centers possess. In particular the National Algorithms Group (NAG) in Britain and the International Mathematical and Statistical Library (IMSL) in the United States are high-quality libraries of routines for scientific computation. To these should be added the more specialized libraries, EISPACK and LINPACK. The reader of this text is highly recommended to

xi

become familiar with the contents of these libraries and the means of using the routines contained therein.

The contents of the text are currently taught in two semester courses at the University of Waterloo. The material forms the basis from which students progress to study a selection of the topics in more detail in more advanced courses. In particular, the linear algebra (Chapter 3) provides the background for a final-year undergraduate/first-year postgraduate course given in one semester on Numerical Linear Algebra. However, the text does not give a comprehensive coverage of all the areas of numerical analysis. The experienced reviewer will notice I have purposely omitted study of partial differential equations, optimization, and mathematical programming and the numerical solution of integral equations. I have also omitted discussion of more specialized topics such as singular value decomposition and fast Fourier transforms. This has been a conscious decision; the text should represent what is possible to cover in the time available and also satisfy the prerequisites being imposed as mentioned earlier. The reader should not conclude that because these topics are not covered in an elementary text that they are either too difficult per se or that they are not important. On the contrary, it is my hope that the reader will be fired with interest to explore further; this material should form the next stage of study when the reader has been introduced to a little more mathematics. Furthermore, there are excellent monographs in these areas which are ideal for the study of these subjects: references may be found at the back of the book.

The text has evolved over a number of years when I have been resident at the Universities of Dundee and Waterloo. During the times spent at these two institutions a great number of faculty and students have aided me in discussions and reactions to the material. In Dundee it is a privilege to acknowledge discussions with R. Fletcher, J. D. Lambert, and G. A. Watson. In Waterloo, R. Bartels, J. A. George, J. D. Lawson, and R. B. Simpson have influenced my thinking considerably. Two close friends who have been of considerable support and encouragement over the years and to whom I extend special thanks are A. R. Gourlay and A. R. Mitchell. Professor D. S. Jones I thank for suggesting the initial text, for comments on the manuscript and patience when, at times, it appeared the text would never be finished.

Finally my gratitude must be expressed to my own family (S, J, R, & J) for their support and understanding—they have made it possible, and worthwhile.

Dundee
September 1981   **J. Ll. Morris**

CHAPTER 1
An Introduction

1 Numerical Analysis

We introduce the reader to the subject matter of this book by pointing out that the subject title is a misnomer! The subject of *numerical analysis* is not, as the title implies, a numerical form of analysis, where by analysis we understand that branch of mathematics concerned with the study of real- and complex-valued functions of one or several variables. A student desiring to know whether a course of *numerical analysis* rather than a course in *topology* (say) is for him wants to know before making his choice what the subject material entails. In a sense, to know the content of a course requires a study of that course, which requires a student to decide to take the course, which in turn requires him to be aware of the course content, ..., an infinite loop. However, things are not as bad as they seem as we can usually make decisions based on a selection of topics and examples from the subject. Consequently, in this introductory chapter we will endeavor to touch upon the aims of the subject of numerical analysis and give some examples of problems which motivate a study of the subject. We hope this alone will lead the reader to study the contents of the remaining ten chapters.

This history of mathematics (see for example Bell [1]†) makes fascinating reading. The reader will often be impressed by the geniuses of mathematics attempting to solve what, at the time, seemed like (and in many cases still are) insoluble problems. Often when one problem is solved it gives rise to new problems which have to be solved, often harder than the original one. At any one time there have always been many problems which defied even the greatest mathematical brains of the time. With the coming of the technological age, problems have required solution which have not yet been solvable by the great (or even the less great) mathematicians, yet for which technology has demanded a solution. The engineers, say, faced with such problems cannot

---

† Numbers in square brackets cite the references which are listed at the end of the book.

wait for a new generation of super-mathematicians—they need the solution to their problems *now*! An everyday example comes to mind. We are all very concerned about tomorrow's weather! The weather on the earth's surface is governed by very complicated mathematical equations. The equations have not, to date, been solved analytically. But we want to know what the weather is going to be like tomorrow. The answer to this seemingly impossible situation is to accept an *approximation* to the required solution rather than demand the *exact* answer. That is, we will be quite happy if the weatherman forecasts rain with winds of 30 miles per hour and daytime temperatures reaching 10 °C and it turns out to be frequent showers, wind 25 miles an hour, and temperature 12 °C (our picnic would be spoiled anyway). How accurate such an approximation will be depends upon the method of approximation proposed. This then is a rôle of numerical analysis: (1) approximate a mathematical model by a simpler one (simpler in the sense that we will be able to get answers); (2) specify how good (in some sense) the approximation will be.

In a broad sense this defines numerical analysis: the derivation and analysis of *approximation methods*.

In the derivation of approximation methods it must always be borne in mind that an approximate solution must be calculated using the method. Therefore, this calculation must be able to be performed in as efficient and convenient a fashion as possible. For example, with reference to our weather forecasting it is no use obtaining an approximate solution which takes so long to calculate that we get today's weather forecast tomorrow! Neither is it any use obtaining today's weather today because it happens to be the first of the month but we cannot expect another weather forecast until the beginning of next month because it costs (the government?) too much to produce the forecast every day.

Thus, recapitulating, we know that practical problems exist which give rise to exceedingly complicated mathematical models for which analytical solutions cannot be obtained, but which demand some form of solution. Numerical analysis provides a means for proposing and analyzing methods which give approximations to the required solution. These approximations should be convenient, efficient and "good" (in some sense).

In contrast to the insoluble problems there often arise problems which *can* be solved analytically. However, the *form* of the solution is not such as to yield a convenient numerical value—which is what an engineer is more likely to want (rather than a sophisticated expression). Consider the following example.

Example 1 Bessel's equation

$$x^2 \frac{d^2 y}{dx^2} + x \frac{dy}{dx} + (x^2 - n^2)y = 0 \tag{1}$$

where n is a given constant, arises in many problems of applied science, particularly in boundary-value problems for right circular cylinders (for example, heat conduction in a circular annulus—see Spiegel [38, p. 358]; or the propagation of electromagnetic waves in circular waveguides—see Jones [26, p. 240]).

It is known that the solution of Eq. (1) may be written as

$$y = \sum_{k=0}^{\infty} \frac{(-1)^k}{k!(n+k)!} \left(\frac{x}{2}\right)^{n+2k}$$

hardly in a form convenient for numerical evaluation!

Thus in such problems we are again faced with deriving some form of approximation to the solution y. This may entail approximating the actual function y or even starting from scratch, approximating the original problem by a simpler one, neglecting the analytical solution. In either case we will require an approximation which is "easy" to evaluate and which is guaranteed to provide a certain accuracy with respect to the original solution.

Finally, we meet a class of problems which gives rise to solutions which can be written down mathematically and which in certain cases can be solved in a straightforward manner using a classical method of solution but which in other cases (which are the ones most often to occur) yields a process which is so time-consuming as to be practically useless. Consider the following example.

Example 2   An electrical network is depicted in Figure 1. The resistances are shown in ohms Let the voltage applied at node i be represented by V_i volts. Let the current flowing between node i and node j be denoted I_{ij} amps. To obtain the voltage existing between the nodes, we apply the two standard laws of electrical networks.

(a) Ohm's law, which states that for a current I_{ij} flowing between nodes i and j and voltages V_i and V_j at the nodes respectively with an effective resistance R_{ij} existing between the nodes, then

$$I_{ij} = \frac{V_i - V_j}{R_{ij}}$$

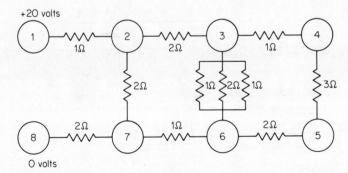

Figure 1

(b) Kirchhoff's current law which states that the sum of the currents entering and leaving any node is algebraically zero. Applying these laws in turn to nodes $2, 3, \ldots, 7$ we obtain

Node 2

$$I_{12} + I_{72} + I_{32} = \frac{V_1 - V_2}{1} + \frac{V_7 - V_2}{2} + \frac{V_3 - V_2}{2} = 0$$

hence $-4V_2 + V_3 + V_7 = -20$

Node 3

$$I_{23} + I_{63} + I_{43} = \frac{V_2 - V_3}{2} + \frac{V_6 - V_3}{R_{63}} + \frac{V_4 - V_3}{1} = 0$$

where R_{63} is found from

$$\frac{1}{R_{63}} = \frac{1}{1} + \frac{1}{2} + \frac{1}{1} = \frac{5}{2}$$

Thus $V_2 - 8V_3 + 2V_4 + 5V_6 = 0$

Node 4

$$I_{34} + I_{54} = \frac{V_3 - V_4}{1} + \frac{V_5 - V_4}{3} = 0$$

i.e. $3V_3 - 4V_4 + V_5 = 0$

Node 5

$$I_{45} + I_{65} = \frac{V_4 - V_5}{3} + \frac{V_6 - V_5}{2} = 0$$

Thus $2V_4 - 5V_5 + 3V_6 = 0$

Node 6

$$I_{36} + I_{56} + I_{76} = \frac{V_3 - V_6}{2/5} + \frac{V_5 - V_6}{2} + \frac{V_7 - V_6}{1} = 0$$

Hence $5V_3 + V_5 - 8V_6 + 2V_7 = 0$

Finally,

Node 7

$$I_{67} + I_{27} + I_{87} = \frac{V_6 - V_7}{1} + \frac{V_2 - V_7}{2} + \frac{V_8 - V_7}{2} = 0$$

so that $V_2 + 2V_6 - 4V_7 + V_8 = 0$

Consequently, we have six linear equations in six unknowns $V_2, \ldots, V_7$ which we write

$$
\begin{aligned}
-4V_2 + V_3 \qquad\qquad\qquad + V_7 &= -20 \\
V_2 - 8V_3 + 2V_4 \qquad + 5V_6 \qquad &= 0 \\
3V_3 - 4V_4 + V_5 \qquad\qquad &= 0 \\
2V_4 - 5V_5 + 3V_6 \qquad &= 0 \\
5V_3 \qquad + V_5 - 8V_6 + 2V_7 &= 0 \\
V_2 \qquad\qquad\qquad + 2V_6 - 4V_7 &= 0
\end{aligned} \tag{2}
$$

The system of equations (2) is an example of the more general system of N linear equations in N unknowns, which we may write

$$
\sum_{j=1}^{N} a_{ij}x_j = b_i \qquad i = 1, 2, \ldots, N \tag{3}
$$

In a standard elementary *mathematics* book on systems of linear equations the student will be told that Cramer's rule [17], which involves ratios of *determinants* (see [40]), can be used to solve systems of the form (3). For N as small as 4 or 5 this is feasible. For $N = 26$, say, the number of multiplications required by this method is about $25 \times 26!$ On a computer performing 100,000 multiplications per second, about 10^{17} years would be necessary to perform the multiplications alone! Thus here we have a situation for which a known method of calculation exists but is so time-consuming as to be useless. The numerical analyst again has a rôle to play: an alternative method must be designed which yields a solution in a reasonable time. Algorithms for this type of problem will be dealt with in Chapter 3.

In conclusion, therefore, numerical analysis is concerned with solving "hard" problems approximately or more efficiently than analytical methods. More than this, it deals with the *statement* of the problem (is there sufficient information available to start solving the problem and to guarantee an answer?). It deals with the derivation of an appropriate method or class of methods by which an answer to the problem may be obtained. It deals with an analysis of the behavior of this class of methods and hence a choice of a "best" one for particular circumstances, i.e. which method works and gives the result in as efficient a manner as possible? The numerical analyst† should

---

† By this we mean anyone who is a student of numerical analysis, a researcher in the field of numerical analysis or, in fact, anyone who seriously wants to use (existing) numerical methods for the solution of a problem and who is not content to simply take an existing program for an appropriate (or perhaps an inappropriate) algorithm to solve his problem without being in possession of the full facts relating to his problem, i.e. without fully understanding all the physical, analytical, and computer science facts relating to his problem.

produce a characterization of the algorithm in computer terms so that a computer programmer can produce an efficient code for a digital computer which takes minimal computer resources. The numerical analyst should analyze the results produced by his algorithm and decide whether these results meet the specifications laid down by the problem formulator. He must decide whether further computer runs should be undertaken, whether the original data for the problem was the correct (type of) data and, if so, if there were sufficient data. If not, perhaps the engineer will have to undertake further experimentation or measurement. (The numerical analyst should decide whether the original mathematical model needs modifying.)

The numerical work is performed by the digital computer so the numerical analyst plays little or no rôle here. (This is in strict contrast to the desk-calculator days!) His rôle is one of analysis of *model*, *method*, and *results*. It is in this sense that an *analyst* is concerned with *numerical* work. Thus, as we see, a numerical analyst must be a broadly based scientist: (s)he must be a capable *analyst*, a competent *physical scientist*, and also a *computer scientist* all rolled into one.

The rôle of physical science is emphasized in the following section where we consider several problems which give rise to mathematical problems which in turn serve to exemplify the classes of method which we will discuss in later chapters. In the description of these examples the reader should not worry if (s)he does not understand the mathematics of any particular example since the examples are designed to cover a number of different disciplines. The rôle of analysis will become clear as we proceed through the chapters. The rôle of computer scientist may be less clear, as in an elementary textbook of this nature we will, in exercises and examples, necessarily keep the discussion as clear and simple as we possibly can. Thus complex problems of data management and efficient computer coding will not arise as a result of the examples and exercises that *we* discuss. However, the exercises and examples have been constructed to make maximal use of the computer. Hence the reader should bear in mind that as (s)he answers the exercise problems, in real life these problems can be on a larger, more complex scale which will demand that (s)he uses the computer resources (both time and storage) in as an efficient manner as possible. Consequently, in answering the problems the reader should get into the habit of programming efficiently even in the situation where, because of the simple problems, the scale of the problem might not seem to warrant it. Acquiring good programming habits from an early stage is worthwhile achievement. We will not give computer programs or flowcharts in examples as it is the author's belief that in a textbook of this nature the reader gains little from such programs. What is important, however, is that the reader program the exercises himself. The programming language used is not important (provided it is a scientific language—COBOL is not really the language for the type of exercises described!), although the text is written in terms of ALGOL (the author's preference) and FORTRAN.

2 Physical Examples

In the previous section we introduced the reader to two physical examples which give rise to mathematical equations which need to be solved numerically. These two examples serve to motivate the material discussed in Chapters 4, 5, 6, and 3, respectively. In this section we will describe some other physical problems which give rise to mathematical equations which will motivate the material to be described in the remaining chapters.

Example 3 A company produces oil from a reservoir by pumping the oil from wells placed in a straight line. For well i placed in position x_i a production of oil p_i is measured each year, the magnitude of p_i being dependent on geological conditions of the reservoir. It is required to ascertain the production which would occur at any new well placed at some intermediate point x, without drilling a large number of wells (they are expensive to sink) on a "try it and see" basis. Let us represent the situation by Figure 2. We assume the

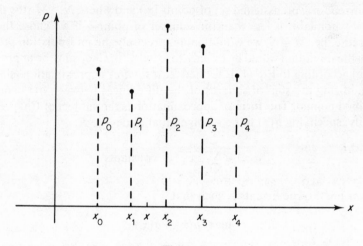

Figure 2

position of the new well is to be x, as shown. We need a method of "filling in" the productions at all the intermediate points. The way in which we "fill in" these values will determine a type of approximation problem. For example, if the actual production curve (if we knew it) is represented by $y = p(x)$ and an approximation to this curve is denoted by $y = q(x)$ (p and q will, in general, be different) then there are many criteria by which we can set about choosing the approximating function $q(x)$. For example, we can demand that $q(x)$ be a polynomial of a specified degree, say N, so that

$$q(x) = \sum_{j=0}^{N} a_j x^j \qquad a_j \text{ constants}$$

and determine the coefficients a_j such that

$$q(x_i) = p(x_i) \qquad i = 0, 1, \ldots, N$$

i.e. the approximation $q(x)$ *exactly* agrees with the unknown (but measurable at the points x_i) function $p(x)$ at preassigned points x_i, $i = 0, 1, \ldots, N$. Such an approximation is called *polynomial interpolation* or *exact approximation*, and is the subject of Chapter 4.

Alternatively, we may ask that the approximating function $q(x)$ be a polynomial of degree M so that

$$q(x) = \sum_{j=0}^{M} b_j x^j \qquad b_j \text{ constants}$$

and determine the coefficients b_j such that

$$\sum_{i=0}^{N} [q(x_i) - p(x_i)]^2 \tag{4}$$

is minimized for a preassigned set of points $\{x_i\}$ and where $N > M$ (the degree of the polynomial q is less than the number of points—if we chose these to be the same, i.e. $M = N$, we would obtain precisely the interpolation problem for then the minimum would be exactly zero (why?)). Choosing the approximation $q(x)$ according to Eq. (4) is termed *least squares* approximation (since we are minimizing the sum of squares of the differences between $q(x)$ and $p(x)$ at specified points). This form of approximation is the subject of Chapter 5.

Finally, specifying $q(x)$ to be a polynomial of degree R,

$$q(x) = \sum_{j=0}^{R} c_j x^j \qquad c_j \text{ constants}$$

and choosing the coefficients c_j such that

$$\max_i |p(x_i) - q(x_i)|$$

is minimized, i.e. the maximum difference between our approximation and the observed value at the specified points is minimized, is called *minimax* approximation and is the subject of Chapter 6.

Example 4 Consider the problem of heat diffusion in a thin bar of length l whose ends are maintained at a temperature of $0\,°C$ and which is initially heated so that the temperature at any point x is given as $g(x)$, $0 \le x \le l$ (see Figure 3). The temperature u at any point x and time t satisfies the *partial differential equation*

$$\frac{\partial u}{\partial t} = \frac{\partial^2 u}{\partial x^2} \tag{5}$$

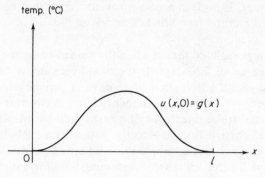

Figure 3

The total problem comprises finding that $u(x, t)$ which satisfies Eq. (5) and which satisfies the *initial condition* $u(x, 0) = g(x)$ and the *boundary condition* $u(0, t) = u(l, t) = 0$. The *solution* u to Eq. (5) may be obtained analytically by the classical method of separation of variables (see Dixon [9]). Carrying out this analysis, it is found that the required solution is

$$u(x, t) = \sum_{n=1}^{\infty} c_n \exp\left(-n^2 \pi^2 / l^2 t\right) \sin n\pi x/l \qquad (6)\dagger$$

where the constants c_n are given by

$$c_n = \frac{2}{l} \int_0^l g(x) \sin \frac{n\pi}{l} x \, dx \qquad n = 1, 2, \ldots \qquad (7)$$

Thus to evaluate the infinite series (6) we first require to evaluate the constants c_n given by Eq. (7). Our success or failure here will depend upon the function $g(x)$. If $g(x)$ is "simple" we may well be able to carry out the integration exactly. However, for "nasty" $g(x)$ we will most likely be unable to carry out the integration exactly. Consequently, we will need an *approximate* method of integration. Such techniques are described in Chapter 7. Note that once the c_n have been obtained we are still left with an enumeration problem in Eq. (6). Here we need to approximate to the infinite series. For example, we may simply decide that 10 or 20 terms of the expansion will be sufficient whence we *truncate* the series at the tenth or twentieth term respectively. However, we may decide that the technicalities are too involved in this approach and start afresh—making an approximation to the original equation, Eq. (5). Such an approximation may be that of *finite differences* where derivatives are replaced by *divided differences* (see Chapter 4) and the resulting methods are known as finite difference methods (see Mitchell and Griffiths

---

† We will consistently use the notation "exp" to denote the exponential function e (see Jones and Jordan [27]).

[32]). Alternatively, Eq. (5) can be approximated by the method known as the finite element method (see Mitchell and Wait [33]).

Example 5   It is postulated that in a healthy human being the red blood cells die with a frequency of 120 days. If the blood cells decay at a much faster rate than this, say with a frequency of 15 days, it may be concluded that the patient is seriously ill (!). A measurement of the decay rate of blood cells, which is postulated to be linear, may be performed by introducing a known quantity of a radioactive isotope into the patient, thus labeling a measured quantity of blood. Each day a sample of blood may be taken and the amount of the radioactive isotope measured. Assuming a uniform distribution of the labeled blood, the amount of radioactive isotope present each day will indicate the quantity of blood cells which have died since the preceding day. From these measurements the gradient (the derivative) is required to determine the decay rate. Clearly we are unable to differentiate a discrete set of data. Consequently, a means of obtaining *numerical differentiation* is required; this is described in Chapter 8.

Example 6   In high-school physics we met the *ideal gas law*

$$P = RT/V$$

where P is the pressure, V is the molar volume, and T is the temperature of the gas. However, nature is not ideal and modifications to the ideal gas law have been proposed. Notably, the Beattie–Bridgeman equation

$$P = RT/V + \beta/V^2 + \gamma/V^3 + \delta/V^4 \tag{8}$$

is such a modification where β, γ and δ are temperature-dependent parameters characteristic of the gas. Equation (8) is an example of a more general modification given by

$$P = \sum_{i=0}^{N} \alpha_i/V^{i+1} \tag{9}$$

where $\alpha_0 = RT$ and $\alpha_1, \alpha_2, \ldots, \alpha_N$ are parameters. To obtain the pressure corresponding to a given temperature T and molar volume V we merely need to substitute the values of the parameters occurring on the right-hand side of Eq. (9). However, to discover what molar volume corresponds to a given pressure P and temperature T is clearly not a simple task. We need to solve a *nonlinear* equation of V. To perform this task a numerical algorithm is required which will obtain an approximation to the appropriate V. Such methods are the subject matter of Chapter 9.

Example 7   In the previous example we described a modification to the ideal gas law which gives rise to a nonlinear equation in the variable V. Such an equation is the special case of the more general situation where we have a *system* of nonlinear equations of more than one variable. Consider the principal reactions in the production of synthesis gas by partial oxidation of methane with oxygen, namely

$$CH_4 + \tfrac{1}{2}O_2 \rightleftharpoons CO + 2H_2$$

$$CH_4 + H_2O \rightleftharpoons CO + 3H_2$$

$$H_2 + CO_2 \rightleftharpoons CO + H_2O$$

It is required to find the reactant ratio of O_2 to CH_4 which will produce an adiabatic equilibrium temperature of 2200 °F at an operating pressure of 20 atmospheres when the reactant gases are preheated to an entering temperature of 1000 °F. A system of nonlinear equations has to be solved.

If x_1 is the mole fraction of CO in the equilibrium mixture;
x_2 the mole fraction of CO_2 in the equilibrium mixture;
x_3 the mole fraction of H_2O in the equilibrium mixture;
x_4 the mole fraction of H_2 in the equilibrium mixture;
x_5 the mole fraction of CH_4 in the equilibrium mixture;
x_6 the number of moles of O_2 per mole of CH_4 in the input gases;
x_7 the number of moles of product gases in the equilibrium
  mixture per mole of CH_4 in the input gases;
then it can be shown that these variables satisfy the seven simultaneous nonlinear equations:

$$
\begin{aligned}
x_1x_7 + 2x_2x_7 + x_3x_7 \qquad\qquad - \quad x_6 &= 0 \\
x_3x_7 + x_4x_7 + 2x_5x_7 &= 2 \\
x_1x_7 + x_2x_7 \qquad\qquad + x_5x_7 &= 1 \\
28{,}837x_1x_7 + 139{,}009x_2x_7 + 78{,}213x_3x_7 - 18{,}927x_4x_7 - 8427x_5x_7 + 10{,}690x_6 &= 13{,}492 \\
x_1 + x_2 + x_3 + x_4 + x_5 &= 1 \\
20x_1x_4^3 - 1.7\times10^5 x_3x_5 & \\
x_1x_3 - 2.6058x_2x_4 &= 0
\end{aligned}
$$

A discussion of methods for solving such nonlinear systems may be found in Chapter 9.

Example 8 Consider four equally spaced particles on a weightless string held under tension T. Let the particles of weights m_1, m_2, m_3, and m_4 be displaced small distances y_1, y_2, y_3, and y_4, respectively, in a direction perpendicular to the rest position of the string as depicted in Figure 4. Neglecting

12

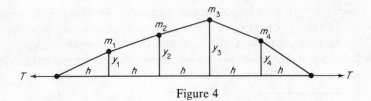

Figure 4

gravity and assuming the resulting vibrations are in the same direction as the initial displacements, the governing equations of motion are given by

$$m_1\frac{d^2y_1}{dt^2} = -\frac{Ty_1}{h} + \frac{T(y_2-y_1)}{h}$$

$$m_2\frac{d^2y_2}{dt^2} = -\frac{T(y_2-y_1)}{h} + \frac{T(y_3-y_2)}{h}$$

$$m_3\frac{d^2y_3}{dt^2} = -\frac{T(y_3-y_2)}{h} - \frac{T(y_3-y_4)}{h}$$

$$m_4\frac{d^4y_4}{dt^2} = \frac{T(y_3-y_4)}{h} - \frac{Ty_4}{h}$$

(10)

Introducing the vector

$$\mathbf{y} = (y_1, y_2, y_3, y_4)^T †$$

and letting $b_i = m_ih/T$, $i = 1, 2, 3, 4$, we may write the system of ordinary differential equations (10) in vector form as

$$B\frac{d^2\mathbf{y}}{dt^2} = C\mathbf{y}$$

where

$$B = \begin{bmatrix} b_1 & 0 & 0 & 0 \\ 0 & b_2 & 0 & 0 \\ 0 & 0 & b_3 & 0 \\ 0 & 0 & 0 & b_4 \end{bmatrix} \quad \text{and} \quad C = \begin{bmatrix} -2 & 1 & 0 & 0 \\ 1 & -2 & 1 & 0 \\ 0 & 1 & -2 & 1 \\ 0 & 0 & 1 & -2 \end{bmatrix}.$$

When the particles vibrate in phase or in direct opposition we say the system vibrates in *normal mode*. In this situation the equation

$$\frac{d^2\mathbf{y}}{dt^2} = -w^2\mathbf{y}$$

(11)

† We will consistently denote *column* vectors by bold type: T will be used to denote the transposed vector.

holds where w is a normal frequency. Hence, combining Eqs. (10) and (11) we obtain

$$Bw^2\mathbf{y} = -C\mathbf{y}$$

Introducing $\mathbf{x} = B^{1/2}\mathbf{y}$, we can write this equation as

$$A\mathbf{x} = -w^2\mathbf{x} \tag{12}$$

where $A = B^{-1/2}CB^{-1/2}$ where

$$B^{1/2} = \begin{bmatrix} \sqrt{b_1} & 0 & 0 & 0 \\ 0 & \sqrt{b_2} & 0 & 0 \\ 0 & 0 & \sqrt{b_3} & 0 \\ 0 & 0 & 0 & \sqrt{b_4} \end{bmatrix} \quad \text{and}$$

$$B^{-1/2} = \begin{bmatrix} 1/\sqrt{b_1} & 0 & 0 & 0 \\ 0 & 1/\sqrt{b_2} & 0 & 0 \\ 0 & 0 & 1/\sqrt{b_3} & 0 \\ 0 & 0 & 0 & 1/\sqrt{b_4} \end{bmatrix}$$

Equation (12) is called an *algebraic eigenvalue problem*. The values of w which result in solving Eq. (12) are called the *eigenvalues*. The corresponding vectors $\mathbf{x}$ are called the *eigenvectors*. We will consider algorithms for special forms of Eq. (12) in Chapter 10.

Example 9 An electrical circuit is described by Figure 5. C denotes a capacitance of C farads, L denotes an inductance of L henries, and R a resistor of R ohms. S is a switch which is to be closed at time $t = 0$. Assume that initially the voltage across C is v_0. The voltage v across the capacitance C when the switch S is closed is governed by differential equation

$$LC\frac{d^2v}{dt^2} + RC\frac{dv}{dt} + v = 0 \tag{13}$$

and subject to the initial conditions

$$v(0) = v_0$$

$$\frac{dv(0)}{dt} = 0$$

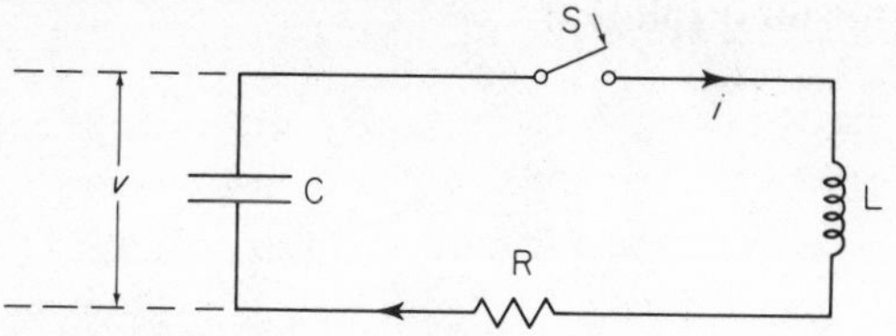

Figure 5

Equation (13) can be written as a system of first-order equations by introducing new variables

$$y_1 = v \qquad y_2 = \frac{dv}{dt}$$

Hence

$$\frac{dy_1}{dt} = y_2 \tag{14}$$

and

$$LC\frac{dy_2}{dt} + RCy_2 + y_1 = 0$$

We can write Eq. (14) in vector form as

$$\begin{bmatrix} 1 & 0 \\ 0 & LC \end{bmatrix} \frac{d}{dt}\begin{bmatrix} y_1 \\ y_2 \end{bmatrix} = \begin{bmatrix} 0 & 1 \\ -1 & -RC \end{bmatrix}\begin{bmatrix} y_1 \\ y_2 \end{bmatrix}$$

or

$$\frac{d}{dt}\begin{bmatrix} y_1 \\ y_2 \end{bmatrix} = \begin{bmatrix} 0 & 1 \\ -1/LC & -R/L \end{bmatrix}\begin{bmatrix} y_1 \\ y_2 \end{bmatrix}$$

Hence

$$\frac{d\mathbf{y}}{dt} = A\mathbf{y} \tag{15}$$

where

$$\mathbf{y} = (y_1, y_2)^{\mathrm{T}} \qquad \text{and} \qquad A = \begin{bmatrix} 0 & 1 \\ -1/LC & -R/L \end{bmatrix} \tag{16}$$

and $y_1(0) = v_0$, $y_2(0) = 0$

Equations (15) and (16) comprise a first-order system of ordinary differential equations. Such equations are considered in Chapter 11.

The examples quoted in this section are, in the main, taken from [4]. This text contains a wealth of interesting physical examples and accounts of numerical methods used for their solution.

CHAPTER 2

Errors in Numerical Computation

1 Introduction

In this chapter we want to discuss the concept of "error" used in the context
of numerical analysis. We first want to emphasize that the error which we
will consider is not one which might equate to a mistake or bug. For example,
in the simple ALGOL program:

begin integer sum, i;

    for $i := 1$ **step** 1 **until** 10 **do**

        sum $:=$ sum $+ 1/i$;

        print sum;

    end

which is supposed to form and print the sum $\sum_{i=1}^{10} 1/i$, a bug is present in that
the identifier *sum* has not been initialized to zero. (We assume that the
compiler does not automatically initialize all declared variables.)

Likewise, in trying to find the values of x and y which satisfy the equations

$$x + 2y = 0.26$$

$$x - y = 1.31$$

we might use the second equation to write

$$x = 1.31 - y \tag{1}$$

(instead of the correct equation $x = 1.31 + y$) and then substitute this
expression for x in the first equation, namely

$$1.31 + y = 0.26$$

15

from which

$$y = -1.05$$

and using Eq. (1),

$$x = 2.36$$

The correct solution to this problem is

$$x = 0.96 \qquad y = -0.35$$

The difference between the *calculated* solution and the *correct* solution is caused by the *mistake* in Eq. (1).

In this text we will assume, whenever a discussion of errors arises, that no mistakes or bugs have been made in the calculations (or programs) which have produced the numbers under discussion. The errors in these numbers will be consequences of our inability to calculate exactly the solution of a given problem. The types of errors that can arise will, in the main, take up the rest of this section.

There are several sources of errors which can arise in a numerical calculation. We list here the most common ones.

(i) *Errors arising from the inadequacy of the mathematical model.* A mathematical problem has usually been derived as the result of the model of a physical situation. For example, in Chapter 1 we assumed the temperature at any point x in a thin bar is given as the solution of the partial differential equation

$$\frac{\partial u}{\partial t} = \frac{\partial^2 u}{\partial x^2} \tag{2}$$

A more realistic model might be one that takes into account that the diffusive coefficient multiplying the term $\partial^2 u / \partial x^2$, the constant one in our case, is dependent upon the temperature of the bar. That is, a more realistic model (but one which is more difficult to solve—hence the choice of simpler model) might be that u is the solution of

$$\frac{\partial u}{\partial t} = D(u) \frac{\partial^2 u}{\partial x^2} \tag{3}$$

where D is the diffusive coefficient (see [5]).

Thus any solution of Eq. (2) will only approximate the solution of Eq. (3) and hence approximate the physical temperature, if it is accepted that Eq. (3) is the correct model. Consequently, even if we could solve Eq. (2) exactly by the infinite series given by Eqs. (6) and (7) of Chapter 1, the resulting temperature will be in error owing to the simplification of the mathematical model.

(ii) *Noise in the input data.* In Example 3 of Chapter 1, to determine the production of a new well at a point in the oil field, measurements of data at existing wells have to be obtained. These measurements are made by reading flow rates at each well. If these readings are not (able to be) made exactly (as is likely to be the case) then errors (sometimes referred to as "noise") in the data will arise. Consequently, any subsequent model which uses this data as input will be suspect, and account will have to be taken of the noise in analyzing the final results (see in particular Chapter 5).

(iii) *Errors due to truncation.* Consider once again Example 4 of Chapter 1. The temperature u in the thin bar is given analytically by the infinite series

$$u(x, t) = \sum_{n=1}^{\infty} c_n \exp \frac{(-n^2\pi^2 t)}{l^2} \sin \frac{(n\pi x)}{l} \tag{4}$$

Even for a simple initial function $g(x)$ which allows the exact calculation of the coefficients c_n given by Eq. (7) of Chapter 1 we will not be able to use all the terms in the infinite expansion (6) of Chapter 1. Consequently we might accept the first ten terms on the basis that the remaining terms do not contribute significantly to the numerical result for u. Thus we would accept as an approximation

$$\tilde{u}(x, t) = \sum_{n=1}^{10} c_n \exp \frac{(-n^2\pi^2 t)}{l^2} \sin \frac{(n\pi x)}{l}$$

The error which arises in neglecting

$$\sum_{n=11}^{\infty} c_n \exp \frac{(-n^2\pi^2 t)}{l^2} \sin \frac{(n\pi x)}{l}$$

is referred to as *truncation error.*

(iv) *Rounding error.* Consider solving the quadratic equation

$$6x^2 - 7x + 2 = 0$$

from which $x = 1/2$ or $2/3$. Expressing these roots as decimal fractions we have

$$x = 0.5 \text{ or } 0.6666^{\cdot}$$

where the 6˙ means recurring, i.e. repeated an infinite number of times. If we require to express these roots to three decimal places we may write

$$x = 0.500$$

without further ado, but $x = 0.6666^{\cdot}$ needs modification. We may *chop* the digits after the third decimal place so that the required root is $x = 0.666$, or we may *round* the digits after the third decimal place so that the required root is $x = 0.667$. The object of rounding is to choose as the representation of the number x that number closest to x. Thus we *round down* for the digits 0, 1, 2, 3, 4 following the third decimal digit and *round up* for the digits 5, 6, 7, 8, 9 following the third digit. In fact the number with digit 5 (and the rest 0) following the third digit is equally distant between two numbers. We adopt the rule of rounding up if the preceding digit is odd and rounding down if the preceding digit is even.

In either case of chopping or rounding as described, the resulting error is called *rounding error*.

To this list we may add:

(v) *Machine (hardware) errors.* Computers are complex structures comprising thousands of electronic parts. On very rare occasions these components can produce hardware errors which turn up as strange results in a user's program or calculation. As in the case of human mistakes, these errors are beyond our control and for the purposes of this text we assume they do not exist.

In the above discussion we have assumed the reader has had an intuitive feeling for the size of the errors we have discussed. We now want to be more precise.

The size of an error may be viewed in one of two ways; namely the size relative to the size of the quantity one is measuring or the size *absolutely*. Let $\tilde{x}$ be an approximate value for a quantity whose exact value is x.

Definition   The absolute error is defined to be $|\tilde{x} - x|$.

Definition   The relative error is defined to be

$$\left| \frac{\tilde{x} - x}{x} \right|$$

where it is assumed $x \neq 0$.

Note in this latter definition the approximate value $\tilde{x}$ is frequently substituted for x in the denominator, assuming $\tilde{x} \neq 0$.

To exemplify what we mean, consider a rocket landing on the moon having traveled all the way from the earth, a distance of approximately 270,000 miles. The personnel on board the landing craft are concerned about the actual point of touch-down on the surface of the moon. They will accept, probably, a landing at most a few hundred feet from the planned position. A mile or two either way will not be acceptable (it may be disastrous if there are craters or mountains nearby!). Consequently the error in the landing position measured in absolute terms must be small per se. The fact that the error is guaranteed to be 1/100,000 of the distance traveled by the moon rocket from earth will not be much good. What we are saying is that the absolute error must be small. (Of course as a consequence the relative error will be small.) Now consider a builder estimating the number of bricks for a new building complex. Let us suppose he estimates that the building will require 270,000 bricks. If a fleet of trucks delivers 270,001 bricks the builder will probably not be alarmed—one brick in 270,000 is not serious. In this case we are concerned with the relative error in the number of bricks being small instead of the absolute error being small, as was the case in the moon landing.

2 Machine-representable Numbers

Consider converting 1/4 to a decimal fraction. We can carry this out exactly, i.e. $1/4 \to 0.25$. Thus 1/4 is a *terminating decimal fraction*. Now consider 1/3 in decimal. We have $1/3 \to 0.333\ldots$, where the dots indicate an infinite number of digits. Thus 1/3 is a *nonterminating decimal fraction*.

In between these two situations we have fractions that require many, but less than an infinite number of decimal digits for their representation. For personal reasons we may restrict the number of decimal digits to be less than or equal to 4.

Thus in this sense the decimal fraction 0.12345 cannot be represented exactly. As we stated in the previous section we round (or chop) such numbers so that in this instance we would represent our number by .1234 and incur rounding error.

Our decision to restrict the number of digits to be less than, or equal to four adds the restriction to the size of the numbers we will be able to represent as well as the accuracy of the numbers we will be able to represent. In the case of four digits we have a maximum number .9999 and the smallest nonzero number will be .0001.

If we adopt the system of working with a constant number of decimal digits we then say that the representation is one of *fixed point*. To circumvent such restriction we can adopt the strategy of using a constant number of digits to represent any number and adjust an exponent to give the given number its correct magnitude. Thus we would represent

$$25.67 \quad \text{as} \quad .2567(2)$$

and

$$0.002\,567 \quad \text{as} \quad .2567(-2)$$

or in general

$$y = \pm .d_1 d_2 d_3 d_4 (e)$$

where $1 \le d_1 \le 9$ and $0 \le d_i \le 9$, $i = 2, 3, 4$, and e, the integer exponent is adjusted to give the number its correct size, and where we add a representation of zero given by $y = +.0000(m)$ where m is a suitably chosen value of e—see below. Such a system of representing numbers is called the *floating point* representation.

Consider a real number

$$r = \pm \left(\sum_{i=1}^{\infty} d_i 10^{-i} \right) 10^e$$

with $1 \le d_1 \le 9, 0 \le d_i \le 9, i = 2, 3, \ldots$, and e an exponent satisfying $m \le e \le M$ with usually $m < 0 < M$. Let us represent r to a finite number, say p, of decimals by

$$y = \pm \left(\sum_{i=1}^{p} \bar{d}_i 10^{-i} \right) 10^e$$

where $\bar{d}_i$ may or may not be d_i, depending on the rounding which takes place in representing r as y. The integer p is called the *precision*. The greater is p the more accurate can real numbers r be represented.

The above discussion has been restricted to decimal arithmetic. However, computers seldom use base 10 arithmetic but rather use base 2, 8, or 16. Consequently, in representing numbers in a computer arithmetic we will be concerned about representing the numbers in some base other than base 10. Let us use the generic base β and assume p such β-digits can be used in the representation of a number. Thus a real number

$$r = \pm \left(\sum_{i=1}^{\infty} d_i \beta^{-i} \right) \beta^e$$

where $1 \le d_i \le \beta - 1$; $0 \le d_i \le \beta - 1$, $i = 2, 3, \ldots$, and $m \le e \le M$, can be represented by

$$y = \pm \left(\sum_{i=1}^{p} \bar{d}_i \beta^{-i} \right) \beta^e \tag{5}$$

For convenience we will omit the bar from the digits d_i and recall that these digits have arisen from a rounding, or chopping, process applied to r.

Clearly if y is to approximate r there will be many numbers r which are represented by the same number y. Let us denote the set of all possible numbers of the form given by Eq. (5) (and include zero) by

$$F = F(\beta, p, m, M)$$

Hence each nonzero computer number $x \in F$ has the β-base representation

$$x = \pm .d_1 d_2 \cdots d_p(e)$$

where, as specified above, the digits d_i, $i = 1, 2, \ldots, p$ in the *mantissa* satisfy $1 \le d_1 \le \beta - 1$; $0 \le d_i \le \beta - 1$, $i = 2, 3, \ldots, p$, and the exponent e satisfies $m \le e \le M$, and the value of $x \in F$ is understood to be

$$x = \pm \left(\sum_{i=1}^{p} d_i \beta^{-i} \right) \beta^e$$

and we choose to represent zero as $+.00 \cdots 0(m)$.

It is clear by assumptions that $\beta \ge 2$ and $p \ge 1$. Hence the only real numbers r which can be exactly represented in $F(\beta, p, m, M)$ are those with not more than p significant digits and with magnitudes in the appropriate range.

It is straightforward to calculate that the largest number in $F(\beta, p, m, M)$ is

$$.d_1 d_2 \cdots d_p(M)$$

with $d_1 = d_2 = \cdots = d_p = \beta - 1$. This number has the value

$$\sum_{i=1}^{p} (\beta - 1)\beta^{-i}\beta^M = (\beta - 1)\beta^M[\beta^{-1} + \beta^{-2} + \beta^{-3} + \cdots + \beta^{-p}]$$

$$= \beta^{M-p}[\beta^p - 1] \tag{6}$$

Likewise, it is simple to show that the smallest number representable in $F(\beta, p, m, M)$ distinguishable from zero is

$$.100\,000\,0 \cdots 00(m) \tag{7}$$

which has the value β^{m-1}.

Any number r which is larger in magnitude than expression (6) is said to *overflow* in the number system $F(\beta, p, m, M)$.

Any smaller (but nonzero) number than (7) is said to *underflow* in the number system $F(\beta, p, m, M)$. In an actual computation, if underflow occurs the computation can usually proceed but a note (error message) of the underflow is given. In contrast, if overflow occurs the computation has to stop, with an associated error message being printed.

The first thing which will strike the reader is that whereas the real number system $\mathbb{R}$ has an infinite number of numbers, our system $F(\beta, p, m, M)$ must have a finite number of numbers. In fact it contains precisely

$$2(\beta - 1)\beta^{p-1}(M - m + 1) + 1 \quad \text{numbers}$$

Example 1 Consider the (binary) base 2 arithmetic, $\beta = 2$. Let us assume a precision $p = 3$ and $m = -1$, $M = 2$. The floating point numbers in this system are then

.100(−1)	.101(−1)	.110(−1)	.111(−1)
(1/4)	(5/16)	(6/16)	(7/16)
.100(0)	.101(0)	.110(0)	.111(0)
(1/2)	(5/8)	(6/8)	(7/8)
.100(1)	.101(1)	.110(1)	.111(1)
(1)	(5/4)	(6/4)	(7/4)
.100(2)	.101(2)	.110(2)	.111(2)
(2)	(5/2)	(6/2)	(7/2)

together with the negatives of the above numbers and zero, which we have assumed to be represented by $0.000(-1)$.

A straightforward count shows that the system $F(2, 3, -1, 2)$ has exactly 33 numbers. We note that there are no numbers in the intervals $(-1/4, 0)$ and $(0, 1/4)$ so that as we calculated, the smallest representable number distinguishable from 0 is $+1/4$. If we denote these numbers graphically

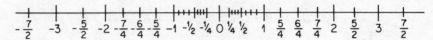

we note that the floating point numbers are *not* evenly distributed. For *common exponents* the numbers are, however, equally spaced. Consequently the absolute error in representing a given real number r in a floating point system $F(\beta, p, m, M)$ will depend upon the magnitude of r. The larger is r in magnitude, the larger can be the rounding error. Graphically we can see that in rounding r we are in fact choosing the nearest number $F(\beta, p, m, M)$ to represent r.

Example 2 The IBM 360/370 computer uses the floating point number system $F(16, 6, -64, 63)$ so that we have $\beta = 16$, $p = 6$, $m = -64$ and $M = 63$. Base 16 arithmetic is commonly called *hexadecimal* arithmetic. The precision $p = 6$ here indicates that each number will have six hexadecimal digits. A hexadecimal digit d_i is one of

$$0, 1, 2, 3, 4, 5, 6, 7, 8, 9, A, B, C, D, E, \text{ or } F$$

where A–F represent ten–fifteen respectively. Hence any number $x \in F(16, 6, -64, 63)$ will have the form

$$x = \pm \left(\sum_{i=1}^{6} d_i 16^{-i} \right) 16^e$$

This number system is utilized as a consequence of the IBM 360/370 series having a computer word comprising 32 bits (binary digits). Pictorially,

1 word =				

1 bit   7 bits   8 bits $\equiv$ 1 byte   8 bits $\equiv$ 1 byte   8 bits $\equiv$ 1 byte

bit 1 = sign bit which is 0 for + and 1 for −.

The last three bytes of the word are used to store the mantissa and consist of six hexadecimal digits. Each hexadecimal digit comprises $1/2$ byte = 4 bits. Each bit can take on one of two values, namely 0 or 1 and hence the number of possible different numbers $= 2^4 = 16$, exactly the number of digits in the hexadecimal system.

The seven digits following the sign bit can take $2^7 = 256$ different values. This gives a degree of flexibility on the lower and consequent upper bound on e. The values of m and M are chosen as -64 and 63 respectively.

The system $F(16, 6, -64, 63)$ contains (using our formula) $15 \times 2^{28} + 1 = 4, 026, 531, 841$ different numbers. The largest representable number is

$$.FFFFFF(63) = \frac{(16^{16} - 1)}{16^6} 16^{63} = 0.72 \times 10^{76}$$

In contrast, the smallest representable number distinguishable from 0 is

$$.100000(-64) = 16^{-65} = 0.54 \times 10^{-78}$$

The representation of zero is $.000000(-64)$.

3 Error Analysis of Floating Point Arithmetic

In the previous section we saw that in general, in representing a decimal fraction in a computer arithmetic we will incur an error, namely the rounding error. This rounding error will occur, however, even for exactly representable numbers, i.e. ones whose computer representation does not require an approximation, in the event that arithmetic (i.e. addition (+), subtraction (−), multiplication (∗), or division (/)) is performed on these numbers. For example, if we consider numbers in $F(10, 4, -1, 2)$ then multiplying

$$u = .1234(1) \quad \text{by} \quad v = .2672(-1)$$

produces in real arithmetic the number

$$.03297248$$

In the number system $F(10, 4, -1, 2)$ this will be represented as

$$.3297(-1)$$

and we have incurred an error $0.248(-5)$. Thus the numbers u and v are exactly representable in $F(10, 4, -1, 2)$ but $u \ast v$ is not.

Since arithmetic operations occur in their thousands, particularly in a large-scale scientific computation, for example in predicting weather forecasts in meteorology, we are concerned about the propagation of these rounding errors as the computation progresses. As we will see, there will be circumstances where these rounding errors will not be significant (for example solving a differential equation by a stable method—see Chapter 11) whereas in other cases these rounding errors can be catastrophic (for example in solving badly scaled systems of linear equations—see Chapter 3). Consequently, in this section we will present a brief introduction to the analysis of errors arising from the floating point arithmetic—the reference text for this subject is Wilkinson [48], to which we direct the interested reader for further details.

Suppose the nonzero real number r has base-β expansion

$$r = \pm \left(\sum_{i=1}^{\infty} d_i \beta^{-i} \right) \beta^e \qquad d_1 \neq 0$$

Let us represent this in the floating point system $F(\beta, p, m, M)$ as

$$y = Fl(r) = \pm .\bar{d}_1 \bar{d}_2 \cdots \bar{d}_p(e)$$

and keeping with the previous section, Eq. (5), we immediately drop the bars.

If we have adopted chopped arithemtic then y has been obtained from r by omitting the final digits (those after digit number p). If we have adopted

rounded arithmetic then y has been obtained from r by "adding"†
$\beta/2 \cdot \beta^{-p-1} \cdot \beta^e$ to r. In either case we define the absolute rounding error as
$|r - Fl(r)|$. In chopped arithemtic therefore

$$|r - Fl(r)| = \left| \sum_{i=p+1}^{\infty} d_i \beta^{-i} \beta^e \right| = \left| \sum_{i=1}^{\infty} d_{p+i} \beta^{-i} \right| \beta^{e-p} \tag{8}$$

In rounded arithmetic,

$$|r - Fl(r)| = \left| \left\{ \sum_{i=p+1}^{\infty} d_i \beta^{-i} - \tfrac{1}{2} \beta^{-p} \right\} \beta^e \right| \tag{9}$$

As we indicated in Example 1 in the previous section, the absolute rounding
error given by Eqs. (8) and (9) can be large when e is large. However, it is
clear that even in this case the rounding error will be smaller for rounded
arithmetic than for chopped arithmetic.

We define the *relative rounding error* as

$$\left| \frac{r - Fl(r)}{r} \right|$$

where we assume $r \neq 0$.

In terms of a given system $F(\beta, p, m, M)$ we can bound the relative error
which occurs by representing r by $y \in F$.

Theorem 1 *If $Fl(r)$ is the chopped representation of $r \neq 0$ in $F(\beta, p, m, M)$ then*

$$\left| \frac{r - Fl(r)}{r} \right| \leq \beta^{1-p}$$

Proof From the definition of relative rounding error we have

$$\left| \frac{r - Fl(r)}{r} \right| = \left| \left(\sum_{i=1}^{\infty} d_{p+i} \beta^{-i} \right) \beta^{e-p} \middle/ \left(\sum_{i=1}^{\infty} d_i \beta^{-i} \right) \beta^e \right|$$

By normalization of the floating point numbers, we have for nonzero r,

$$0 < \sum_{i=1}^{\infty} d_{p+i} \beta^{-i} \leq 1$$

Since $d_1 \neq 0$, $1/\beta \leq d_1/\beta + d_2/\beta^2 + \cdots \leq 1$. Hence

$$\left| \frac{r - Fl(r)}{r} \right| \leq \frac{\beta^{-p}}{1/\beta} = \beta^{1-p}$$

which is independent of the exponent e.

Thus from Theorem 1 the relative rounding error is a function of the base
β and the precision p. For a given base β, the higher is the precision, the

† "adding" because we strictly add if r is positive and subtract if r is negative.

smaller will be the relative rounding error. Consequently, if a standard calculation is carried out with a precision p and suffers too greatly from the relative rounding error, the calculation may be repeated with a greater precision, say $2p$, whence the relative rounding error will be reduced. Carrying out computations in such a manner is commonly called calculating in *extended precision* or *multiple precision*. Most computations are usually carried out in single precision and only in special circumstances, by the use of appropriate computer instructions, is extended precision used.

Example 3 For the IBM 360/370, $\beta = 16$, $p = 6$, and the relative rounding error is 16^{-5} when single precision arithmetic is used. For the Honeywell 6660, $\beta = 2$, $p = 27$, and the relative rounding error is $2^{-26} = 2^{-2}16^{-6}$ when single precision is used.

An analogous theorem exists for rounded arithmetic.

Theorem 2 *If $Fl(r)$ is the rounded representation of $r \neq 0$ in $F(\beta, p, m, M)$ then*

$$\left| \frac{r - Fl(r)}{r} \right| \leq \tfrac{1}{2} \beta^{1-p}$$

The proof is left as an exercise to the reader (Exercise 1) but the details are essentially the same as the proof of Theorem 1.

The bounds given by Theorem 1 and 2 are fundamental to the analysis of floating point arithmetic for the following reason. If x and $y \in F(\beta, p, m, M)$ and if $\circ$ denotes one of the four basic operations $+, -, *, /$, the computer result $Fl(x \circ y)$ of performing $x \circ y$ is the correct representation of the exact result $x \circ y$. That is, in forming $x \circ y$ where x and y are already machine representable, it is assumed that all the figures which can significantly affect the exact value of $x \circ y$ before the number $x \circ y$ is represented as a number in $F(\beta, p, m, M)$ are taken into account. Only after $x \circ y$ has been formed is the rounding (chopping) process applied.

On an actual computer, floating point arithmetic is made to behave like this model by using, for example, a double-precision accumulator.

With this basic model, then, Theorems 1 and 2 state that

$$\left| \frac{(x \circ y) - Fl(x \circ y)}{(x \circ y)} \right| \leq \begin{cases} \beta^{1-p} & \text{if the computer chops} \\ \tfrac{1}{2} \beta^{1-p} & \text{if the computer rounds} \end{cases} \tag{10}$$

Definition In the floating point number system $F(\beta, p, m, M)$ the *unit rounding error u* is defined by

$$u = \begin{cases} \beta^{1-p} & \text{for chopped arithmetic} \\ \tfrac{1}{2} \beta^{1-p} & \text{for rounded arithmetic} \end{cases}$$

Using this definition and rearranging Eq. (10), we have:

Lemma 1 *For numbers* $x, y \in F(\beta, p, m, M)$

$$Fl(x \circ y) = (x \circ y)(1 + \delta) \quad where \ |\delta| < u$$

In words, Lemma 1 states that the floating point representation of $(x \circ y)$ is the exact result to within a unit rounding error for $\circ$ equated to any one of the basic operations $+, -, *, /$ when x and y are exactly representable in $F(\beta, p, m, M)$.

Note this lemma is *not* stating that

$$Fl(w \circ v) = (w \circ v)(1 + \delta)$$

in general for $w, v \notin F$, i.e., this result is not true for numbers which are not exactly representable in F. So if $x = Fl(w)$ and $y = Fl(v)$, then all we can say is that

$$Fl(x \circ y) = (x \circ y)(1 + \delta)$$

which does not give us information about $Fl(w \circ v)$ but of

$$Fl(Fl(w) \circ Fl(v)).$$

It is important to realize this fact, for example, when considering the subtraction of two nearly equal numbers.

Consider $w = 0.53214$ and $v = 0.53221$ and let us assume a number system F with four decimal digits. Now $w - v = -0.00007$ in exact arithmetic. So $Fl(w - v) = -.7000(-4)$. To evaluate $w - v$ in our number system we must first represent w and v as four-decimal-digit numbers. Let us assume rounded arithmetic so that

$$x = Fl(w) = .5321(0) \quad and \quad y = Fl(v) = 0.5322(0)$$

So $x \circ y = x - y$ and hence

$$Fl(x \circ y) = Fl(-0.0001) = -.1000(-3)$$

Hence if we calculate

$$\frac{(x \circ y) - Fl(x \circ y)}{(x \circ y)} \quad \text{we obtain } 0$$

That is, the subtraction of y from x is obtained without error. But what we are endeavoring to calculate is $w \circ v = w - v$. So calculating

$$\frac{(w - v) - Fl(x \circ y)}{(w - v)} \quad \text{we obtain } -0.42857$$

That is, error which arises in representing w and v in the number system F and then performing the subtraction has produced an answer which has a significant relative error.

This example, is of course, not in conflict with Lemma 1 and the reader should realize that we have performed more than one floating point operation while the lemma is strictly only applicable to a single operation.

Let us therefore generalize our discussion and consider a sequence of arithmetic operations. When we discuss a sequence of operations we must be precise about the order in which operations are performed. hence we will assume that the precedence of the arithmetic operators is precisely that given by the rules of FORTRAN and ALGOL.†

Theorem 3 *If x, y, $z \in F$ then*

$$\frac{|(x+y+z)-Fl(x+y+z)|}{|x+y+z|} \leq \frac{(|x|+|y|+|z|)(2u+u^2)}{|x+y+z|}$$

Proof By direct computation using our rules of precedence,

$$Fl(x+y+z) = Fl(Fl(x+y)+z)$$

Now using Lemma 1, $Fl(x+y) = (x+y)(1+\delta_1)$, where $|\delta_1| \leq u$, so $Fl(x+y+z)$ $= Fl((x+y)(1+\delta_1)+z)$. Again using Lemma 1, we have

$$Fl(x+y+z) = ((x+y)(1+\delta_1)+z)(1+\delta_2) \quad \text{where } |\delta_2| \leq u$$

$$= (x+y+z)+(x+y)(\delta_1+\delta_2+\delta_1\delta_2)+z\delta_2$$

Hence

$$\frac{|(x+y+z)-Fl(x+y+z)|}{|x+y+z|} = \frac{|(x+y)(\delta_1+\delta_2+\delta_1\delta_2)+z\delta_2|}{|x+y+z|}$$

Now we use the triangular inequality so that

$$\frac{|(x+y+z)-Fl(x+y+z)|}{|x+y+z|} \leq \frac{|x+y||\delta_1+\delta_2+\delta_1\delta_2|+|z||\delta_2|}{|x+y+z|}$$

$$\leq \frac{|x+y|[|\delta_1|+|\delta_2|+|\delta_1\delta_2|]+|z||\delta_2|}{|x+y+z|}$$

Now using $|\delta_1| \leq u$ and $|\delta_2| \leq u$, we have

$$\frac{|(x+y+z)-Fl(x+y+z)|}{|x+y+z|} \leq \frac{(|x|+|y|+|z|)(2u+u^2)}{|x+y+z|}$$

as required.

† The only difference which occurs between precedences of operators in ALGOL and FORTRAN concerns the exponentiation operator which we have not included in our arithmetic operators. The reason for this exclusion is that exponentiation is not an elementary arithmetic operator and is evaluated using the natural logarithm and the exponential functions: see Ekman and Fröberg [10].

The effect of the two operations is now apparent. The bound can be large if $|x+y+z|$ is small while $|x|+|y|+|z|$ is not, so that the small factor $(2u+u^2)$ does not compensate for the large one. Such a situation will arise when the members of the sum have opposite signs so that cancellation occurs.

Example 4   Consider the set of floating point numbers $F(10, 7, m, M)$ with chopping.

(a) $Fl(10^5 + 3.141592 + 10^5) = 100003.1 + 100000.0 = 200003.1$
$$\rightarrow .2000031(6).$$

$$\text{The relative error} = \left| \frac{20003.141592 - 200003.1}{200003.141592} \right| = \frac{0.041592}{200003.141592}$$

$$= 0.21 \times 10^{-6}$$

(b) $Fl(10^5 + 3.141592 - 10^5) = 100003.1 - 100000 = 3.1 \rightarrow .3100000(1)$

and the relative error is

$$\frac{3.141592 - 3.1}{3.141592} = \frac{0.041592}{3.141592} = 0.13 \times 10^{-1}$$

In (b) we once again see the effect of subtracting two nearly equal numbers—the subtraction is carried out exactly but the error made in forming the first of the operands 100003.1 now becomes much more significant in the final result.

Theorem 4   *If* x, y, $z \in F$ *then*

$$\frac{|x*y*z - Fl(x*y*z)|}{|x*y*z|} \leq 2u + u^2$$

Proof   This may be proved in an analogous manner to the proof given in Theorem 3 and is left to the reader as an exercise—see Exercise 2. A similar result also applies for division—see Exercise 3.

To extend the analysis of rounding error involving a general number of operators we will need the result of the following theorem.

Theorem 5   *If* $|\delta_i| \leq u$, $1 \leq i \leq k$, u *the unit rounding error, then*

$$\prod_{i=1}^{k} (1+\delta_i) \leq 1 + 1.01ku \quad \text{if } ku \leq 0.01$$

Proof   The numbers $\delta_i 1 \leq i \leq k$ may be positive, negative, or zero. In all such cases, however, we have trivially

$$\prod_{i=1}^{k} (1+\delta_i) \leq (1+u)^k$$

Consider $\exp(u) = 1 + u + \frac{1}{2}u^2 \exp(\xi)$, $\xi \in (0, u)$ obtained using Taylor's series with the Lagrange form of the remainder (see Jones and Jordan [27]).

Now since $1/2u^2 \exp(\xi) \geq 0$

$$\exp(u) > 1 + u \tag{11}$$

Now consider

$$\exp(ku) = 1 + ku + 1/2k^2u^2 \exp(k\xi) \qquad \xi \in (0, u)$$

obtained using Taylor series. By assumption in the statement of the theorem, $ku \leq 0.01$ and so

$$\exp(ku) \leq 1 + ku(1 + 1/2(0.01)\exp(0.01))$$

$$\leq 1 + ku(1.01) \tag{12}$$

Combining Eqs. (11) and (12), we find that

$$\prod_{i=1}^{k}(1+\delta_i) \leq (1+u)^k \leq \exp(ku) \leq 1 + 1.01ku$$

which is the required result.

Remark The number 0.01 appearing as the bound on ku in Theorem 5 is somewhat arbitrary since we may obtain the inequality (12) for any ku such that $\exp(ku) \leq 2$, namely $ku \leq 0.693$. However, we want to choose ku small enough so that the restriction on ku is not too severe; $ku \leq 0.01$ is commonly used to this end.

Thus the restriction states that the number of terms is bounded by $0.01u^{-1}$, u the unit rounding error. Typically, on the IBM 360 this produces a limit of 10,486 terms.

We will make use of Theorem 5 in a slightly modified form, namely we write

$$\sum_{i=1}^{k}(1+\delta_i) = 1 + 1.01k\delta \quad \text{for some } \delta \text{ such that } |\delta| \leq u \tag{13}$$

This result follows simply from Theorem 5—see Exercise 4.

Theorem 6 *If $x_i \in F(\beta, p, m, M)$, $1 \leq i \leq n$, then*

$$\frac{\left|\sum_{i=1}^{n} x_i - Fl\left(\sum_{i=1}^{n} x_i\right)\right|}{\left|\sum_{i=1}^{n} x_i\right|} \leq \frac{\left(\sum_{i=1}^{n} |x_i|\right)1.01nu}{\left|\sum_{i=1}^{n} x_i\right|}$$

Proof

$$Fl\left(\sum_{i=1}^{n} x_i\right) = (\cdots((x_1 + x_2)(1+\delta_2) + x_3)(1+\delta_3) + \cdots + x_n)(1+\delta_n)$$

where δ_i is the rounding error associated with adding x_i to the previously formed sum of x_j, $j = 1, 2, \ldots, i-1$. In each case $|\delta_i| \le u$, $i = 1, 2, \ldots, n$. Hence

$$Fl\left(\sum_{i=1}^{n} x_i\right) = \sum_{i=1}^{n} x_i \prod_{j=1}^{n} (1 + \delta_j)$$

where we have introduced a $\delta_1 = 0$ for convenience.

Hence using result (13), we have

$$Fl\left(\sum_{i=1}^{n} x_i\right) = \sum_{i=1}^{n} x_i[1 + 1.01(n - i + 1)u_i] \quad \text{where } |u_i| \le u.$$

Hence

$$\left|\left(\sum_{i=1}^{n} x_i\right) - Fl\left(\sum_{i=1}^{n} x_i\right)\right| = \left|\sum_{i=1}^{n} x_i 1.01(n - i + 1)u_i\right|$$

We now employ the triangle inequality on the right-hand side so that

$$\left|\left(\sum_{i=1}^{n} x_i\right) - Fl\left(\sum_{i=1}^{n} x_i\right)\right| \le \sum_{i=1}^{n} |x_i| 1.01 n u$$

and we have used $|u_i| \le u$, $i = 1, 2, \ldots, n$. Hence the result follows.

Theorem 7 *If $x_i \in F(\beta, p, m, M)$, $1 \le i \le n$, then*

$$\frac{\left|\left(\prod_{i=1}^{n} x_i\right) - Fl\left(\prod_{i=1}^{n} x_i\right)\right|}{\left|\prod_{i=1}^{n} x_i\right|} \le 1.01 n u$$

The proof is left to the reader as an easy exercise—see Exercise 5. A similar result also holds for division—see Exercise 6.

Remarks We note that if all of the numbers x_i are of the same sign then the rounding error bound given by Theorem 6 will be small and we may predict, with confidence, the accuracy of our computer result. However, if the value of $\sum_{i=1}^{n} |x_i|$ is large relative to $|\sum_{i=1}^{n} x_i|$, which may occur when the numbers x_i comprise negative and positive numbers, then the error bound will be large and we can only say that the relative rounding error can be large.

Example 5 Obtain relative error bounds for the computations:

$$\text{(i)} \quad r_1 = Fl(x^2 - y^2)$$

$$\text{(ii)} \quad r_2 = Fl((x - y)(x + y))$$

Applying our rules of precedence we evaluate expression (i) as

$$r_1 = Fl(Fl(x * x) - Fl(y * y))$$

$$= Fl(x * x(1 + \delta_1) - y * y(1 + \delta_2))$$

using Lemma 1. So

$$r_1 = ((x*x)(1+\delta_1) - (y*y)(1+\delta_2))(1+\delta_3)$$

again using Lemma 1, and where $|\delta_i| \le u$. Hence

$$\frac{|(x^2-y^2)-Fl(x^2-y^2)|}{|(x^2-y^2)|} = \frac{|x^2(\delta_1+\delta_3+\delta_1\delta_3) - y^2(\delta_2+\delta_3+\delta_2\delta_3)|}{|x^2-y^2|}$$

Using the triangle inequality we may write this as

$$\frac{|(x^2-y^2)-Fl(x^2-y^2)|}{|x^2-y^2|} \le \frac{x^2|\delta_1+\delta_3+\delta_1\delta_3| + y^2|\delta_2+\delta_3+\delta_2\delta_3|}{|x^2-y^2|}$$

$$\le \frac{[|\delta_1|+|\delta_3|+|\delta_1\delta_3|]x^2 + [|\delta_2|+|\delta_3|+|\delta_2\delta_3|]y^2}{|x^2-y^2|}$$

$$\le (2u+u^2)\frac{(x^2+y^2)}{|x^2-y^2|} \tag{14}$$

which may not be small if $|x^2-y^2|$ is small relative to x^2+y^2.

Now consider (ii):

$$r_2 = Fl(Fl(x-y)*Fl(x+y))$$

$$= Fl((x-y)(1+\delta_1)*(x+y)(1+\delta_2))$$

$$= (x-y)(x+y)(1+\delta_1)(1+\delta_2)(1+\delta_3) \qquad \text{where } |\delta_i| \le u$$

Hence

$$\frac{|(x-y)(x+y)-Fl((x-y)(x+y))|}{|(x-y)(x+y)|}$$

$$= \frac{|(x-y)(x+y)[\delta_1+\delta_2+\delta_3+\delta_1\delta_2\delta_3+\delta_3\delta_1+\delta_1\delta_2\delta_3]|}{|(x-y)(x+y)|}$$

$$\le \frac{|(x-y)(x+y)|(3u+3u^2+u^3)}{|(x-y)(x+y)|}$$

$$= (3u+3u^2+u^3) \tag{15}$$

which is small for x and y.

Consequently, the error bounds (14) and (15) indicate that if we calculate the difference of two squares by procedure (ii) we can guarantee an accuracy of our result which we cannot do for procedure (i).

Example 6   An efficient algorithm for evaluating a polynomial $p_N(x)$ of degree N is commonly called Horner's rule, and is defined by

$$p_N(x) = \sum_{i=0}^{N} a_i x^i = [\cdots\{(a_N x + a_{N-1})x + a_{N-2}\}x + \cdots]x + a_0$$

Investigate the error bound for the rounding error arising in this calculation assuming the coefficients $\{a_i\}$ and the variable x are machine representable numbers and assuming $p_N(x)$ is nonzero.

Using the assumed rules of precedence, let us denote by θ_i the rounding error which arises when a multiplication is performed, and by ζ_i the rounding error which arises when an addition is performed. Hence

$$Fl(p_N(x)) = [(\cdots\{a_N x(1+\theta_N) + a_{N-1}\}(1+\zeta_{N-1})x(1+\theta_{N-1})$$
$$+ (a_{N-2}\cdots + a_1)(1+\zeta_1)x(1+\theta_1) + a_0](1+\theta_0)$$

So

$$Fl(p_N(x)) = \sum_{i=0}^{N} a_i x^i \prod_{j=0}^{i}(1+\zeta_j)\prod_{j=1}^{i}(1+\theta_j)$$

where we have introduced for convenience $\theta_N = 0$.

Let us write $\delta_j = \theta_j$, $j = 1, \ldots, N$, and $\delta_{j+1+N} \equiv \zeta_j$, $j = 0, 1, \ldots, N$. Then

$$Fl(p_N(x)) = \sum_{i=0}^{N} a_i x^i \prod_{j=1}^{2i+1}(1+\delta_j)$$

We now use Eq. (13) so that

$$Fl(p_N(x)) = \sum_{i=0}^{N} a_i x^i (1 + (2i+1)1.01 u_i) \qquad \text{where } |u_i| \le u$$

Hence

$$\frac{|p_N(x) - Fl(p_N(x))|}{|p_N(x)|} = \frac{\left|-\sum_{i=0}^{N} a_i x^i (2i+1)1.01 u_i\right|}{|p_N(x)|}$$

$$\le \frac{\sum_{i=0}^{n} |a_i||x^i|(2i+1)1.01|u_i|}{|p_N(x)|}$$

$$\le \frac{2(N+1)1.01u \sum_{i=0}^{N} |a_i||x^i|}{|p_N(x)|}$$

which is the required error bound.

Exercises

1. Prove Theorem 2.
2. Prove Theorem 4.
3. Prove Theorem 4 for division.
4. Prove that $\prod_{i=1}^{k}(1+\delta_i) = 1 + 1.01 k\delta$ for some δ such that $|\delta| \le u$.
5. Prove Theorem 7.
6. Obtain a bound for the floating point calculation of $\prod_{i=1}^{n}(1/x_i)$ where $x_i \in F(\beta, p, m, M)$.

4 The Effects of Computational Error: An Example†

In this final section we wish to consider the effects of the occurrence of computational error, i.e. rounding error and truncation error, in an actual computation. The effects of rounding error are well documented. For example, calculation of zeros of a quadratic equation subject to rounding error; the solution of a system of linear equations; the calculation of zeros of polynomials; etc. For a readable account of these problems the reader is referred to Forsythe [11]. Consider the following problem.

It is required to construct a routine to calculate sinh (x) for a real argument x, assuming one does not exist in our computer library. To construct this routine we need to determine the "best" ‡ way of approximating the hyperbolic sine function taking into account the occurrence of truncation error and rounding error. From our high-school mathematics we know that

$$\sinh (x) = 0.5(\exp (x) - \exp (-x)) \tag{16}$$

where

$$\exp (x) = 1 + x + \frac{x^2}{2!} + \frac{x^3}{3!} + \cdots \tag{17}$$

Thus we can perform the calculation of the sinh function in one of the following ways:

(1) Calculate $\exp (x)$ using the series (17) for the given value of x. Use the same series for $-x$ and then subtract the latter result from the former.

(ii) Use the series for $\exp (x)$ for positive x. Calculate $\exp (-x)$ as $1/\exp (x)$, this denominator being given by the first step. Subtract the results.

(iii) Use the Taylor series expansion of sinh (x) in which the alternate terms are cancelled formally, i.e.

$$\sinh (x) = x + \frac{x^3}{3!} + \frac{x^5}{5!} + \cdots \tag{18}$$

Before analyzing these possible algorithms let us run the calculations in single precision (8 decimal digits) and compare the results. Consider first (i) and (ii).

To calculate $\exp (x)$ we need to decide on the number of terms to be used in the expansion (17). Let us add terms until the truncation error is less in magnitude than $5u$ (say), u being the unit rounding error for our computer. The truncation error is given by

$$\frac{x^{n+1}}{(n+1)!} \exp (\xi) \qquad \xi \in (0, x)$$

† This section may be omitted on first reading.
‡ For the purposes of this discussion we assume the only tool at our disposal for approximation is the Taylor series. Better approximations are certainly possible and will be discussed in Chapter 6.

when n terms are retained in the expansion (see Jones and Jordan [27]). Hence we require to choose n such that

$$\frac{x^{n+1}}{(n+1)!} \exp(\xi) \leq 5u \tag{19}$$

for x in the interval of interest. Let us assume initially $-10 \leq x \leq 10$. Let us in fact evaluate $\exp(-9.1)$. The terms of the series obtained in performing the summation of the series (17) for $x = -9.1$ are tabulated in Table 1, column 2. The reader will note that the alternate terms in the series are of opposite signs and consequently, in the light of our discussion in section 3, we expect significant cancellation of significant digits to occur. The number of terms taken in the series before the sum remained constant (i.e. the sums agreed to within the unit rounding error) was 44 on our machine (a Honeywell 6660). Substituting $n = 44$ and $x = -9.1$ into Eq. (19) indicates that this certainly is a sufficient number of terms to ensure Eq. (19) is satisfied (it is in fact satisfied for smaller n—the discrepancy is caused by the occurrence of rounding error). The value of $\exp(-9.1)$ using the standard function from our computer indicates that the computed result given in column 1, Table 2 has no significant digits. In contrast, the computer result given by calculating $1/\exp(9.1)$ from the series of terms in column 3 of Table 1 and reported in column 2 of Table 2 is extremely accurate so that rounding eror is not a problem. Hence the example indicates that using a large number of terms of alternating signs in a series can produce numerical values which have no significant figures owing to the build-up of rounding error. We would therefore surmise that the safest way of accumulating the sum would be to use the series (17) for $|x|$ and use the reciprocal of the result if $x < 0$.

The numerical results are now to be substituted into (16). The resulting computation produces the numbers reported in Table 3. The first column contains the result obtained using the (inaccurate) result in column 1, Table 2 for $\exp(-9.1)$. From this is subtracted $\exp(9.1)$ obtained by the Taylor series (17) with all positive terms. Column 2 of Table 3 contains the result of using the reciprocal of $\exp(9.1)$ and subtracting from it the numerical value of $\exp(9.1)$ again obtained from (17) with positive x. As can be seen the results in columns 1 and 2 are nearly identical: the effect of the poor result for $\exp(-9.1)$ does not affect the result (the reason for this is that for $x = -9.1$ the dominant term in $\sinh(x)$ is $\exp(9.1)$ and consequently an error in $\exp(-9.1)$ is (more or less) insignificant. We have also reported the result of using (18) to calculate $\sinh(-9.1)$ in column 3 of Table 3 and the correct result in the final column of Table 3.

Let us now investigate the situation for $x = -0.123E-03$ and $x = -0.678E-06$. The results of evaluating the series in a manner similar to that above are displayed in Table 4.

In Table 5 we have calculated the value of $\sinh(x)$ using the three different strategies. These values are reported in Table 5 together with the correct results.

Table 1

term number	term in series (17) $x = -9.1$	term in series (17) $x = 9.1$
1	$-0.91000000E$†$+01$	$.91000000E+01$
2	$+.41405000E+02$	$.41405000E+02$
3	$-.12559517E+03$	$.12559517E+03$
4	$+.28572901E+03$	$.28572901E+03$
5	$-.52002680E+03$	$+.52002680E+03$
6	$+.78870731E+03$	$+.78870731E+03$
7	$-.10253195E+04$	$+.10253195E+04$
8	$+.11663009E+04$	$+.11663009E+04$
9	$-.11792599E+04$	$+.11792599E+04$
10	$+.10731265E+04$	$+.10731265E+04$
11	$-.88776826E+03$	$+.88776826E+03$
12	$+.67322427E+03$	$+.67322427E+03$
13	$-.47125699E+03$	$+.47125699E+03$
14	$+.30631704E+03$	$+.30631704E+03$
15	$-.18583234E+03$	$+.18583234E+03$
16	$+.10569214E+03$	$+.10569214E+03$
17	$-.56576383E+02$	$+.56576383E+02$
18	$+.28602505E+02$	$+.28602505E+02$
19	$-.13699094E+02$	$+.13699094E+02$
20	$+.62330880E+01$	$+.62330880E+01$
21	$-.27010048E+01$	$+.27010048E+01$
22	$+.11172338E+01$	$+.11172338E+01$
23	$-.44203598E+00$	$+.44203598E+00$
24	$+.16760531E+00$	$+.16760531E+00$
25	$-.61008334E-01$	$+.61008334E-01$
26	$+.21352917E-01$	$+.21352917E-01$
27	$-.71967238E-02$	$+.71967238E-02$
28	$+.23389352E-02$	$+.23389352E-02$
29	$-.73394176E-03$	$+.73394176E-03$
30	$+.22262900E-03$	$+.22262900E-03$
31	$-.65352385E-04$	$+.65352385E-04$
32	$+.18584584E-04$	
33	$-.51248400E-05$	
34	$+.13716483E-05$	
35	$-.35662857E-06$	
36	$+.90147778E-07$	
37	$-.22171481E-07$	
38	$+.53094861E-08$	
39	$-.12388801E-08$	
40	$+.28184522E-09$	
41	$-.62555892E-10$	
42	$+.13553777E-10$	
43	$-.28683574E-11$	
44	$+.59322846E-12$	

† Throughout the text we will use the E-format to indicate that the mantissa is to be scaled by the appropriate power of 10; $E+n$ means mutliply by 10^n; $E-n$ means multiply by 10^{-n}; En (without a sign) is taken as $E+n$.

Table 2

exp (−9.1) from column 2 (Table 1)	1/exp (9.1) from column 3 (Table 1)	correct value
0.81754984E+04	0.11166583E−03	0.11166581E−03

Table 3

1/2(exp (−9.1) −exp (9.1))	1/2(1/exp (9.1) −exp (9.1)	series (18)	correct value
−0.44776456E04	−0.44776457E04	−0.44776476E04	−0.44776464E04

Table 4

term	computer terms in series (17)			
	$x = -0.123E-03$	$x = 0.123E03$	$x = -0.678E-06$	$x = 0.678E06$
1.	−0.123000000E−03	0.123000000E−03	−0.678000000E−06	0.678000000E−06
2.	0.75644999E−08	0.75644999E−08	0.22984200E−12	0.22984200E−12
3.	−0.31014450E−12	0.31014450E−12		
4.	0.95369433E−17	0.95369433E−17		
calculated value	0.99987700E00	0.9987702E00	0.99999932E00	0.99999933E00
correct value		0.99987701E00		0.99999932E00

Table 5

$x =$	(i) & Eq. (17)	(ii) & Eq. (17)	Eq. (18)	correct value
−0.123E−03	−0.12299710E−03	−0.12298663E−03	−0.12300000E−03	−0.12300000E−03
−0.678E−06	−0.67427754E−06	−0.67055225E−06	−0.67800001E−06	−0.67800000E−06

A comparison of the computations of the exponentials in both cases described by Table 4 indicates that a very small rounding error occurs. However, when we come to use these values in the computation of the sinh function it is clear that the subtraction of nearly equal numbers produces results which have poor accuracy—only two significant figures are correct using either of strategies (i) or (ii). In contrast, the calculation of sinh (x) via the Taylor series (18) produces an extremely accurate result as compared with the entries in the fourth column of Table 5.

In the remainder of this section we will use the tools developed in section 3 to analyze the effect of the truncation error and rounding error which can arise when we compute an approximation to sinh (x). In the light of the calculations just carried out, let us assume we are required to design an algorithm for computing $\sinh(x)$ for $-5 \le x \le 5$ according to the following rules:

if $|x| < 1/2$ **then** calculate $\sinh(x)$ via its Taylor series

      else   compute $\exp(x)$ via its Taylor series

               compute $\sinh(x)$ by the formula

               $\sinh(x) = (\exp(x) - \exp(-x)) * 0.5$

We will consider the **else** part of the algorithm first. Let us first require that a sufficient number of terms be taken in the Taylor series for $\exp(x)$ so that the truncation error is less than $5u$ for $x \in [-5, 5]$.

The truncation error is $x^{n+1}/(n+1)! \exp(\xi) \xi \in (-5, 5)$ when $(n+1)$ terms are taken in the Taylor series.

Hence we require

$$\frac{5^{n+1}}{(n+1)!} \le 5u \tag{20}$$

For the particular computer under consideration we assume an n large enough is chosen to ensure that (20) is satisfied. So

$$|\exp(x) - p_n(x)| \le 5u \qquad -5 \le x \le 5 \tag{21}$$

By the algorithm we are to use $p_n(x)$ for $x \ge 0$ and $1/p_n(x)$ for $x < 0$. The explicit representation of $p_n(x)$, for a given value of n, is

$$p_n(x) = \sum_{j=0}^{n} \frac{x^j}{j!}$$

If the coefficient $1/j!$ is to be calculated by forming the reciprocal of $j!$ for large values of n ($n = 40, 41$, etc.) the evaluation of $j!$ can exceed the machine representable numbers, i.e. overflow quickly occurs if evaluation of the coefficients proceeds in this manner. Consequently, we will use an alternative procedure to Horner's rule, given by the following algorithm:

     term $(0) \leftarrow 1$; sum $(0) \leftarrow 1$;

     for $j := 1$ **step** 1 **until** n **do**

     term $(j) := $ term $(j-1) * x/j$;

     sum $(j) := $ sum $(j-1) + $ term (j)†;   (22)

In this way $j!$ is never explicitly evaluated so overflow from this source is no longer a problem. Using the analysis of section 3 we now require to estimate a bound on the relative rounding error given the method of calculation and assuming the given value of x satisfies $Fl(x) = x(1 + \lambda)$ where λ is the associated rounding error.

† There is, of course, no need to use subscribed variables in the actual computation. However, we have use subscripted variables here to facilitate the discussion of the rounding errors.

Consider algorithm (22) written in terms of floating point arithmetic. Hence

$Fl(\text{term }(0)) := 1;$ $Fl(\text{sum }(0)) := 1;$

for $j := 1$ **step** 1 **until** n **do**

$$Fl(\text{term }(j)) := [Fl(\text{term }(j-1)) * Fl(x/j)](1 + \theta_j); \tag{23}$$

$$Fl(\text{sum }(j)) := [Fl(\text{sum }(j-1)) + Fl(\text{term }(j))](1 + \phi_j); \tag{24}$$

where θ_j is the rounding error associated with the floating point multiplication in Eq. (23) and ϕ_j is the rounding error associated with the floating point addition in Eq. (24). Eq. (23) can be further expanded to

$$Fl(\text{term }(j)) := [Fl(\text{term }(j-1)) * Fl(x)/j(1 + \psi_j)](1 + \theta_j) \tag{25}$$

where ψ_j is the rounding error associated with floating point division in Eq. (25).

Consider Eq. (24) written for $j = 1, 2, \ldots, n$. Then

$$Fl(\text{sum }(n)) := [Fl(\text{sum }(n-1)) + Fl(\text{term }(n))](1 + \phi_n)$$

$$= [\{Fl(\text{sum }(n-2)) + Fl(\text{term }(n-1))\}(1 + \phi_{n-1}) + Fl(\text{term }(n))]$$

$$\times (1 + \phi_n)$$

$$= \cdots$$

$$= ((((\cdots (Fl(\text{sum }(0)) + Fl(\text{term }(1)))(1 + \phi_1) + Fl(\text{term }(2)))$$

$$\times (1 + \phi_2) + \cdots$$

$$+ \cdots Fl(\text{term }(n-1)))(1 + \phi_{n-1}) + Fl(\text{term }(n)))(1 + \phi_n)$$

$$= \sum_{j=0}^{n} Fl(\text{term }(j)) \prod_{i=j}^{n} (1 + \phi_i) \tag{26}$$

where we have used $Fl(\text{sum }(0)) = 1 = Fl(\text{term }(0))$ and, for convenience, we have introduced $\phi_0(=0)$.

We now substitute $Fl(\text{term }(j))$ from Eq. (25) into Eq. (26). So

$$Fl(\text{sum }(n)) = \sum_{j=0}^{n} \left[\prod_{k=1}^{j} \frac{Fl(x)}{k} (1 + \psi_k)(1 + \theta_k) \prod_{i=j}^{n} (1 + \theta_i) \right]$$

$$\times \sum_{j=0}^{n} \frac{x^j}{j!} \prod_{k=1}^{j} (1 + \psi_k)(1 + \theta_k)(1 + \lambda) \prod_{i=j}^{n} (1 + \phi_i)$$

$$= \sum_{j=0}^{n} \frac{x^j}{j!} \prod_{k=0}^{n+1+2j} (1 + \delta_k)$$

where we have introduced a convenient redefinition of $\{\delta_k\}$ in terms of the $\{\psi_k\}, \{\phi_k\}, \{\theta_i\}$, and λ.

We are now able to estimate the relative error in computing the polynomial $p_n(x)$; so

$$\frac{\left| \sum_{j=0}^{n} \frac{x^j}{j!} - \sum_{j=0}^{n} \frac{x^j}{j!} \prod_{k=0}^{n+2j+1} (1+\delta_k) \right|}{\left| \sum_{j=0}^{n} \frac{x^j}{j!} \right|} \leq \frac{\sum_{j=0}^{n} \frac{|x^j|}{j!} (n+2j+2)1.01|u_j|}{\left| \sum_{j=0}^{n} \frac{x^j}{j!} \right|}$$

where we have used Eq. (13) and $|u_j| \leq u$, $j = 0, 1, \ldots, n$.
Hence

$$\frac{|p_n(x) - Fl(p_n(x))|}{|p_n(x)|} \leq \frac{\left\{ \sum_{j=0}^{n} \frac{|x^j|}{j!} \right\} 3(n+1)1.01u}{\left| \sum_{j=0}^{n} \frac{x^j}{j!} \right|}$$

But the polynomial $p_n(x)$ is used for nonnegative x only. Hence

$$\frac{|p_n(x) - Fl(p_n(x))|}{|p_n(x)|} \leq 3(n+1)1.01u \tag{27}$$

and we have a bound on the relative rounding error in terms of n, the number of terms taken. So from Eq. (27) we have

$$|p_n(x) - Fl(p_n(x))| \leq 3(n+1)1.01u|p_n(x)|$$

and hence

$$\frac{|p_n(x) - Fl(p_n(x))|}{|\exp(x)|} \leq 3(n+1)1.01u \frac{|p_n(x)|}{|\exp(x)|} \tag{28}$$

Now from Eq. (21) we have

$$p_n(x) = \exp(x) + 5\delta \qquad \text{where } |\delta| \leq u$$

So

$$\frac{p_n(x)}{\exp(x)} = \frac{1 + 5\delta}{\exp(x)}$$

and hence

$$\frac{|p_n(x)|}{\exp(x)} \leq 1 + \max_{0 \leq x \leq 5} \frac{5|\delta|}{\exp(x)} \leq 1 + 5u$$

Hence substitute this result into (28) and we obtain

$$\frac{|p_n(x) - Fl(p_n(x))|}{|\exp(x)|} \leq 3(n+1)1.01u(1 + 5u) \tag{29}$$

We may now use the result (29) to estimate the total relative error arising in approximating $\exp(x)$ by $p_n(x)$, since

$$\frac{|\exp(x) - Fl(p_n(x))|}{|\exp(x)|} \leq \frac{|\exp(x) - p_n(x)|}{|\exp(x)|} + \frac{|p_n(x) - Fl(p_n(x))|}{|\exp(x)|} \qquad (30)\dagger$$

Substituting the results (21) and (29) into the right-hand side of the inequality (30) we obtain

$$\frac{|\exp(x) - Fl(p_n(x))|}{|\exp(x)|} \leq \max_{0 \leq x \leq 5} \frac{5u}{|\exp(x)|} + 3(n+1)1.01u(1+5u)$$

$$= (3.03n + 8.03)u + 1.01(n+1)u^2 \qquad (31)$$

For $x \in [0, 5]$ the value of n for inequality (20) to be satisfied is typically 30 or less. Let us assume $n = 30$. Then Eq. (31) becomes

$$\frac{|\exp(x) - Fl(p_n(x))|}{|\exp(x)|} \leq 100u \text{ (approximately)} \qquad (32)$$

i.e. $Fl(p_n(x)) \leq \exp(x)(1 + 100\delta)|\delta| \leq u$

Using this estimate of $\exp(x)$, we require to evaluate

$$Fl(\sinh(x)) = Fl\left(\frac{\exp(x) - \exp(-x)}{2}\right)$$

$$= Fl\left(\frac{Fl(\exp(x)) - Fl(\exp(-x))}{2}\right)$$

$$= 1/2(\exp(x)(1 + 100\delta_1) - \exp(-x)(1 + 100\delta_2))(1 + \delta_3)$$

where $|\delta_i| \leq u$, $i = 1, 2, 3$ and where we have incorporated the rounding error arising from the division $1/\exp(x)$ to produce $\exp(-x)$ into the approximation (32) and, further, we assume division of 2 here produces no rounding error. So

$$Fl(\sinh(x)) = \tfrac{1}{2}[\exp(x)(1 + 100\delta_1 + \delta_3 + 100\delta_1\delta_3) - \exp(-x)$$

$$\times (1 + 100\delta_2 + \delta_3 + 100\delta_2\delta_3)]$$

Hence by direct computation we find

$$Fl\left(\frac{\exp(x) - \exp(-x)}{2}\right) \leq \frac{(|\exp(x)| + |\exp(-x)|)}{|\exp(x) - \exp(-x)|}(101u + 100u^2) \qquad (33)$$

The right-hand side of (33) is simply

$$\frac{(\exp(x) + \exp(-x))}{|\exp(x) - \exp(-x)|}(101u + 100u^2) = |\coth(x)|[101u + 100u^2]$$

---

† Note this inequality is stating that the total relative error is less than or equal to the sum of the relative truncation error and the relative rounding error.

From the description of the algorithm this approximation is to be used for $|x| > 1/2$. Hence a bound on $|\coth(x)|$ for $|x| > 1/2$ is easily found, namely $|\coth(x)| \le 2.164$. So

$$Fl \frac{(\exp(x) - \exp(-x))}{2} \le 219u + 217u^2 \le 300u \qquad (34)$$

We have left to examine the error bound for the **then** part of the algorithm on page 37.

The Taylor series for $\sinh(x)$ is

$$\sinh(x) = x + \frac{x^3}{3!} + \frac{x^5}{5!} + \cdots + \frac{x^{2m+1}}{(2m+1)!} + \cdots$$

$$p_{2m+1}(x) = x + \frac{x^3}{3!} + \cdots + \frac{x^{2m+1}}{(2m+1)!}$$

with a truncation error

$$\frac{f^{(2m+3)}(\xi)}{(2m+3)!} x^{2m+3} \qquad \xi \in (-\tfrac{1}{2}, \tfrac{1}{2})$$

where $m + 1$ terms are taken in the series. Hence the relative truncation error is

$$\frac{|\sinh(x) - p_{2m+1}(x)|}{|(\sinh(x)|} = \frac{f^{(2m+3)}(\xi)}{(2m+3)!} \frac{|x^{2m+3}|}{|(\sinh(x)|}$$

But

$$\frac{|x^{2m+3}|}{|\sinh(x)|} = \frac{|x^{2m+2}|}{|\sinh(x)/x|} \le |x^{2m+2}|$$

since $|\sinh(x)| > |x|$. Now

$$f^{(2m+3)}(x) = \cosh(x)$$

So

$$|f^{(2m+3)}(x)| \le \cosh(\tfrac{1}{2}) \text{ for } |x| \le \tfrac{1}{2}$$

Hence

$$\frac{|\sinh(x) - p_{2m+1}(x)|}{|\sinh(x)|} \le \frac{\cosh(\tfrac{1}{2})(\tfrac{1}{2})^{2m+3}}{(2m+3)!}$$

Now m is determined by the requirement that

$$|\sinh(x) - p_{2m+1}(x)| \le 5u$$

Typically this requires $m = 2$ or 3 for $|x| \le \tfrac{1}{2}$. Let us assume $m = 3$. Hence

$$p_7(x) = x + \frac{x^3}{3!} + \frac{x^5}{5!} + \frac{x^7}{7!}$$

For this polynomial we can use our algorithm (22) with the additional starting value term $(-1) = 0$. We can then use the bound in the rounding error we have derived in (29) where $n = 2m + 1 = 7$. So

$$\frac{|\sinh (x) - Fl(p_7))|}{|\sinh (x)|} \leq 24(1.01)u \frac{|p_7(x)|}{|\sinh (x)|}$$

$$\leq 24(1.01)u \frac{|p_7(x)/x|}{|\sinh (x)/x|} \leq 24(1.01)u|p_7(x)/x|$$

Now it is easy to show that $|p_7(x)/x|$ achieves its maximum at $x = \frac{1}{2}$. So

$$\frac{|\sinh (x) - Fl(p_7(x))|}{|\sinh (x)|} \leq 24(1.01)u1.04$$

which is clearly less than $300u$ given by (34). We therefore conclude, under the assumption $n = 30$ in the **else** portion of the algorithm and $m = 3$ in the **then** clause that

$$\max_{-5 \leq x \leq 5} \frac{|\sinh (x) - Fl(p_7(x)|}{|\sinh (x)|} \leq 300u$$

where $p_7(x)$ is the polynomial approximation defined by our algorithm for approximating $\sinh (x)$.

CHAPTER 3
Systems of Linear Equations

1 Introduction

In Example 2 of Chapter 1 we described an electrical network in which it was required to determine certain voltages between nodes in a network. The mathematical model for this problem gives rise to a system of linear equations; see Eq. (2) of Chapter 1. In the present chapter we want to consider the problem of obtaining, numerically, solutions to such problems.

Let us write the system of equations in the form

$$a_{11}x_1 + a_{12}x_2 + \cdots + a_{1N}x_N = b_1$$
$$a_{21}x_1 + a_{22}x_2 + \cdots + a_{2N}x_N = b_2$$
$$\vdots$$
$$a_{M1}x_1 + a_{M2}x_2 + \cdots + a_{MN}x_N = b_M$$

(1)

where the coefficients a_{ij}, $i = 1, 2, \ldots, M$, $j = 1, 2, \ldots, N$ and the constants b_i, $i = 1, 2, \ldots, M$ are known. We want to find the unknowns x_j, $j = 1, 2, \ldots, N$. In general, the system (1) which arises in practice is of two distinct types. First, there is the system in which all or nearly all of the coefficients a_{ij} are *nonzero*. In this case, the system is said to be *full*. Second, there is the system in which a great many of the coefficients are *zero*. In this case the system is said to be *sparse*. Full systems in general arise out of physical problems. Sparse systems arise a great deal in numerical analysis, in particular, in the solution of ordinary and partial differential equations; and also in perturbation theory, for example, quantum mechanics. In this case, the values of M and N are large (greater than 100).

In order to understand the principles involved in the solution of linear systems of equations we will assume the reader is familiar with the more elementary aspects of matrices. The actual properties of matrices which we will require are described in Strang [40]. We will adopt a standard notation in this chapter where we represent *column* vectors by bold type, e.g.

$$\mathbf{x} \equiv \begin{bmatrix} x_1 \\ x_2 \\ \vdots \\ x_N \end{bmatrix},$$

is a column vector with N elements $x_1, \ldots, x_N$. We will represent *row* vectors as transposed column vectors denoted by $\mathbf{x}^\mathrm{T}$. We will denote matrices by capital letters, $A, B, \ldots$, e.g.

$$A \equiv \begin{bmatrix} 1 & 3 & 6 \\ 7 & 1 & 5 \end{bmatrix}$$

is a (2×3) matrix with elements 1, 3, 6, 7, 1, and 5. We will denote the transpose of a matrix by appending $^\mathrm{T}$ to the capital letter representing the matrix, e.g.

$$A^\mathrm{T} \equiv \begin{bmatrix} 1 & 7 \\ 3 & 1 \\ 6 & 5 \end{bmatrix}$$

We will denote the *inverse* of A, where it exists, by A^{-1}, and the *adjoint* of A by $\tilde{A}$. The identity (or unit) matrix will be denoted by I. We will denote the determinant of a matrix A by det A. We note the following results from linear algebra.

Result 1   Given an $N \times N$ matrix A with a determinant det A which is nonzero, the inverse of A can be represented by

$$A^{-1} = \frac{1}{\det A} \tilde{A}$$

Result 2   The necessary and sufficient condition for a matrix to be singular is that det $A = 0$.

Using the notation described above, we may now write system (1) as

$$A\mathbf{x} = \mathbf{b} \tag{2}$$

where

$$A = \begin{bmatrix} a_{11} & \cdots & a_{1N} \\ \vdots & & \\ a_{M1} & \cdots & a_{MN} \end{bmatrix} \qquad \mathbf{x} = \begin{bmatrix} x_1 \\ \vdots \\ x_M \end{bmatrix} \qquad \mathbf{b} = \begin{bmatrix} b_1 \\ \vdots \\ b_M \end{bmatrix}$$

If we assume $M = N$, then we have the following important theorem.

Theorem 1 *If a solution to Eq. (2) exists, it is unique if and only if the matrix A is nonsingular. In this event, the solution is given by*

$$\mathbf{x} = A^{-1}\mathbf{b}$$

Further, if $\mathbf{b} = \mathbf{0}$ *(i.e. all the elements of* $\mathbf{b}$ *are zero), then the homogeneous system has a solution other than the trivial one* $\mathbf{x} = 0$, *if and only if A is singular.*

A classical method of obtaining the solution x, where it exists, is *Cramer's* rule. This method is a simple implementation of the result of Theorem 1. Let D_i, $i = 1, 2, \ldots, N$ be the N determinants formed by deleting the ith column from the determinant of A and replacing it by $\mathbf{b}$, namely

$$D_1 = \det \begin{bmatrix} b_1 & a_{12} & \cdots & a_{1N} \\ \vdots & & & \\ b_N & a_{N2} & \cdots & a_{NN} \end{bmatrix}$$

$$D_2 = \det \begin{bmatrix} a_{11} & b_1 & \cdots & a_{1N} \\ \vdots & & & \\ a_{N1} & b_N & \cdots & a_{NN} \end{bmatrix}$$

$$D_N = \det \begin{bmatrix} a_{11} & a_{12} & \cdots & b_1 \\ \vdots & & & \\ a_{N1} & a_{N2} & \cdots & b_N \end{bmatrix}$$

Then Cramer's rule gives the solution to Eq. (2) in terms of determinants, namely

$$x_i = \frac{D_i}{\det A} \qquad i = 1, 2, \ldots, N$$

where x_i, $i = 1, 2, \ldots, N$, are the elements of $\mathbf{x}$. For small values of N ($N = 2$ or 3) this would be a feasible method for the solution of the system (2). However, for values of N as small as 10, over 3,000,000 operations are required to evaluate the corresponding determinants and there are $N = 10$ such expressions to evaluate for the x_i. The arithmetic thus becomes prohibitive.† It is for this reason that numerical procedures are required as an alternative to Cramer's rule.

Exercises

1. Calculate the solution of the system of linear equations

$$3x_1 + 2x_2 - 3x_3 = -4$$

$$x_1 + x_2 - x_3 = 0$$

$$2x_1 - x_2 - x_3 = -7$$

using Cramer's rule.

---

† Forsythe [11] quotes the interesting example of the time required to solve a 25×25 system of linear equations using Cramer's rule programmed on an IBM 360: it would take 10^{17} years to perform just the multiplications alone!

2. Calculate the solution of the system of linear equations

$$3x_1 + 4x_2 - 2x_3 + x_4 = 6$$
$$2x_1 - 2x_2 + 3x_3 + x_4 = 4$$
$$-x_1 + x_2 - x_3 + x_4 = 0$$
$$2x_1 - 2x_2 + 3x_3 + 3x_4 = 6$$

using Cramer's rule.

2 Direct Methods of Solution for Full Systems

To consider a method of solution of a system of linear equations, let us first consider a simple example and follow the steps which arise in determining the solution. Consider the system of equations

$$x_1 + 2x_2 - 3x_3 = 2 \tag{3}$$

$$2x_1 - 2x_2 - x_3 = 1 \tag{4}$$

$$x_1 - x_2 - 2x_3 = -1 \tag{5}$$

Using Eq. (3), eliminate x_1 from equations (4) and (5). By subtracting $2 \times$ Eq. (3) from Eq. (4), we obtain

$$-6x_2 + 5x_3 = -5 \tag{6}$$

By subtracting Eq. (3) from Eq. (5) we obtain

$$-3x_2 + x_3 = -3 \tag{7}$$

Thus our original system has been reduced to the *equivalent* system

$$x_1 + 2x_2 - 3x_3 = 2$$
$$-6x_2 + 5x_3 = -3$$
$$-3x_2 + x_3 = -3$$

x_2 can now be eliminated from Eqs. (6) and (7) by subtracting $\frac{1}{2} \times$ Eq. (6) from Eq. (7). We then obtain

$$\frac{-3x_3}{2} = \frac{-3}{2} \tag{8}$$

Thus the original system has been *reduced* to the three equations

$$x_1 + 2x_2 - 3x_3 = 2$$
$$-6x_2 + 5x_3 = -3$$
$$\frac{-3x_3}{2} = \frac{-3}{2}$$

It is now a straightforward procedure to *back-substitute*. That is, from Eq. (8) we obtain $x_3 = 1$. This is substituted into Eq. (7) to give $x_2 = 4/3$. Finally, x_3 and x_2 can be substituted into Eq. (3) to obtain $x_1 = 7/3$.

This example has been solved by the procedure of systematic elimination, more commonly known as Gaussian elimination. It is clear that the same process can be applied to a more general matrix defining a system of equations which has more than three unknowns. To define this method for a system of N equations in N unknowns we proceed as before. Consider the matrix

$$A = \begin{bmatrix} a_{11} & \cdots & a_{1N} \\ \vdots & & \vdots \\ a_{N1} & \cdots & a_{NN} \end{bmatrix}$$

and the vectors

$$\mathbf{x} = \begin{bmatrix} x_1 \\ \vdots \\ x_N \end{bmatrix} \qquad \mathbf{b} = \begin{bmatrix} b_1 \\ \vdots \\ b_N \end{bmatrix}$$

Let us augment the matrix A by the vector $\mathbf{b}$ to give a new matrix A_b given by

$$A_b = \begin{bmatrix} a_{11} & \cdots & a_{1N} & b_1 \\ \vdots & & \vdots & \\ a_{N1} & \cdots & a_{NN} & b_N \end{bmatrix}$$

which is $(N \times (N+1))$. The first step of Gaussian elimination is to eliminate all the elements in the first column of A_b except a_{11}. This is effected by defining multipliers $m_{i1} = a_{i1}/a_{11}$, $i = 2, 3, \ldots, N$, and subtracting m_{i1} times the first row from row i, for $i = 2, 3, \ldots, N$. (Compare with the worked example above.) The resulting matrix is then of the form

$$\begin{bmatrix} a_{11} & & \cdots & a_{1N} & b_1 \\ 0 & a_{22} - \dfrac{a_{21}}{a_{11}} a_{12} & \cdots & a_{2N} - \dfrac{a_{21}}{a_{11}} a_{1N} & b_2 - \dfrac{a_{21}}{a_{11}} b_1 \\ \vdots & & & & \\ 0 & a_{N2} - \dfrac{a_{N1}}{a_{11}} a_{12} & \cdots & a_{NN} - \dfrac{a_{N1}}{a_{11}} a_{1N} & b_N - \dfrac{a_{N1}}{a_{11}} b_1 \end{bmatrix}$$

To save space we write this new matrix as

$$A_b^{(1)} = \begin{bmatrix} a_{11} & a_{12} & \cdots & a_{1N} & b_1 \\ 0 & a_{22}^{(1)} & \cdots & a_{2N}^{(1)} & b_2^{(1)} \\ \vdots & & & & \\ 0 & a_{N2}^{(1)} & \cdots & a_{NN}^{(1)} & b_N^{(1)} \end{bmatrix}$$

where the superscript (1) indicates that one elimination has been carried out.

We then repeat the process, now starting with the second row so that we further reduce the matrix $A_b^{(1)}$ by defining multipliers $m_{i2} = a_{i2}^{(1)}/a_{22}^{(1)}$, $i = 3$, $4, \ldots, N$, and by subtracting $m_{i2} \times$ second row from the ith row, $i = 3, 4, \ldots, N$. Following the above elimination we obtain the new matrix:

$$A_b^{(2)} = \begin{bmatrix} a_{11} & a_{12} & \cdots & & a_{1N} & b_1 \\ 0 & a_{22}^{(1)} & \cdots & & a_{2N}^{(1)} & b_2^{(1)} \\ 0 & 0 & a_{33}^{(2)} & \cdots & a_{3N}^{(2)} & b_3^{(2)} \\ \vdots & \vdots & \vdots & & \vdots & \vdots \\ 0 & 0 & a_{N3}^{(2)} & \cdots & a_{NN}^{(2)} & b_N^{(2)} \end{bmatrix}$$

where

$$a_{ij}^{(2)} = a_{ij}^{(1)} - m_{i2}a_{2j}^{(1)}, \qquad i = 3, 4, \ldots, N, \qquad j = 3, 4, \ldots, N \qquad (9)$$

and $b_i^{(2)}$ are similarly defined. We can repeat the process until we obtain a matrix of the form

$$A_b^{(N-1)} = \begin{bmatrix} a_{11} & a_{12} & \cdots & a_{1N} & b_1 \\ 0 & a_{22}^{(1)} & \cdots & a_{2N}^{(1)} & b_2^{(1)} \\ \vdots & \vdots & & \vdots & \vdots \\ 0 & 0 & & a_{NN}^{(N-1)} & b_N^{(N-1)} \end{bmatrix} \qquad (10)$$

provided that at no stage any of the *pivots* $a_{ii}^{(i-1)}$, $i = 1, 2, 3, \ldots, N$, vanish, and where at any stage,

$$a_{ij}^{(k)} = a_{ij}^{(k-1)} - m_{ik}a_{kj}^{(k-1)}$$

$$k = 1, 2, \ldots, N-1 \quad j = k+1, \ldots, N \quad i = k+1, \ldots, N \qquad (11)$$

$$\text{where } a_{ij}^{(0)} = a_{ij}$$

and

$$b_i^{(k)} = b_i^{(k-1)} - m_{ik}b_k^{(k-1)},$$

$$k = 1, 2, \ldots, N-1 \quad i = k+1, \ldots, N \quad \text{and where } b_i^{(0)} = b_i$$

where

$$m_{ik} = \frac{a_{ik}^{(k-1)}}{a_{kk}^{(k-1)}}$$

Having obtained this final form of the reduced matrix we can now back-substitute to obtain the solution vector **x**. Thus we find

$$x_N = \frac{b_N^{(N-1)}}{a_{NN}^{(N-1)}} \qquad (12)$$

which is then substituted into the penultimate row of Eq. (10) to evaluate

$$x_{N-1} = [b_{N-1}^{(N-2)} - a_{N-1N}^{(N-2)}x_N]/a_{N-1N-1}^{(N-2)}$$

or in general

$$x_i = \left[b_i^{(i-1)} - \sum_{j=i+1}^{N} a_{ij}^{(i-1)} x_j \right] \Big/ a_{ii}^{(i-1)} \qquad i = N-1, \ldots, 2, 1 \qquad (13)$$

Here x_N is given by Eq. (12).

The method described above is known as Gaussian elimination *without pivoting*. This is because at each stage we proceed hoping (or assuming) that the pivotal element, i.e. the $a_{ii}^{(i-1)}$, is nonzero. It is simple to construct examples which give such zero elements so that using this method for these examples the procedure would break down. To ensure such zero pivots are not used to calculate the multipliers at stage k, a search is made in column k for the element of maximum absolute value from rows $k, k+1, \ldots, N$. The row which contains the maximum-modulus column element is interchanged if necessary with row k. Following this criterion we are assured of obtaining a nonzero pivotal element unless all the elements of the column under consideration are zero, in which case the matrix is singular (why?). When the row interchanges have taken place, the algorithm continues as for the case of Gaussian elimination without pivoting. This method, employing the element of maximum modulus as pivot in a given column of A, is termed Gaussian elimination with *partial* pivoting.

Example 1 Solve the following system using Gaussian elimination with partial pivoting.

$$\begin{bmatrix} 1 & 0 & 2 & 3 \\ -1 & 2 & 2 & -3 \\ 0 & 1 & 1 & 4 \\ 6 & 2 & 2 & 4 \end{bmatrix} \begin{bmatrix} x_1 \\ x_2 \\ x_3 \\ x_4 \end{bmatrix} = \begin{bmatrix} 1 \\ -1 \\ 2 \\ 1 \end{bmatrix}$$

In the first column, 6 is the maximum (modulus) element and hence this element is used as pivot. We therefore interchange rows 1 and 4, remembering to interchange the elements of the right-hand side vector too. Thus

$$A_b = \begin{bmatrix} 1 & 0 & 2 & 3 & 1 \\ -1 & 2 & 2 & -3 & -1 \\ 0 & 1 & 1 & 4 & 2 \\ 6 & 2 & 2 & 4 & 1 \end{bmatrix} \quad \text{becomes} \quad \begin{bmatrix} 6 & 2 & 2 & 4 & 1 \\ -1 & 2 & 2 & -3 & -1 \\ 0 & 1 & 1 & 4 & 2 \\ 1 & 0 & 2 & 3 & 1 \end{bmatrix}$$

We then proceed to eliminate the element x_1 from rows 2, 3, and 4 by reducing to zero the elements -1, 0, 1 by subtracting the appropriate multiples of the first row of A_b; the reduced matrix is then

$$\begin{bmatrix} 6 & 2 & 2 & 4 & 1 \\ 0 & 7/3 & 7/3 & -7/3 & -5/6 \\ 0 & 1 & 1 & 4 & 2 \\ 0 & -1/3 & 5/3 & 7/3 & 5/6 \end{bmatrix}$$

To proceed further we note that the maximum (modulus) element in the second column is 7/3 which already occupies the pivotal position; hence no interchange of rows is required. To eliminate x_2 from rows 3 and 4 we subtract the appropriate multiples of the second row from rows 3 and 4. The resulting reduced matrix then becomes

$$\begin{bmatrix} 6 & 2 & 2 & 4 & 1 \\ 0 & 7/3 & 7/3 & -7/3 & -5/6 \\ 0 & 0 & 0 & 5 & 33/14 \\ 0 & 0 & 2 & 2 & 5/7 \end{bmatrix}$$

Finally, to eliminate x_3 using the last two rows of this matrix requires the interchange (using our present rules for pivoting) of the last two rows. This yields the reduced matrix

$$\begin{bmatrix} 6 & 2 & 2 & 4 & 1 \\ 0 & 7/3 & 7/3 & -7/3 & -5/6 \\ 0 & 0 & 2 & 2 & 5/7 \\ 0 & 0 & 0 & 5 & 33/14 \end{bmatrix}$$

We obviously need go no further since we have the required format for our matrix and back-substitution can proceed (see Exercise 3).

The third variant of Gaussian elimination is termed Gaussian elimination with *total* pivoting. This extension of partial pivoting is a simple idea; we determine the maximum-modulus element existing in both rows and columns. This element is then used as the pivot to eliminate the variable corresponding to this element of A_b. The detailed steps proceed as follows.

Determine the maximum-modulus element of A; this is used as the pivot. If the pivot corresponds to variable x_i, interchange the row in which the pivot is currently situated with row i; the pivot then becomes the diagonal element of row i. Subtract suitable multiples of the pivotal row from *each* of the remaining rows so as to reduce to zero each element in column i as described by Eq. (9). Repeat the process, where the pivot is chosen from the elements omitting those elements previously used as pivots. The resulting system on completion will be diagonal; no back-substitution will be necessary.

To clarify this algorithm with total pivoting, let us recompute the solution of Example 1 using the Gaussian elimination with total pivoting.

Example 2 Compute the solution of the system of linear equations given in Example 1 using Gaussian elimination with total pivoting. As before,

$$A_b = \begin{bmatrix} 1 & 0 & 2 & 3 & 1 \\ -1 & 2 & 2 & -3 & -1 \\ 0 & 1 & 1 & 4 & 2 \\ 6 & 2 & 2 & 4 & 1 \end{bmatrix} \quad \text{becomes} \quad \begin{bmatrix} 6 & 2 & 2 & 4 & 1 \\ -1 & 2 & 2 & -3 & -1 \\ 0 & 1 & 1 & 4 & 2 \\ 1 & 0 & 2 & 3 & 1 \end{bmatrix}$$

since 6 is the maximum (modulus) element. Using Eq. (9) we reduce the matrix as before to

$$\begin{bmatrix} 6 & 2 & 2 & 4 & 1 \\ 0 & 7/3 & 7/3 & -7/3 & -5/6 \\ 0 & 1 & 1 & 4 & 2 \\ 0 & -1/3 & 5/3 & 7/3 & 5/6 \end{bmatrix}$$

The maximum element now is 4, which corresponds to x_4. Hence, we interchange rows 3 and 4. Hence our matrix has become

$$\begin{bmatrix} 6 & 2 & 2 & 4 & 1 \\ 0 & 7/3 & 7/3 & -7/3 & -5/6 \\ 0 & -1/3 & 5/3 & 7/3 & 5/6 \\ 0 & 1 & 1 & 4 & 2 \end{bmatrix}$$

We now subtract suitable multiples of the last row from each of the other rows. We then obtain

$$\begin{bmatrix} 6 & 1 & 1 & 0 & -1 \\ 0 & 35/12 & 35/12 & 0 & 2/6 \\ 0 & -11/12 & 13/12 & 0 & -2/6 \\ 0 & 1 & 1 & 4 & 2 \end{bmatrix}$$

Our pivot is now 35/12 which already occupies the diagonal position, so that no interchange of rows is necessary. The reduced matrix now becomes

$$\begin{bmatrix} 6 & 0 & 0 & 0 & -39/35 \\ 0 & 35/12 & 35/12 & 0 & 2/6 \\ 0 & 0 & 2 & 0 & -8/35 \\ 0 & 0 & 0 & 4 & 66/35 \end{bmatrix}$$

The remaining element 2 is clearly the last pivot; the final matrix thus becomes

$$\begin{bmatrix} 6 & 0 & 0 & 0 & -39/35 \\ 0 & 35/12 & 0 & 0 & 2/3 \\ 0 & 0 & 2 & 0 & -8/35 \\ 0 & 0 & 0 & 4 & 66/35 \end{bmatrix}$$

The diagonal system is therefore obtained. The solution x can be obtained explicitly, namely

$$x_1 = -13/70 \qquad x_2 = 8/35 \qquad x_3 = -4/35 \qquad x_4 = 33/70$$

The algorithm we have just described is a variant of a method known as the Gauss–Jordan elimination. At first sight this algorithm might seem to be more efficient than the method with partial pivoting since we now do not have to implement the back-substitution. However, to offset this, we have had to perform additional operations in reducing to zero the elements situated above the diagonal. As we will see, this in fact entails more work, in general, than that associated with the back-substitution and consequently the Gauss–Jordan algorithm is not often used in favor of the partial pivoting algorithm.

Exercises

3. Calculate the solution of Example 1 using the reduced matrix as a basis for back-substitution.
4. Determine using three decimal arithmetic x_1 and x_2 using (a) Gaussian elimination without pivoting, (b) Gaussian elimination with partial pivoting in the following systems. How well do the solutions satisfy the equations?

$$\text{(i)} \quad \begin{bmatrix} 1 & 6 \\ 2 & 1 \end{bmatrix} \begin{bmatrix} x_1 \\ x_2 \end{bmatrix} = \begin{bmatrix} 3 \\ 1 \end{bmatrix}$$

$$\text{(ii)} \quad \begin{bmatrix} 0.43 & 1.1 \\ 0.48 & 2.3 \end{bmatrix} \begin{bmatrix} x_1 \\ x_2 \end{bmatrix} = \begin{bmatrix} 5.128 \\ 3.601 \end{bmatrix}$$

$$\text{(iii)} \quad \begin{bmatrix} -2.1 & 1.01 \\ 3.675 & 2.3 \end{bmatrix} \begin{bmatrix} x_1 \\ x_2 \end{bmatrix} = \begin{bmatrix} -2.1 \\ 3.1 \end{bmatrix}$$

5. Develop a flow diagram for Gaussian elimination without pivoting.
6. Modify the flow diagram of Exercise 5 to allow partial pivoting.
7. Develop the flow diagram for Gaussian elimination with total pivoting.
8. Construct a computer program for the flow diagram of Exercise 5.
9. Incorporate, as subroutines (FORTRAN) or procedures (ALGOL), the changes to the program of Exercise 8 to allow for partial pivoting.
10. Construct a computer program for the flow diagram of Exercise 7.
11. Use your programs of Exercises 8, 9, and 10 (for the purpose of comparison) to compute, if possible, the solutions of the following systems of equations. Comment on your results.

$$\text{(i)} \quad \begin{bmatrix} 1 & 1 & 1 & 1 \\ 1 & -1 & -1 & 1 \\ 2 & 2 & -2 & 1 \\ -1 & 2 & 3 & -4 \end{bmatrix} \begin{bmatrix} x_1 \\ x_2 \\ x_3 \\ x_4 \end{bmatrix} = \begin{bmatrix} 10 \\ 0 \\ 4 \\ -4 \end{bmatrix}$$

$$\text{(ii)} \quad \begin{bmatrix} 1.1 & -1.01 & -1 & -1.1 \\ 1 & -1 & -1 & -1 \\ 0.91 & 0.88 & 0.88 & 0.91 \\ 2 & 3 & -4 & 1 \end{bmatrix} \begin{bmatrix} x_1 \\ x_2 \\ x_3 \\ x_4 \end{bmatrix} = \begin{bmatrix} 2.19 \\ 2.00 \\ 0.00 \\ 8.00 \end{bmatrix}$$

$$\text{(iii)} \quad \begin{bmatrix} 2.2 & 3.2 & 4.5 & 6.0 & 4.5 \\ -1 & 1 & -1 & 1 & 1 \\ -2 & 3 & -4 & 5 & -1 \\ 5 & 3 & -1 & 2 & -3 \\ -4 & 8 & -3 & 2 & 2 \end{bmatrix} \begin{bmatrix} x_1 \\ x_2 \\ x_3 \\ x_4 \\ x_5 \end{bmatrix} = \begin{bmatrix} 16 \\ 0 \\ -4 \\ 4 \\ 3 \end{bmatrix}$$

$$\text{(iv)} \quad \begin{bmatrix} 1.000 & -1.633 & 0.9870 & 4.3211 \\ -0.0120 & 1.3400 & 1.6850 & 3.2810 \\ -5.6314 & 1.6840 & 0.0012 & -0.1430 \\ 0.0100 & 0.2000 & 0.3000 & -1.0000 \end{bmatrix} \begin{bmatrix} x_1 \\ x_2 \\ x_3 \\ x_4 \end{bmatrix} = \begin{bmatrix} 1.4130 \\ 2.6180 \\ -6.1341 \\ 0.3012 \end{bmatrix}$$

12. Use your programs of Exercises 9 and 10 to determine the solution of the system of linear equations of Example 2 of Chapter 1.

In the use of pivoting strategies advocated so far, we have suggested choosing the pivot on the basis of the largest-modulus element. If we assume this element corresponds to the diagonal element of the ith row of A_b, then the corresponding linear equation

$$\sum_{j=1}^{N} a_{ij}x_j = b_i$$

may be divided throughout by a (nonzero) constant α without essentially affecting the equation. However, by choosing α large enough we may now ensure that $|a_{ii}|$ is no longer large, i.e. we have changed the pivot without changing the equations. Thus by dividing by suitable constants we can in fact arrange to choose any nonzero element as pivot. The question therefore naturally arises: "Is there perhaps some best choice of pivot independent of divisors like α above?" Consider the following example.

Example 3 Consider determining the solution of the (2×2) system of equations

$$x_1 - x_2 = 2.0 \tag{14}$$

$$0.01x_1 + 20x_2 = -1.00051 \tag{15}$$

which has a true solution $x_1 = 1.949$; $x_2 = -0.051$. If we use two decimal arithmetic, with rounding, then using the first row for the pivot we find we obtain the solution

$$x_1 = 2.0, \quad x_2 = -0.05, \quad \text{"close" to the exact result}$$

If instead we consider Eq. (14) divided by 10^6, we then have

$$10^{-6}x_1 - 10^{-6}x_2 = 10^{-6} \times 2.0$$

The exact solution is of course unchanged. Using our previous strategy for choosing a pivot we would now favor Eq. (15). (If we still use Eq. (14) for the pivot we again obtain the previous result.) If we use Eq. (15) as the pivotal equation we obtain $x_1 = 0$, $x_2 = 0.05$. Multiplying Eq. (15) by 10^6 does not help. Thus the effect of round-off error, which arises in the calculations irrespective of the equation used as pivotal equation, has a disastrous effect (x_1 has no correct significant figures when Eq. (15) is used as pivotal equation), whereas when using Eq. (14) as pivotal equation the rounding error remains insignificant. Thus in this example it is not the absolute magnitude of the pivotal element which is important. Rather, we note that the coefficients of x_1 and x_2 in Eq. (14) are of the same magnitude, whereas in Eq. (15) they differ in magnitude by a factor of 200.

It is this relative size that is crucial in the prevention of the propagation of round-off error and it is this relative size we use to choose the pivotal element at any stage. Thus the choice of pivot is important in the prevention of the build-up of round-off error as well as the obvious prevention of zero pivots from being used.

In choosing a pivotal strategy, therefore, we must decide on the element which is large relative to the other elements in the row under consideration. Whereas there is not a unique choice of determining the maximum *relative* element,[†] we advocate choosing at the kth stage the pivotal element in row $i \geq k$ such that $|a_{ik}|/\|A_i\|_\infty$[‡] is a maximum, where A_i is the ith row of the matrix A. This method is then sometimes referred to as Gaussian elimination with *scaled* (partial/total) pivoting.

Exercises

13. Incorporate the alternative pivotal strategy given in Example 3 into your program of Exercise 9.
14. Repeat Exercise 13 for the program of Exercise 10.
15. Repeat Exercises 11 and 12 using the programs of Exercises 13 and 14. Compare your solutions with the earlier exercises.

Before continuing to describe other alternative direct methods, let us estimate the number of operations (additions, multiplications, etc.) necessary to compute the solution vector $\mathbf{x}$ by one of the methods advocated so far. In what follows we assume that the arithmetic unit of the computer deals with additions and subtractions in exactly the same time and likewise, multiplications and divisions are performed in the same time. Further since, approximately, a multiplication/division will always be accompanied by an addition/subtraction, we will count multiplications/divisions and refer to these as "operations."

To estimate the number of operations, we consider the elimination process (9) and the back-substitution process (12) and (13) for Gaussian elimination with partial pivoting.

For the first stage we see, using Eq. (9) that, leaving the first row unchanged, we have one division per row to obtain the multipliers $m_{i1} = a_{i1}/a_{11}$. Thus the total number of divisions at the first stage is $(N-1)$. In each row, m_{i1}, $i = 2, 3, \ldots, N$, multiplies each element for $j = 1, 2, \ldots, N$. Thus, there are $N(N-1)$ operations. Thus the total number of operations for the first stage of the elimination is $N(N-1) + (N-1)$ operations. Repeating this process, we see that at the kth stage of the elimination process we will have $(N-k+1)(N-k) + (N-k)$ operations. Thus since we have, in general, $(N-1)$ stages to complete in order to reduce the original matrix to its

---

[†] I.e. relative in what sense: relative to the next largest element; relative to the $\|\cdot\|_\infty$ of the elements of the row? etc.

[‡] $\|A_i\|_\infty$ is a notation, read as "norm ∞ of A_i," used to represent the maximum absolute value in the vector A_i. We will introduce and discuss these "norms" in section 7.

triangular form, the total number of operations will be

$$\sum_{k=1}^{N-1} (N-k+2)(N-k) \text{ operations}$$

Using the well-known formulas for summations, we find that we have

$$\tfrac{1}{3}N(N^2-1)+\tfrac{1}{2}N(N-1)$$

operations during the elimination process.

For the back-substitution we find that we need N divisions and $\tfrac{1}{2}N(N-1)$ multiplications. Hence the total number of operations is

$$\tfrac{1}{3}N^3+N^2 \text{ operations} \tag{16}$$

where we have omitted terms linear in N. For large N, the predominant term in this expression is $\tfrac{1}{3}N^3$ and hence we say that the Gaussian elimination method with partial pivoting requires approximately $\tfrac{1}{3}N^3$ operations to evaluate the solution of a system of N equations in N unknowns.

By a similar analysis it can be shown that the number of operations required by the Gauss–Jordan method is $\tfrac{1}{2}N^3$. Clearly when N is large and the N^3 terms predominate, the Gauss–Jordan method is *less* efficient than the Gaussian elimination with partial (or without) pivoting.

Exercises

16. Verify that approximately $\tfrac{1}{2}N^3$ operations are required by the Gauss–Jordan algorithm when used to solve a system containing N equations in N unknowns.
17. For what value of N would the Gauss–Jordan method be more efficient than the Gaussian elimination when including the lower powers of N in the above analysis?

3 *LU* Factorization of a Nonsingular Matrix

Definition A *triangular matrix* is a square matrix with either all elements above the main diagonal equal to zero or all elements below the main diagonal equal to zero. The former triangular matrix is called a lower triangular matrix; the latter an upper triangular matrix. Thus a lower triangular matrix L has the form

$$L = \begin{bmatrix} a_{11} & & & \\ a_{21} & a_{22} & & 0 \\ \vdots & & \ddots & \\ a_{N1} & a_{N2} & \cdots & a_{NN} \end{bmatrix}$$

0 denotes zero everywhere above the main diagonal and an upper triangular matrix U has the form

$$U = \begin{bmatrix} a_{11} & a_{12} & \cdots & a_{1N} \\ & a_{22} & \cdots & a_{2N} \\ & & \ddots & \vdots \\ 0 & & & a_{NN} \end{bmatrix}$$

0 denotes zero everywhere below the main diagonal.

56

Consider the 4×4 nonsingular matrix

$$A = \begin{bmatrix} a_{11} & a_{12} & a_{13} & a_{14} \\ a_{21} & a_{22} & a_{23} & a_{24} \\ a_{31} & a_{32} & a_{33} & a_{34} \\ a_{41} & a_{42} & a_{43} & a_{44} \end{bmatrix}$$

Let us *partition* A into

$$A = \begin{bmatrix} d_1 & \mathbf{a}_{12}^{\mathrm{T}} \\ \mathbf{a}_{21} & \bar{H}_1 \end{bmatrix}$$

where $d_1 = a_{11}$; $\mathbf{a}_{12}^{\mathrm{T}} = (a_{12}, a_{13}, a_{14})$; $\mathbf{a}_{21} = (a_{21}, a_{31}, a_{41})^{\mathrm{T}}$ and

$$\bar{H}_1 = \begin{bmatrix} a_{22} & a_{23} & a_{24} \\ a_{32} & a_{33} & a_{34} \\ a_{42} & a_{43} & a_{44} \end{bmatrix}$$

We may *factorize* A as

$$\begin{bmatrix} 1 & \mathbf{0}_3^{\mathrm{T}} \\ \mathbf{m}_{21} & I_3 \end{bmatrix} \begin{bmatrix} 1 & \mathbf{0}_3^{\mathrm{T}} \\ \mathbf{0}_3 & H_1 \end{bmatrix} \begin{bmatrix} d_1 & \mathbf{a}_{12}^{\mathrm{T}} \\ \mathbf{0}_3 & I_3 \end{bmatrix} \tag{17}$$

where $\mathbf{0}_3 = (0, 0, 0)^{\mathrm{T}}$, I_3 is the 3×3 identity matrix, where H_1 is to be determined, and $\mathbf{m}_{21} = \mathbf{a}_{21}/d_1$, where we assume $d_1 \neq 0$. Denoting the matrices in order left to right in Eq. (17) by L_1, A_1, U_1 respectively, where L_1 is a unit lower triangular matrix (the main diagonal elements are all unity) and U_1 is upper triangular.

Then $A = L_1 A_1 U_1$ if $H_1 = \bar{H}_1 - \mathbf{m}_{21} \mathbf{a}_{12}^{\mathrm{T}}$.

Let us write

$$H_1 = \begin{bmatrix} d_2 & \mathbf{a}_{23}^{\mathrm{T}} \\ \mathbf{a}_{32} & \bar{H}_2 \end{bmatrix}$$

where we assume $d_2 \neq 0$, and note that the elements of $\mathbf{a}_{23}$ and $\mathbf{a}_{32}$ are elements of H_1 (not A). $\bar{H}_2$ is a 2×2 matrix.

In a similar fashion to Eq. (17), we may factorize A_1 as

$$\begin{bmatrix} 1 & 0 & 0 & 0 \\ 0 & 1 & 0 & 0 \\ 0 & \mathbf{m}_{32} & & I_2 \\ 0 & & & \end{bmatrix} \begin{bmatrix} 1 & 0 & 0 & 0 \\ 0 & 1 & 0 & 0 \\ 0 & 0 & & H_2 \\ 0 & 0 & & \end{bmatrix} \begin{bmatrix} 1 & 0 & 0 & 0 \\ 0 & d_2 & \mathbf{a}_{23}^{\mathrm{T}} \\ 0 & 0 & & I_2 \\ 0 & 0 & & \end{bmatrix} \tag{18}$$

where $\mathbf{m}_{32} = \mathbf{a}_{32}/d$ and $H_2 = \bar{H}_2 - \mathbf{m}_{32} \mathbf{a}_{23}^{\mathrm{T}}$, which we write as

$$H_2 = \begin{bmatrix} d_3 & \mathbf{a}_{34}^{\mathrm{T}} \\ \mathbf{a}_{43} & \bar{H}_3 \end{bmatrix}$$

where $\mathbf{a}_{43}$ and $\mathbf{a}_{34}^{\mathrm{T}}$ are 1-vectors (scalars), as is $\bar{H}_3$. Denoting the matrices in order left to right in Eq. (18) by L_2, A_2, U_2 respectively, we see that

$$A = L_1 L_2 A_2 U_2 U_1$$

where L_2 is unit lower triangular and U_2 is upper triangular. Carrying this factorization one stage further,

$$A_2 = \begin{bmatrix} 1 & 0 & 0 & 0 \\ 0 & 1 & 0 & 0 \\ 0 & 0 & 1 & 0 \\ 0 & 0 & \mathbf{m}_{43} & 1 \end{bmatrix} \begin{bmatrix} 1 & 0 & 0 & 0 \\ 0 & 1 & 0 & 0 \\ 0 & 0 & 1 & 0 \\ 0 & 0 & 0 & H_3 \end{bmatrix} \begin{bmatrix} 1 & 0 & 0 & 0 \\ 0 & 1 & 0 & 0 \\ 0 & 0 & d_3 & \mathbf{a}_{34}^{\mathrm{T}} \\ 0 & 0 & 0 & 1 \end{bmatrix} \tag{19}$$

where $\mathbf{m}_{43} = \mathbf{a}_{43}/d_3$ and $H_3 = \bar{H}_3 - \mathbf{m}_{43}\mathbf{a}_{34}^{\mathrm{T}}$. Writing Eq. (19) as

$$A_2 = L_3 A_3 U_3$$

we see that A_3 is diagonal. We write

$$A_3 = IU_4 \quad \text{where} \quad U_4 = \begin{bmatrix} 1 & 0 & 0 & 0 \\ 0 & 1 & 0 & 0 \\ 0 & 0 & 1 & 0 \\ 0 & 0 & 0 & H_3 \end{bmatrix}$$

Hence we have *factorized* A so that

$$A = L_1 L_2 L_3 L_4 U_4 U_3 U_2 U_1$$

(where $L_4 = I$). Now consider the multiplication of

$$L_1 L_2 L_3 L_4$$

It is easily seen that

$$L = L_1 L_2 L_3 L_4 = \begin{bmatrix} 1 & 0 & 0 & 0 \\ m_{21} & 1 & 0 & 0 \\ m_{31} & m_{32} & 1 & 0 \\ m_{41} & m_{42} & m_{43} & 0 \end{bmatrix}$$

where we have assumed

$$\mathbf{m}_{21} = (m_{21}, m_{31}, m_{32})^{\mathrm{T}} \quad \mathbf{m}_{32} = (m_{32}, m_{42})^{\mathrm{T}} \quad \text{and} \quad \mathbf{m}_{43} = m_{43}$$

Similarly,

$$U = U_4 U_3 U_2 U_1 = \begin{bmatrix} d_1 & \mathbf{a}_{12} & & \\ & d_2 & \mathbf{a}_{23}^{\mathrm{T}} & \\ & & d_3 & \mathbf{a}_{34}^{\mathrm{T}} \\ & & & d_4 \end{bmatrix} \doteq \begin{bmatrix} a_{11} & a_{12} & a_{13} & a_{14} \\ 0 & a_{22}^{(1)} & a_{23}^{(1)} & a_{24}^{(1)} \\ 0 & 0 & a_{33}^{(2)} & a_{34}^{(2)} \\ 0 & 0 & 0 & a_{44}^{(3)} \end{bmatrix}$$

using the notation of Eq. (11).

That is, $A = LU$, where L is the unit lower triangular matrix comprising the multipliers arising in the Gaussian elimination algorithm described in the previous section, and U is the upper triangular matrix comprising the coefficients modified by the elimination process Eq. (11). Although we have derived only the factorization of the 4×4 matrix A, it is straightforward to see, assuming that none of the diagonal elements (d_i in our derivation) is zero, that

$$A = LU$$

where L is unit lower triangular and whose elements are the multipliers m_{ij} and U is upper triangular whose elements are given by Eq. (11). Hence in algorithmic form the factorization may be represented as follows:

comment LU decomposition of nonsingular matrix A assuming no pivoting;

> **for** $K \leftarrow 1$ **step** 1 **until** $N - 1$ **do**
>> **for** $i \leftarrow K + 1$ **step** 1 **until** N **do**
>> **begin** $m_{ik} \leftarrow a_{ik}^{(K-1)} / a_{kk}^{(K-1)}$; $a_{ik}^{(K)} \leftarrow m_{ik}$;
>>> **for** $j \leftarrow K + 1$ **step** 1 **until** N **do**
>>> $$a_{ij}^{(K)} \leftarrow a_{ij}^{(K-1)} - m_{iK} a_{Kj}^{(K-1)}$$
>> **end**;

Having obtained this factorization, the solution of the system of equations

$$A\mathbf{x} = \mathbf{b}$$

can be carried out by writing

$$LU\mathbf{x} = \mathbf{b}$$

and introducing $\mathbf{y} = U\mathbf{x}$ so that $L\mathbf{y} = \mathbf{b}$ and $U\mathbf{x} = \mathbf{y}$ are two triangular systems of equations which can be solved quite simply. The actual algorithm is easily found:

comment algorithms for the triangular systems

$$L\mathbf{y} = \mathbf{b}$$

$$U\mathbf{x} = \mathbf{y}$$

using the LU factorization of A;

$$y_1 = b_1;$$

> **for** $K \leftarrow 2$ **step** 1 **until** N **do**

$$y_K \leftarrow b_K - \sum_{j=1}^{K-1} m_{Kj} y_j;$$

$$x_N \leftarrow y_N;$$

> **for** $K \leftarrow N - 1$ **step** -1 **until** 1 **do**

$$x_K \leftarrow \left(y_K - \sum_{j=K+1}^{N} a_{Kj}^{(K-1)} x_j \right) \bigg/ a_{KK}^{(K-1)};$$

The *LU* factorization of *A* and the subsequent solving for **x** is exactly equivalent to the Gaussian elimination algorithm described in section 2. The main difference in implementation is that we here propose the *storing* of the multipliers $\{m_{ij}\}$. The importance of the *LU* factorization of a matrix comes about when we consider the following problem.

If the solution of the problem

$$AX = B \tag{20}$$

is required where

$$X \text{ is } N \times M, \qquad B = N \times M$$

and *A* is again our nonsingular $N \times N$ matrix whose factors are *L* and *U*, then Eq. (20) is just a convenient *notation* for the *M* systems of linear equations

$$A\mathbf{x}_i = \mathbf{b}_i \qquad i = 1, 2, \ldots, M \tag{21}$$

where $\mathbf{x}_i$ and $\mathbf{b}_i$ are the *i*th columns of *X* and *B* respectively.

Using the factorization of *A*, Eq. (21) is hence equivalent to solving

$$LU\mathbf{x}_i = \mathbf{b}_i \qquad i = 1, 2, \ldots, M \tag{22}$$

The number of operations needed to produce the factors *L* and *U* is easily seen to be $\frac{1}{3}N^3 + O(N)$. Equation (22) can be solved by writing

$$U\mathbf{x}_i = \mathbf{y}_i$$

and solving

$$\left.\begin{array}{l} L\mathbf{y}_i = \mathbf{b}_i \\ U\mathbf{x}_i = \mathbf{y}_i \end{array}\right\} \qquad i = 1, 2, \ldots, M \tag{23}$$

The number of operations required for solving for each $\mathbf{y}_i$ is $\frac{1}{2}N^2 + O(N)$, and similarly for $\mathbf{x}_i$ is $\frac{1}{2}N^2 + O(N)$.

Hence the number of operations to solve for *X*, having computed the *LU* factorization of *A*, will be

$$M \times N^2 + O(MN) \text{ operations}$$

and therefore the total number of operations is

$$\tfrac{1}{3}N^3 + MN^2 + O(MN) \tag{24}$$

If we had solved

$$A\mathbf{x}_i = \boldsymbol{b}_i \qquad i = 1, 2, \ldots, N$$

using the Gaussian elimination algorithm of section 2 we would have required

$$\frac{MN^3}{3} + MN^2 \tag{25}$$

operations; a considerable increase for any $M > 1$. Hence in the event that several systems of linear equations are required to be solved with the *same* coefficient matrix A, the LU factorization should be performed once and the solutions obtained through the relatively inexpensive Eq. (23).

The LU factorization of A has the added advantage, if the matrix A is not required further, of being efficient with regard to computer storage—the multipliers m_{ij}, as they are formed, may be stored in the (i, j) position of A. Hence the lower triangular matrix L occupies the lower triangle of A, and U the diagonal and upper triangle of A (no storage is required for the unit diagonal).

Example 4   Decompose the 4×4 matrix

$$A = \begin{bmatrix} 1 & 1 & 2 & -4 \\ 2 & -1 & 3 & 1 \\ 3 & 1 & -1 & 2 \\ 1 & -1 & -1 & -1 \end{bmatrix}$$

into its L and U factors and hence solve the system

$$A\mathbf{x} = (0, 5, 5, 0)^{\mathrm{T}}$$

For $K = 1$ evaluate

$$m_{21} = 2$$

$$m_{31} = 3$$

$$m_{41} = 1$$

so that at the first stage, A is reduced to

$$
\begin{array}{r}
A \rightarrow \\
m_{21} \\
m_{31} \\
m_{41}
\end{array}
\begin{bmatrix}
1 & 1 & 2 & -4 \\
\textcircled{2} & -3 & -1 & 9 \\
\textcircled{3} & -2 & -7 & 14 \\
\textcircled{1} & -2 & -3 & 5
\end{bmatrix}
$$

Then for $K = 2$, $m_{32} = 2/3$

$$m_{42} = 2/3$$

so that at the second stage

$$
\begin{array}{r}
A \rightarrow \\
\\
m_{32} \\
m_{42}
\end{array}
\begin{bmatrix}
1 & 1 & 2 & -4 \\
\textcircled{2} & -3 & -1 & 9 \\
\textcircled{3} & \textcircled{2/3} & -19/3 & 8 \\
\textcircled{1} & \textcircled{2/3} & -7/3 & -1
\end{bmatrix}
$$

Finally, $m_{43} = 7/19$

$$A \rightarrow \begin{bmatrix} 1 & 1 & 2 & -4 \\ ② & -3 & -1 & 9 \\ ③ & ②/③ & -19/3 & 8 \\ ① & ②/③ & ⑦/⑲ & -75/19 \end{bmatrix}$$

$$\underset{m_{43}}{\uparrow}$$

Hence we have

$$L = \begin{bmatrix} 1 & 0 & 0 & 0 \\ 2 & 1 & 0 & 0 \\ 3 & 2/3 & 1 & 0 \\ 1 & 2/3 & 7/19 & 1 \end{bmatrix} \quad \text{and} \quad U = \begin{bmatrix} 1 & 1 & 2 & -4 \\ 0 & -3 & -1 & 9 \\ 0 & 0 & -19/3 & 8 \\ 0 & 0 & 0 & -75/19 \end{bmatrix}$$

The problem is completed by solving the two triangular systems (23). For $L\mathbf{y} = \mathbf{b}$ we obtain, using the algorithm on page (58) for $K = 4$,

$$y_1 = 0$$

$$y_2 = b_2 - m_{21}y_1 = 5$$

$$y_3 = b_3 - m_{31}y_1 - m_{32}y_2 = \tfrac{5}{3}$$

$$y_4 = b_4 - m_{41}y_1 - m_{42}y_2 - m_{43}y_3 = \frac{-225}{57}$$

Similarly, for $U\mathbf{x} = \mathbf{y}$ we find

$$x_4 = \frac{-225/57}{-75/19} = 1$$

$$x_3 = \frac{y_3 - a_{34}^{(2)}x_4}{a_{33}^{(2)}} = 1$$

$$x_2 = \frac{y_2 - a_{24}^{(1)}x_4 - a_{34}^{(1)}x_3}{a_{22}^{(2)}} = 1$$

and

$$x_1 = \frac{y_1 - a_{14}x_4 - a_{13}x_3 - a_{12}x_2}{a_{11}} = 1$$

The required solution is $\mathbf{x}^T = (1, 1, 1, 1)$.

4 *PLU* Factorization of a Nonsingular Matrix

In this section we will generalize the *LU* factorization of the previous section in the direction of taking account of the pivotal strategy we described in section 2. We recall that at the kth stage (i.e. the elimination of x_k) we searched in column k for that row which contains the element a_{ik} such that $a_{ik}/\|A_i\|_\infty$ is largest. If the row i for which this ratio is largest is not the kth row, we advocated interchanging the ith row and the kth row. However, the simple copying of elements of arrays in high-level languages like FORTRAN and ALGOL is not achieved without cost. Consequently we advocate *accounting* which rows have been used as pivotal rows rather than physically interchanging elements of the coefficient matrix. We do this in the following manner.

Define an integer vector $\mathbf{p} = (p_1, p_2, \ldots, p_N)^T$ which is initialized to be $(1, 2, 3, \ldots, N)^T$.

Starting with row p_1, determine in column 1 the pivotal element. Assume this is situated in row p_l. Interchange the elements (i.e. their values) p_1 and p_l. At this first stage because of the initialization of $\mathbf{p}$, after this interchange,

$$\mathbf{p} = (l, 2, 3, \ldots, l-1, 1, l+1, \ldots, N)^T$$

Using row p_1 $(=l)$ as pivotal row, multipliers m_{i1}, $i = 1, 2, \ldots, N$, $i \neq p_1$, are formed and the elements in rows $i = 1, 2, \ldots, N$, $i \neq p_1$ are transformed by subtracting $m_{i1} \times a_{p_1 j}$ from each a_{ij}, $i = 1, 2, \ldots, N$, $i \neq p_1$, $j = 2, 3, \ldots, N$. We now repeat the process. In column 2 we search for that row $i = 1, 2, 3, \ldots, N$, $i \neq p_1$, with a_{i2} such that $a_{i2}/\|A_i\|_\infty$ is largest. Say this occurs in row p_r; then interchange p_2 and p_r $(=r$ say$)$.

i.e. $\qquad \mathbf{p} = (l, r, 2, 3, \ldots, l-1, 1, l+1, \ldots, r-1, 2, r+1, \ldots, N)^T$

Recall the values of $p_1 = l$

$$p_2 = r \quad \text{etc.}$$

i.e. the *names* of the elements in p are unchanged; just the values are interchanged.

Now define multipliers m_{i2}, $i = 1, 2, \ldots, N$, $i \neq p_1$, $i \neq p_2$ and subtract $m_{i2} \times a_{p_2 j}$ from each a_{ij}, $i = 1, 2, \ldots, N$; $i \neq p_1$; $i \neq p_2$; $j = 3, 4, \ldots, N$. Thus at stage k, the vector $\mathbf{p}$ contains

$$\mathbf{p} = (p_1, p_2, \ldots, p_k, p_{k+1}, \ldots, p_N)$$

where the values of the first $k-1$ elements indicate the indices of the rows which have been previously used as pivotal rows. Hence in column k, rows $p_k, \ldots, p_N$ search for the pivotal row; interchanging (if necessary) the elements of $\mathbf{p}$. Form the multipliers m_{ik}, $i = p_{k+1}, \ldots, p_N$, and transform the elements a_{ij}, $i = p_{k+1}, \ldots, p_N$, $j = k+1, \ldots, N$ according to the equation

$$a_{ij} \leftarrow a_{ij} - m_{ik} a_{p_k j}.$$

Hence in algorithmic form we have the following:

comment initialize $\mathbf{p}$;
$\qquad$ **for** $i \leftarrow 1$ **step** 1 **until** N **do**
$\qquad\qquad p(i) \leftarrow i$;

comment initialize vector **d** to contain $\|A_i\|_\infty$;
    for $i \leftarrow 1$ **step** 1 **until** N **do**
    begin amax $\leftarrow 0.0$;
        for $j \leftarrow 1$ **step** 1 **until** N **do**
        if amax $< |a(i, j)|$ **then** amax $\leftarrow |a(i, j)|$;
        if amax $= 0.0$ **then begin** print singular matrix; **goto** finish
                end

        $d(i) \leftarrow$ amax
    end;

comment LU factorization
    for $k \leftarrow 1$ **step** 1 **until** $N - 1$ **do**

"Find the smallest $j \geq k$ such that

$$\frac{|a_{p_j k}^{(k-1)}|}{d_{p_j}} > \frac{|a_{ik}^{(k-1)}|}{d_i} \qquad i = p_k, p_{k+1}, \ldots, p_N$$

Interchange p_k and p_j in **p**";

    for $i \leftarrow k + 1$ **step** 1 **until** N **do**
    begin

$$m_{p_i k} \leftarrow a_{p_i k}^{(k-1)} / a_{p_k k}^{(k-1)};$$

        for $j \leftarrow k + 1$ **step** 1 **until** N **do**

$$a_{p_i j}^{(k)} \leftarrow a_{p_i j}^{(k-1)} - m_{p_i k} a_{p_k j}^{(k-1)}$$

    end;

Example 5 Using the above algorithm, let us obtain the decomposition of the matrix A of Example 4.

Initially $\mathbf{p} = (1, 2, 3, 4)^T \mathbf{d} = (4, 3, 3, 1)^T$
For $k = 1$: Find the pivot; in the first column of A the appropriate ratios are

$$\frac{|a_{11}|}{d_1} = \frac{1}{4} \qquad \frac{|a_{21}|}{d_2} = \frac{2}{3} \qquad \frac{|a_{31}|}{d_3} = \frac{3}{3} \qquad \frac{|a_{41}|}{d_4} = \frac{1}{1}$$

The smallest row index is $i = 3$.

Hence interchange p_1 and p_3 so that

$$\mathbf{p} = (3, 2, 1, 4)^T$$

Form the multipliers:

$$m_{11} = \frac{a_{11}}{a_{31}} = \frac{1}{3} \qquad m_{21} = \frac{a_{21}}{a_{31}} = \frac{2}{3} \qquad m_{41} = \frac{a_{41}}{a_{31}} = \frac{1}{3}$$

Calculate the elements $a_{p_{i}j}^{(1)}$;

$$A \rightarrow \begin{bmatrix} \boxed{1/3} & 2/3 & 7/3 & -14/3 \\ \boxed{2/3} & -5/3 & 11/3 & -1/3 \\ 3 & 1 & -1 & 2 \\ \boxed{1/3} & -4/3 & -2/3 & 1/3 \end{bmatrix}$$

where the circled entries are the multipliers. For $k = 2$: Find the pivot in the second column from rows p_2, p_3, p_4, namely from rows 2, 1, and 4.

The appropriate ratios are

$$\frac{|a_{12}^{(1)}|}{d_1} = \frac{2/3}{4} \qquad \frac{|a_{22}^{(1)}|}{d_2} = \frac{5/3}{3} \qquad \frac{|a_{42}^{(1)}|}{d_4} = \frac{4/3}{1}$$

Hence the pivotal row is row $p_4(=4)$, and hence interchange p_2 and p_4 so that $\mathbf{p} = (3, 4, 1, 2)^{\mathrm{T}}$.

Form the multipliers

$$m_{12} = \frac{a_{12}^{(1)}}{a_{42}^{(1)}} = -\frac{1}{2} \qquad m_{22} = \frac{a_{22}^{(1)}}{a_{42}^{(1)}} = \frac{5}{4}$$

Calculate the elements $a_{p_{i}j}^{(2)}$: Hence

$$A \rightarrow \begin{bmatrix} \boxed{1/3} & \boxed{-1/2} & 2 & -9/2 \\ \boxed{2/3} & \boxed{5/4} & 9/2 & -3/4 \\ 3 & 1 & -1 & 2 \\ \boxed{1/3} & -4/3 & -2/3 & 1/3 \end{bmatrix}$$

For $k = 3$: Find the pivot in the third column from rows p_3 and p_4, namely rows 1 and 2.

The appropriate ratios are

$$\frac{|a_{13}^{(2)}|}{d_1} = \frac{1}{2} \qquad \frac{|a_{23}^{(2)}|}{d_2} = \frac{3}{2}$$

Hence row 2 ecomes the pivotal row. Interchange p_3 and p_4 so that the *final* $\mathbf{p}$ becomes

$$\mathbf{p} = (3, 4, 2, 1)^{\mathrm{T}}$$

The multiplier $m_{13} = 4/9$ and then

$$A \rightarrow \begin{bmatrix} \boxed{1/3} & \boxed{-1/2} & \boxed{4/9} & -25/6 \\ \boxed{2/3} & \boxed{5/4} & 9/2 & -3/4 \\ 3 & 1 & -1 & 2 \\ \boxed{1/3} & -4/3 & -2/3 & 1/3 \end{bmatrix}$$

The decomposition is complete.

An inspection of the final matrix in Example 5 indicates clearly that no obvious lower triangular matrix L of multipliers or upper triangular matrix U are evident in the final storage locations of the original A. However, note that if we extract the rows in the order 3, 4, 2, 1, and write the lower triangular matrix L as that matrix comprising the circled multipliers, and 1's on the diagonal and U the upper triangular matrix of what is left, we find

$$L = \begin{bmatrix} 1 & 0 & 0 & 0 \\ 1/3 & 1 & 0 & 0 \\ 2/3 & 5/4 & 1 & 0 \\ 1/3 & -1/2 & 4/9 & 1 \end{bmatrix} \quad \text{and} \quad U = \begin{bmatrix} 3 & 1 & -1 & 2 \\ 0 & -4/3 & -2/3 & 1/3 \\ 0 & 0 & 9/2 & -3/4 \\ 0 & 0 & 0 & -25/6 \end{bmatrix}$$

Multiplying L and U we find

$$LU = \begin{bmatrix} 3 & 1 & -1 & 2 \\ 1 & -1 & -1 & 1 \\ 2 & -1 & 3 & 1 \\ 1 & 1 & 2 & -4 \end{bmatrix}$$

which is the matrix A written with its rows rearranged in order 3, 4, 2, and 1. That is, the factors L and U produce a matrix B, say, which is a *permutation* of the rows of the original matrix A. If we define a *permutation matrix*

$$P^{\text{T}} = \begin{bmatrix} 0 & 0 & 1 & 0 \\ 0 & 0 & 0 & 1 \\ 0 & 1 & 0 & 0 \\ 1 & 0 & 0 & 0 \end{bmatrix} \equiv \begin{bmatrix} \mathbf{e}_3^{\text{T}} \\ \mathbf{e}_4^{\text{T}} \\ \mathbf{e}_2^{\text{T}} \\ \mathbf{e}_1^{\text{T}} \end{bmatrix}$$

where $\mathbf{e}_i$ is the unit vector with 1 in the ith position and 0 in each remaining position, then

$$P^{\text{T}}A = \begin{bmatrix} 0 & 0 & 1 & 0 \\ 0 & 0 & 0 & 1 \\ 0 & 1 & 0 & 0 \\ 1 & 0 & 0 & 0 \end{bmatrix} \begin{bmatrix} 1 & 1 & 2 & -4 \\ 2 & -1 & 3 & 1 \\ 3 & 1 & -1 & 2 \\ 1 & -1 & -1 & 1 \end{bmatrix} = \begin{bmatrix} 3 & 1 & -1 & 2 \\ 1 & -1 & -1 & 1 \\ 2 & -1 & 3 & 1 \\ 1 & 1 & 2 & -4 \end{bmatrix}$$

$$= LU$$

Hence we have factored the matrix A into a factorization which has the form

$$A = P^{-\text{T}}LU$$

where $P^{-\text{T}}$ denotes the inverse of P^{T}. But the permutation matrix P^{T} has the property that

$$PP^{\text{T}} = I$$

so that $P^{-\text{T}} = P$. Thus $A = PLU$.

Thus the effect of the pivotal strategy is to produce the triangular factors L and U with their rows "out of order". The rows of L and U may be accessed according to the information contained in the vector $\mathbf{p}$. Namely, since

$$A\mathbf{x} = \mathbf{b}$$

we have $PLU\mathbf{x} = \mathbf{b}$ or $LU\mathbf{x} = P^T\mathbf{b}$. We therefore require to solve

$$L\mathbf{y} = P^T\mathbf{b}$$

$$U\mathbf{x} = \mathbf{y}$$

where

$$P^T\mathbf{b} = \begin{bmatrix} b_3 \\ b_4 \\ b_2 \\ b_1 \end{bmatrix}$$

produces the vector whose elements are the elements of $\mathbf{b}$ accessed in the order 3, 4, 2, 1, i.e. p_1, p_2, p_3, p_4. The elements of

$$L = \begin{bmatrix} 1 & 0 & 0 & 0 \\ 1/3 & 1 & 0 & 0 \\ 2/3 & 5/4 & 1 & 0 \\ 1/3 & -1/2 & 4/9 & 1 \end{bmatrix}$$

are obtained by accessing the elements in the "storage area" A in rows in order p_1, p_2, p_3, and p_4. Likewise, the rows of

$$U = \begin{bmatrix} 3 & 1 & -1 & 2 \\ 0 & -4/3 & -2/3 & 1/3 \\ 0 & 0 & 9/2 & -3/4 \\ 0 & 0 & 0 & -25/6 \end{bmatrix}$$

are obtained from the storage area A in the order p_1, p_2, p_3, p_4. Hence the solution $\mathbf{x}$ is obtained by adding the following algorithmic statements to the algorithm on page 58.

comment forward substitution for $L\mathbf{y} = P^T\mathbf{b}$;

$$y_1 = b_{p_1};$$

for $k \leftarrow 2$ **step** 1 **until** N **do**

$$y_k \leftarrow b_{p_k} - \sum_{j=1}^{k-1} a_{p_k j}^{(k-1)} y_j;$$

comment back substitution for $U\mathbf{x} = \mathbf{y}$

$$x_N = \frac{y_N}{a_{p_N N}^{(N-1)}};$$

for $k \leftarrow N-1$ **step** -1 **until** 1 **do**

$$x_k \leftarrow \frac{y_k - \sum\limits_{j=k+1}^{N} a_{p_k j}^{(k-1)} x_j}{a_{p_k k}^{(k-1)}} \; ;$$

Exercises

18. Implement the algorithm for Gaussian elimination as described on pages 62 and 66.
19. Verify your program by computing the solution to the problem of Example 4.
20. An engineer asks you to compute the inverse of the matrix of Example 4, using your program of Exercise 18. You set about the problem in the following way:
    Decompose A into its PLU factorization. Then solve

    $$PLU\mathbf{x}_i = \mathbf{e}_i \qquad i = 1, 2, \ldots, 4;$$

    the ith vector $\mathbf{x}_i$ being the ith column of the inverse of A.
    Modify your program of Exercise 18 (in a minor way) to accommodate the engineer.
    How many operations are needed to find A^{-1} in this manner? If you had used the Gaussian elimination program of Exercise 9, how many operations would have been required to calculate A^{-1}?
21. When solving $A\mathbf{x} = \mathbf{b}$ why is it (much) better to *solve* the system of linear equations than to find A^{-1} and then form $\mathbf{x}$ as

    $$\mathbf{x} = A^{-1}\mathbf{b}?$$

5 Band Matrices

In certain circumstances the structure of the given matrix A allows a special form of Gaussian elimination to be developed. A case which arises often in numerical analysis is the problem of solving systems of linear equations whose coefficient matrix is *tridiagonal*.

Definition A square matrix A with elements a_{ij} is said to be *tridiagonal* if

$$a_{ij} = 0 \quad \text{for } |i - j| > 1$$

We note that in a tridiagonal matrix all elements of the matrix are zero except possibly for those situated on the *main* diagonal, the *sub*diagonal, and the *super*diagonal.

Example 6 The following matrices are tridiagonal:

$$\begin{bmatrix} 1 & 2 & 0 & 0 \\ 4 & 1 & 1 & 0 \\ 0 & 1 & 1 & 1 \\ 0 & 0 & 2 & 1 \end{bmatrix} \quad \begin{bmatrix} 1 & 0 & 0 & 0 \\ 2 & 0 & 4 & 0 \\ 0 & 0 & 1 & 1 \\ 0 & 0 & 3 & 0 \end{bmatrix}$$

In the case where our matrix is of this special form we will find that we can take into account the special structure of A so as to produce an efficient implementation of the Gaussian elimination whereby time is not spent (as would be the case with the standard algorithms described so far) in eliminating elements which are already zero.

For the moment let us assume pivoting is not necessary. Let us write the tridiagonal matrix A as

$$A = \begin{bmatrix} a_{11} & a_{12} & & & \\ a_{21} & a_{12} & a_{23} & & \\ & \ddots & & \ddots & a_{N-1\,N} \\ & & a_{NN-1} & a_{NN} \end{bmatrix}$$

Assuming no pivoting, the LU factorization of A requires at each stage that just *one* multiplier be evaluated and that only the diagonal element be transformed. Hence only the following simple modification to the LU factorization algorithm is necessary:

for $i \leftarrow k + 1$ **do**
  begin

$$m_{p_i k} \leftarrow a_{p_i k}^{(k-1)} / a_{p_k k}^{(k-1)};$$

  for $j \leftarrow k + 1$ **do**

$$a_{p_i j}^{(k)} \leftarrow a_{p_i j}^{(k-1)} - m_{p_i k} a_{p_k j}^{(k-1)}$$

  end;

The result of this factorization is to produce factors L and U of the form

$$L = \begin{bmatrix} 1 & & & & \\ m_{21} & 1 & & & \\ & m_{32} & 1 & & \\ & & \ddots & \ddots & \\ & & & m_{NN-1} & 1 \end{bmatrix} \qquad U = \begin{bmatrix} a_{11} & a_{12} & & & \\ & a_{22}^{(1)} & a_{23} & & \\ & & \ddots & \ddots & a_{N-1\,N} \\ & & & & a_{NN}^{(N-1)} \end{bmatrix}$$

The introduction of pivoting does not complicate matters too greatly. To see the effect of pivoting, consider the first stage of the factorization. Between rows 1 and 2 we pick the pivotal row. Assume the second row is the required pivotal row. Physically interchange these two rows.† Then the matrix A becomes

$$\begin{bmatrix} a_{21} & a_{22} & a_{23} & & \\ a_{11} & a_{12} & 0 & & \\ & a_{32} & a_{33} & a_{34} & \\ & & \ddots & \ddots & \ddots \end{bmatrix}$$

---

† We know how to avoid the physical interchange don't we? However, for description purposes we will assume we actually interchange the row.

That is, we now have a matrix with two upper diagonals. If we performed this interchange at every stage we would in effect be factorizing a matrix A with *one* subdiagonal and *two* superdiagonals. Hence the simple modification to our LU factorization would be:

for $i \leftarrow k + 1$ **do**
   begin

$$m_{p_i k} \leftarrow a_{p_i k}^{(k-1)} / a_{p_k k}^{(k-1)};$$

   for $j \leftarrow k + 1,\ k + 2$ **do**

$$a_{p_i j}^{(k)} \leftarrow a_{p_i j}^{(k-1)} - m_{p_i k} a_{p_k j}^{(k-1)};$$

We can now generalize the case of a tridiagonal matrix to a general *band* matrix.

Consider a nonsingular matrix A to possess q lower diagonals and r upper diagonals and a main diagonal as shown below:

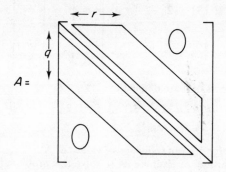

Assume that pivoting is necessary. Then the algorithm for forming the PLU factorization of A is:

   for $k \leftarrow 1$ **step** 1 **until** $N - 1$ **do**

"Find the smallest $j \geq k$ such that

$$\frac{|a_{p_j k}^{(k-1)}|}{d_{p_j}} > \frac{|a_{ik}^{(k-1)}|}{d_i} \qquad i = p_k, p_{k+1}, \ldots, p_{k+q}$$

interchange p_k and p_j in **p**"

   for $i \leftarrow k + 1$ **step** 1 **until** $\min\{k + q, N\}$ **do**
      begin

$$m_{p_i k} \leftarrow a_{p_i k}^{(k-1)} / a_{p_k k}^{(k-1)};$$

      for $j \leftarrow k + 1$ **step** 1 **until** $\min\{k + q + r, N\}$ **do**

$$a_{p_i j}^{(k)} \leftarrow a_{p_i j}^{(k-1)} - m_{p_i k} a_{p_k j}^{(k-1)}$$

     end;

70

We leave to an exercise the minor changes needed for the "solution algorithm."

Exercises

22. Modify the "solution-algorithm" on page 66 for the special case of band matrices with q subdiagonals and r superdiagonals.
23. Assuming the worst-case situation where at each stage pivoting indicates that the current row is to be interchanged with the row placed r rows away, find the number of operations necessary to factorize and solve the system

$$A\mathbf{x} = \mathbf{b}$$

in terms of the order N and the band widths q and r of A.
24. Implement your algorithm of Exercise 22 and the factorization algorithm given on page 69.
25. It is required to solve the tridiagonal system

$$\begin{bmatrix} 2 & -1 & & & \\ -1 & 2 & -1 & & \\ & \ddots & \ddots & \ddots & \\ & & -1 & 2 & -1 \\ & & & -1 & 1 \end{bmatrix} \begin{bmatrix} x_1 \\ x_2 \\ \vdots \\ x_{N-1} \\ x_N \end{bmatrix} = \begin{bmatrix} b_1 \\ b_2 \\ \vdots \\ b_{N-1} \\ b_N \end{bmatrix}$$

where the elements of $\mathbf{b}$ are defined by

$$b_i = 2h^3(2N - 3i) \qquad i = 1, 2, \ldots, N-1$$
$$b_N = (1 - N)h^3$$

where $h = \pi/N$ for a given sequence of values of N.

Use your algorithm of Exercise 24 to solve this problem for $N = 5, 10, 20,$ and 40.

If your computer center has a *timing routine*, does the behavior with N of the time taken to solve these systems agree with your estimate of the number of operations required given in Exercise 23?
26. Use your algorithm of Exercise 24 to solve the following system of equations for $N = 10, 20, 40$:

$$\begin{bmatrix} 6 & -4 & 1 & & & \\ -4 & 6 & -4 & 1 & & \\ 1 & -4 & 6 & -4 & 1 & \\ & \ddots & \ddots & \ddots & \ddots & 1 \\ & & & & & -4 \\ & & & 1 & -4 & 6 \end{bmatrix} \begin{bmatrix} x_1 \\ x_2 \\ \vdots \\ \\ x_N \end{bmatrix} = 24/N^4 \begin{bmatrix} 1 \\ 1 \\ \vdots \\ \\ 1 \end{bmatrix}$$

6 Symmetric Positive-Definite Matrices

In many applications, matrices occur which are symmetric, i.e.

$$A^T = A$$

In such cases the storage required to represent the matrix is reduced from the N^2 locations needed for a general $N \times N$ matrix to the $\frac{1}{2}N(N+1)$ locations needed to store *either* the lower triangle *or* the upper triangle of A, and the main diagonal. For large N this storage saving can be significant.† In the LU (or PLU) decomposition of a symmetric matrix A we have (ignoring pivoting)

$$A = LU$$

Since L is unit diagonal, and, in general U is not, it is clear that $U \neq L^T$. That is, if the LU decomposition algorithm is applied to a general symmetric matrix A, then the triangular factors L and U will not be the transposes of one another. Consequently, the full N^2 storage locations are necessary for the storage of L and U. The question naturally arises: can we determine a factorization of $A = LL^T$ (the L here is not the *same* lower triangular matrix as occurs in the LU factorization). If this *is* possible then the decomposition of A will also only need $\frac{1}{2}N(N+1)$ storage locations. Assume we have performed the LU decomposition of A using the algorithm of section 3, i.e.

$$A = LU$$

Denote the diagonal of U as D and then write

$$U = D\tilde{U}$$

where the element $\tilde{u}_{ij}$ of $\tilde{U}$ are u_{ij}/u_{ii}, u_{ij} being the elements of U. Hence

$$A = LD\tilde{U}$$

Consider a 4×4 symmetric matric

$$A = \begin{bmatrix} a_{11} & a_{12} & a_{13} & a_{14} \\ a_{12} & a_{22} & a_{23} & a_{24} \\ a_{13} & a_{23} & a_{33} & a_{34} \\ a_{14} & a_{24} & a_{34} & a_{44} \end{bmatrix}$$

Partition A in a manner similar to that employed in the derivation of the LU factorization in section 3. Namely,

$$A = \begin{bmatrix} d_1 & \mathbf{a}_{12}^T \\ \mathbf{a}_{12} & \bar{H}_1 \end{bmatrix}$$

where $d_1 = a_{11}$ and

$$\bar{H}_1 = \begin{bmatrix} a_{22} & a_{23} & a_{24} \\ a_{23} & a_{33} & a_{34} \\ a_{24} & a_{34} & a_{44} \end{bmatrix}$$

---

† In many computer centers computer costs to the user are related to the amount of main memory storage used during a computation, so that reducing main memory requirements actually reduces the real costs to the user.

is symmetric. Assuming pivoting is not necessary and following section 3,

$$A = \begin{bmatrix} 1 & 0 & 0 & 0 \\ \mathbf{m}_{21} & & I_3 & \end{bmatrix} \begin{bmatrix} d_1 & 0 & 0 & 0 \\ 0 & & & \\ 0 & & H_1 & \\ 0 & & & \end{bmatrix} \begin{bmatrix} 1 & & \mathbf{m}_{21}^{T} \\ 0 & & & \\ 0 & & I_3 & \\ 0 & & & \end{bmatrix}$$

$$\equiv \qquad L_1 \qquad\qquad A_1 \qquad\qquad L_1^{T}$$

where

$$\mathbf{m}_{21} = (m_{21}, m_{31}, m_{41})^{T} = (a_{12}/d_1, a_{13}/d_1, a_{14}/d_1)^{T}$$

and $H_1 = \bar{H}_1 - \mathbf{a}_{12}\mathbf{a}_{12}^{T}/d_1$ is symmetric. Write

$$H_1 = \begin{bmatrix} d_2 & \mathbf{a}_{23}^{T} \\ \mathbf{a}_{23} & \bar{H}_2 \end{bmatrix}$$

where $\bar{H}_2$ is symmetric. We may therefore decompose H_1 in a similar manner to A. Thus

$$A_1 = \begin{bmatrix} 1 & 0 & 0 & 0 \\ 0 & 1 & 0 & 0 \\ 0 & \mathbf{m}_{32} & I_2 & \\ 0 & & & \end{bmatrix} \begin{bmatrix} d_1 & 0 & 0 & 0 \\ 0 & d_2 & 0 & 0 \\ 0 & 0 & H_2 & \\ 0 & 0 & & \end{bmatrix} \begin{bmatrix} 1 & 0 & 0 & 0 \\ 0 & 1 & \mathbf{m}_{32}^{T} & \\ 0 & 0 & I_2 & \\ 0 & 0 & & \end{bmatrix}$$

$$= \qquad L_2 \qquad\qquad A_2 \qquad\qquad L_2^{T}$$

where $\mathbf{m}_{32} = \mathbf{a}_{23}/d_2$ and $H_2 = \bar{H}_2 - \mathbf{a}_{23}\mathbf{a}_{23}^{T}/d_2$ is symmetric. Once again, write

$$H_2 = \begin{bmatrix} d_3 & \mathbf{a}_{34}^{T} \\ \mathbf{a}_{34} & \bar{H}_3 \end{bmatrix}$$

where (in fact) $\mathbf{a}_{34}$ and $\bar{H}_3$ are scalars. Then

$$A_2 = \begin{bmatrix} 1 & 0 & 0 & 0 \\ 0 & 1 & 0 & 0 \\ 0 & 0 & 1 & 0 \\ 0 & 0 & m_{43} & 1 \end{bmatrix} \begin{bmatrix} d_1 & 0 & 0 & 0 \\ 0 & d_2 & 0 & 0 \\ 0 & 0 & d_3 & 0 \\ 0 & 0 & 0 & d_4 \end{bmatrix} \begin{bmatrix} 1 & 0 & 0 & 0 \\ 0 & 1 & 0 & 0 \\ 0 & 0 & 1 & m_{43} \\ 0 & 0 & 0 & 1 \end{bmatrix}$$

$$= L_3 \qquad\qquad = A_3 \qquad\qquad = L_3^{T}$$

where $d_4 = \bar{H}_3 - \mathbf{a}_{34}\mathbf{a}_{34}^{T}/d_3$. Thus we have factorized A so that

$$A = L_1 L_2 L_3 D L_3^{T} L_2^{T} L_1^{T}$$

where $D \equiv A_3$ is a diagonal matrix. Performing the products $L_1 L_2 L_3$ (and $L_3^{T} L_2^{T} L_1^{T}$), we have

$$A = LDL^{T} \qquad\qquad\qquad (26)$$

where

$$L = \begin{bmatrix} 1 & 0 & 0 & 0 \\ m_{21} & 1 & 0 & 0 \\ m_{31} & m_{32} & 1 & 0 \\ m_{41} & m_{42} & m_{43} & 1 \end{bmatrix}$$

Here L is precisely the matrix L arising in the LU factorization of section 3. Hence $U = DL^{\mathrm{T}}$.

This result tells us that, *if* we had carried out the LU factorization of a symmetric A then we could a posteori obtain the symmetric factorization by factoring out the diagonal matrix and redefining the elements of U ($\rightarrow \tilde{U}$). However, this is not what we require—the upper triangle will have been used to store U! If we make the further assumption of A that

$$\text{the diagonal elements of } D \text{ are positive} \qquad (27)$$

then we can write

$$D = D^{1/2} D^{1/2}$$

where

$$D^{1/2} \equiv \begin{bmatrix} d_1^{1/2} & & & \mathbf{0} \\ & d_2^{1/2} & & \\ & & d_3^{1/2} & \\ \mathbf{0} & & & d_4^{1/2} \end{bmatrix}$$

and hence

$$A = (LD^{1/2})(D^{1/2}L^{\mathrm{T}})$$
$$= \tilde{L}\tilde{L}^{\mathrm{T}}$$

Matrices which are symmetric and have the property (27) belong to a special class of matrices called *positive definite* matrices.

Definition   An $N \times N$ (real) matrix A is positive definite if

$$\mathbf{x}^{\mathrm{T}} A \mathbf{x} > 0$$

for all N-vectors $\mathbf{x}$ ($\neq \mathbf{0}$).

Positive definite matrices occur very frequently in practice; in particular in the numerical solution of differential equations. Such matrices can be recognized by the following properties:

Definition   A matrix is said to be *strictly diagonally dominant* if

$$|a_{ii}| > \sum_{\substack{j=1 \\ j \neq i}}^{N} |a_{ij}| \quad i = 1, 2, \ldots, N$$

If a real matrix A is symmetric, strictly diagonally
dominant, and has positive diagonal elements, $\qquad$ (28)
then A is positive definite.

Hence for such matrices we want to determine a lower triangular matrix $\tilde{L}$ such that

$$A = \tilde{L}\tilde{L}^{\mathrm{T}} \qquad (29)$$

and so that only the lower triangular matrix $\tilde{L}$ need be computed and stored (for which we have shown our previous factorization algorithm is not satisfactory). That such a factorization exists follows from our discussion above provided the diagonal elements d_i are nonzero. It can be proved in a straightforward manner (see Forsythe and Moler [13]) that positive definite matrices require no pivoting. Further, it can be shown that pivoting is not necessary to control round-off error growth (see Wilkinson [49]). Hence we are endeavoring to determine $\tilde{L}$ which satisfies (29), with no permutation matrices to complicate matters! The matrix $\tilde{L}$ can be determined in a straightforward, constructive, manner. Let the matrix

$$\tilde{L} = \begin{bmatrix} l_{11} & & & 0 \\ l_{21} & l_{22} & & \\ l_{31} & l_{32} & l_{33} & \\ l_{41} & l_{42} & l_{43} & l_{44} \end{bmatrix}$$

Then

$$A = \tilde{L}\tilde{L}^{\mathrm{T}} = \begin{bmatrix} l_{11} & & & 0 \\ l_{21} & l_{22} & & \\ l_{31} & l_{32} & l_{33} & \\ l_{41} & l_{42} & l_{43} & l_{44} \end{bmatrix} \begin{bmatrix} l_{11} & l_{21} & l_{31} & l_{41} \\ & l_{22} & l_{32} & l_{33} \\ & & l_{33} & l_{43} \\ 0 & & & l_{44} \end{bmatrix}$$

Form the product column by column, forming only the elements in the lower triangle.

Column 1

$$l_{11}^2 = a_{11} \rightarrow l_{11} = \sqrt{a_{11}}$$

$$l_{11}l_{21} = a_{21} \rightarrow l_{21} = a_{21}/l_{11}$$

$$l_{11}l_{31} = a_{31} \rightarrow l_{31} = a_{31}/l_{11}$$

$$l_{11}l_{41} = a_{41} \rightarrow l_{41} = a_{41}/l_{11}$$

Column 2

$$l_{21}^2 + l_{22}^2 = a_{22} \rightarrow l_{22} = \sqrt{a_{22} - l_{21}^2}$$

$$l_{31}l_{21} + l_{32}l_{22} = a_{32} \rightarrow l_{32} = (a_{32} - l_{31}l_{21})/l_{22}$$

$$l_{41}l_{21} + l_{42}l_{22} = a_{42} \rightarrow l_{42} = (a_{42} - l_{41}l_{21})/l_{22}$$

Column 3

$$l_{31}^2 + l_{32}^2 + l_{33}^2 = a_{33} \rightarrow l_{33} = \sqrt{a_{33} - l_{31}^2 - l_{32}^2}$$

$$l_{41}l_{31} + l_{42}l_{32} + l_{43}l_{33} = a_{43} \rightarrow l_{43} = \frac{a_{43} - l_{41}l_{31} - l_{42}l_{32}}{l_{33}}$$

Finally, *Column 4*

$$l_{41}^2 + l_{42}^2 + l_{43}^2 + l_{44}^2 = a_{44} \rightarrow l_{44} = \sqrt{a_{44} - l_{41}^2 - l_{42}^2 - l_{43}^2}$$

Note that each stage we are able to obtain *explicitly* the required elements l_{ij}. For a general $N \times N$ matrix the above steps generalize into the following algorithm:

comment Cholesky decomposition of positive definite matrix ;

for $j \leftarrow 1$ **step** 1 **until** N **do**
  begin

$$l_{jj} \leftarrow \sqrt{\left(a_{jj} - \sum_{k=1}^{j-1} l_{jk}^2 \right)} ;$$

  for $i \leftarrow j + 1$ **step** 1 **until** N **do**

$$l_{ij} \leftarrow \left(a_{ij} - \sum_{k=1}^{j-1} l_{ik}l_{jk} \right) \Big/ l_{jj}$$

  end ;

The success of the algorithm in producing *real* factors $\tilde{L}$ and $\tilde{L}^{\mathrm{T}}$ depends on the expression under the square root signs. It is a simple matter to show that if the matrix A is positive definite then no square roots of negative numbers occur, so that the arithmetic is always real. It may be noted that positive definite matrices are not exclusively of the form dictated by (28). Consequently, if a matrix is real symmetric and is suspected of being positive definite, the fact can be checked (and verified) by using the Cholesky factorization algorithm to obtain, if possible, real symmetric factors $\tilde{L}$ and $\tilde{L}^{\mathrm{T}}$. If at any stage a square root of a negative number occurs then the original matrix A was not positive definite.

Exercises

27. By hand, using three decimal arithmetic, obtain the LU factorization and the $\tilde{L}\tilde{L}^{\mathrm{T}}$ factorization of

$$A = \begin{bmatrix} 4 & -1 & 0 & 0 \\ -1 & 4 & -1 & 0 \\ 0 & -1 & 4 & -1 \\ 0 & 0 & -1 & 4 \end{bmatrix}$$

and verify that the Cholesky factorization can be obtained by manipulation of the LU factors.

28. We have verified that the storage requirements of the Cholesky algorithm is approximately half that of the LU factorization for the same positive definite matrix A. How does the number of operations compare between the two algorithms?

29. Implement the Cholesky algorithm and hence determine if the coefficient matrix of Exercise 26 is positive definite.

30. Let A be a symmetric positive band matrix with $q = r = \beta$. Let B and C be $N \times N$ full nonsingular matrices. Let T be a symmetric positive definite tridiagonal matrix. All the matrices are $N \times N$ and you may assume $N \gg \beta > 1$. Let $f(\mathbf{x})$ be defined as follows:

$$f(\mathbf{x}) = \mathbf{x}^T B^{-1} C T^{-1} A^{-1} \mathbf{x} + \mathbf{b}^T B^{-T} \mathbf{x}$$

where $\mathbf{x}$ and $\mathbf{b}$ are given N-vectors.

The function $f(\mathbf{x})$ is to be evaluated for K different vectors $\mathbf{x}$.

   (i) Describe carefully how you would design a subroutine (procedure) or set of subroutines (procedures) to evaluate $f(\mathbf{x})$.
   (Consider the cases $K \ll N$ and $K \gg N$.)
   What are the total operation counts and storage requirements necessary to implement this problem?

   (ii) How would you find $\mathbf{x} \neq \mathbf{0}$ such that $f(\mathbf{x}) = 0$?

31. You are given the nonsingular $2N \times 2N$ linear system $A\mathbf{x} = \mathbf{b}$ to solve partitioned as shown below:

$$\begin{pmatrix} B & C \\ C^T & E \end{pmatrix} \begin{pmatrix} \mathbf{x}_1 \\ \mathbf{x}_2 \end{pmatrix} = \begin{pmatrix} \mathbf{b}_1 \\ \mathbf{b}_2 \end{pmatrix}$$

where B is symmetric and positive definite and E has no special features. All the matrices are $N \times N$ and full. Explain carefully how you would solve this problem. How many operations would be required to find $\mathbf{x}$? What storage requirements does your solution require?

32. The $N \times N$ positive definite matrix A is partitioned into n blocks, each block being an $n \times n$ matrix so that

$$A = \begin{bmatrix} B_1 & C_1 & & \\ A_2 & B_2 & C_2 & \\ & \ddots & \ddots & \cdot \cdot C_n \\ & & A_n & B_n \end{bmatrix}$$

The $n \times n$ matrices A_i, B_i, and C_i are all full matrices. Derive an algorithm which will determine the *block LU* factorization of A so that the matrices L and U have the form

$$L = \begin{bmatrix} I & & & \\ M_2 & I & & \\ & M_3 & \ddots & \\ & & \ddots & M_n & I \end{bmatrix} \qquad U = \begin{bmatrix} D_1 & C_1 & & \\ & D_2 & C_2 & \\ & & \ddots & \ddots & C_{n-1} \\ & & & D_n \end{bmatrix}$$

where all the matrices I, M_i, D_i, and C_i are $n \times n$.

How would you solve the matrix equation $AX = G$
where X is an $N \times M$ matrix and G is a given $N \times M$ matrix? Indicate the number of operations necessary to solve for X.

If you had proposed decomposing the matrix A using the Cholesky factorization method of section 6 modified for a band matrix with $p = q = 2n - 1$, how many operations would have been necessary to solve for X using this algorithm?

7 Ill Conditioning

The pivotal strategy described in section 2 endeavors to pick that pivot which will restrict any growth in rounding error to be as small as possible. In many situations this choice will ensure a solution correct to p significant digits in a floating point number system with precision p. However, there are (unfortunately common) cases where this is not so. Consider the following example.

Example 7

$$2x_1 + 3x_2 = 5$$

$$2x_1 + 3.1x_2 = 5.1$$

which has the solution $x_1 = x_2 = 1$. If we change the coefficients in the equations by a small amount so that a new system of equations is obtained, namely,

$$2x_1 + 3x_2 = 5$$

$$1.999x_1 + 3x_2 = 4.99$$

it can be easily shown that this system has the solution $x_1 = 10$, $x_2 = -5$.

Thus a very small change in the coefficients gives rise to a *large change* in the solution. Such behavior characterizes what is called an *ill-conditioned* system. Note that this behavior is a feature of the problem, *not* of the method used to solve the problem. If we were able to represent coefficients and carry out the computations of the Gaussian elimination algorithm with infinite precision, then such ill-conditioned systems would cause no problem. However, we deal with a real world and rounding errors will occur at every stage of the algorithm; in particular, at the stage of representing the coefficients. Thus in the floating point arithmetic we may not be solving the given system of equations but one which is a small perturbation of it. If these rounding errors arise *only* at the stage of representing the coefficient matrix and we were able to conduct the arithmetic in the Gaussian elimination algorithm exactly, then as indicated by Example 7 we still cannot conclude that the computed solution is necessarily close to the theoretical solution. Without knowing the theoretical solution how are we to check? Let the theoretical solution of the system of linear equations

$$A\mathbf{x} = \mathbf{b} \tag{2}$$

be denoted by $\mathbf{x}_T$. Let the computed solution, in the presence of rounding error, of Eq. (2) be denoted by $\mathbf{x}_C$. If $\mathbf{x}_C = \mathbf{x}_T$ then Eq. (2) will be identically satisfied when $\mathbf{x}_C$ is replaced by $\mathbf{x}_T$. That is, when the vector $A\mathbf{x}$ is evaluated it should produce $\mathbf{b}$. Calculate

$$A\mathbf{x}_C - \mathbf{b}$$

and denote the result by $\mathbf{r}$, i.e.

$$\mathbf{r} = A\mathbf{x}_C - \mathbf{b} \tag{30}$$

Clearly if $\mathbf{r} = \mathbf{0}$ we have the exact solution. In general we will find $\mathbf{r} \neq \mathbf{0}$. However, provided $\mathbf{r}$ was small we would surmise that the solution $\mathbf{x}_C$ would be "close" to the theoretical solution $\mathbf{x}_T$. Unfortunately, this need not be true, as the following shows. Subtract Eq. (30) from Eq. (2). Then

$$A(\mathbf{x}_T - \mathbf{x}_C) = -\mathbf{r}$$

or

$$\mathbf{x}_T - \mathbf{x}_C = -A^{-1}\mathbf{r} \tag{31}$$

Thus if the residual vector $\mathbf{r}$ is "small" but the elements of A^{-1} are large, the product of the right-hand side of Eq. (31) can be large. To show this, consider the solution of the first system in Example 7 to be substituted into the second system. Computing the residual vector, we find

$$r_1 = 2(1) + 3(1) - 5 = 0$$

$$r_2 = 1.999(1) + 3(1) - 4.990 = 0.009.$$

That is, the residual $\mathbf{r} = (r_1, r_2)^T$ has elements r_1, r_2 which are small although we know the theoretical solution $(10, 5)$, which is very different from the "computed" solution $(1, 1)$. It may be easily shown that the inverse of the matrix A defined by

$$A = \begin{bmatrix} 2 & 3 \\ 2 & 3 \end{bmatrix} \tag{32}$$

does not exist since the determinant of A is zero. It is clear that both the matrices of Example 7 are perturbations of the matrix A in Eq. (32). It is fair to surmise therefore that these matrices have inverses whose elements are large, as is easily verified:

$$\begin{bmatrix} 2 & 3 \\ 2 & 3.1 \end{bmatrix}^{-1} = \begin{bmatrix} 15.5 & -15 \\ -10 & 10 \end{bmatrix}$$

and

$$\begin{bmatrix} 2 & 3 \\ 1.999 & 3 \end{bmatrix}^{-1} = \begin{bmatrix} 1000 & -1000 \\ -666\frac{1}{3} & 666\frac{2}{3} \end{bmatrix}$$

The reference to "close-to," "large," "small," etc. has been intuitive so far. We want to be more precise and further discover what we mean when we talk of "the vector $\mathbf{x}_T$ being close to the vector $\mathbf{x}_C$." Namely, we want to describe a means of measuring "how big" is a given vector. Consequently we will digress here and describe briefly the concept of the vector and matrix norms. The reader is directed to Strang [40] for a fuller account.

Definition The generalization of the modulus (size) of a scalar function to a vector-valued function is called a *vector norm*. The associated quantity associated with a matrix-valued function is called a *matrix norm*.

Properties of vector norms

A vector norm of the vector $\mathbf{x}$ denoted by $\|\mathbf{x}\|$ must satisfy the following conditions. If $x \in V$, a linear N-dimensional vector space, then:

(i) $\|\mathbf{x}\| \geq 0 \quad \forall \mathbf{x} \in V \quad$ and $\quad \|\mathbf{x}\| = 0 \quad$ iff $\mathbf{x} = \mathbf{0}$

(ii) $\|\alpha \mathbf{x}\| = |\alpha| \cdot \|\mathbf{x}\| \quad \forall$ scalars α and $\forall \mathbf{x} \in V$

(iii) $\|\mathbf{x} + \mathbf{y}\| \leq \|\mathbf{x}\| + \|\mathbf{y}\| \quad \forall \mathbf{x}, \mathbf{y} \in V$

Condition (iii) is known as the *triangular inequality*.

Example 8 The following examples satisfy conditions (i)–(iii) of vector norms. To distinguish between them it is convenient to append the subscript indicated:

(a) $\|\mathbf{x}\|_1 = \sum_{i=1}^{N} |x_i|$

(b) $\|\mathbf{x}\|_2 = \left\{ \sum_{i=1}^{N} |x_i|^2 \right\}^{1/2}$

(c) $\|\mathbf{x}\|_\infty = \max_i |x_i|$

(d) $\|\mathbf{x}\|_p = \left\{ \sum_{i=1}^{N} |x_i|^p \right\}^{1/p} \qquad p \geq 1$

Examples (a)–(c) are special cases of p, where we note for (c),

$$\lim_{p \to \infty} \|\mathbf{x}\|_p = \|x\|_\infty$$

The vector norm $\|\mathbf{x}\|_2$ is frequently referred to as the *Euclidean vector norm* since it corresponds to the generalization to N dimensions of the quantity we usually associate with distance or length in two-dimensional Euclidean space. The vector norm $\|\mathbf{x}\|_\infty$ is often called the *maximum norm* or *uniform norm*. The vector norm (d) for $p \geq 1$ is generally referred to as the *p-norm*.

Exercises

33. Prove that $\|\mathbf{x}\|_1$ satisfies the conditions (i)–(iii) for a vector norm.
34. Prove that $\|\mathbf{x}\|_2$ satisfies the conditions (i)–(iii) for a vector norm.
35. Prove that $\|\mathbf{x}\|_\infty$ satisfies conditions (i)–(iii) for a vector norm, (Reference Isaacson and Keller [24]).
36. Calculate $\|\mathbf{x}\|_p$ for $p = 1, 2, \infty$ for the following vectors:
    (a) $(1, 2, 3)$
    (b) $(0, 0, 0)$
    (c) $(-1, -2, -3, -4)$
    (d) (x_1, x_2, x_3) with $x_1 = x_2 = 1 + i\sqrt{3}, x_3 = 1 - i\sqrt{3}; i = \sqrt{-1}$.

Properties of matrix norms

A matrix norm of the $(N \times N)$ matrix A denoted by $\|A\|$† must satisfy the following conditions:

$$\text{(i)} \quad \|A\| \geq 0 \quad \text{and} \quad \|A\| = 0 \quad \text{iff } A = 0$$

$$\text{(ii)} \quad \|\alpha A\| = |\alpha| \, \|A\| \quad \forall \text{ scalars } \alpha$$

$$\text{(iii)} \quad \|A + B\| \leq \|A\| + \|B\|$$

$$\text{(iv)} \quad \|AB\| \leq \|A\| \, \|B\|$$

There are again many ways of constructing expressions which satisfy the conditions for the matrix norms (i)–(iv). However, since matrices and vectors appear jointly, it is felicitous to choose the matrix norm *induced* by the vector norms. That is, if $\|\mathbf{x}\|$ is a vector norm of $\mathbf{x}$ then since $A\mathbf{x}$ is also a vector we can determine its norm $\|A\mathbf{x}\|$. Hence we define

$$\|A\| = \frac{\sup_{\mathbf{x} \neq \mathbf{0}} \|A\mathbf{x}\|}{\|\mathbf{x}\|} \tag{33}$$

We may rewrite Eq. (33) since if $\mathbf{x} \neq \mathbf{0}$ we can define $\mathbf{y} = \mathbf{x}/\|\mathbf{x}\|$ so that $\|\mathbf{y}\| = 1$ and hence

$$\|A\| = \max_{\|\mathbf{y}\|=1} \|A\mathbf{y}\| \tag{34}$$

By Eq. (33) we note that this definition implies that

$$\|A\mathbf{x}\| \leq \|A\| \, \|\mathbf{x}\| \quad \text{for any } \mathbf{x} \tag{35}$$

A matrix norm satisfying (35) is said to be *compatible* with the corresponding vector norm.

Example 9 The following examples satisfy the conditions (i)–(iv) for a matrix norm and are compatible with the corresponding vector norms. We assume the $(N \times N)$ matrix $A = \{a_{ij}\}$ has elements a_{ij}, $i, j = 1, 2, \ldots, N$. We again append a subscript to distinguish the various norms.

$$\text{(a)} \qquad \|A\|_1 = \max_j \sum_{i=1}^{N} |a_{ij}|$$

is the matrix norm induced by the vector norm (a) of Example 8.

(b) Define the conjugate transpose of the matrix A by $A^* = \{\bar{a}_{ji}\}$ where the bar denotes complex conjugate.‡ Further define the spectral radius of A,

---

† $\| \quad \|$ is used for the matrix norm as well as for the vector norm; there will be no confusion in notation since the context in which the norm is used will make clear which norm is being used.

‡ For our present application the elements of the matrix A are real so that it is sufficient to consider just the usual transpose of A. However, we retain the possibility of complex elements of A at this stage since it will introduce to the reader the more general setting for matrix norms.

which we denote by $\rho(A)$, as the maximum eigenvalue of A, i.e., if λ_k is an eigenvalue of A then

$$\rho(A) = \max_k |\lambda_k|$$

Then we define

$$\|A\|_2 = \{\rho(A^*A)\}^{1/2}$$

This matrix norm is induced by the Euclidean vector norm. It is frequently referred to as the spectral (matrix) norm.

(c) The matrix norm induced by the maximum vector norm (c) of Example 8 is

$$\|A\|_\infty = \max_i \sum_{j=1}^{N} |a_{ij}|$$

Exercises

37. Prove that $\|A\|_1$ satisfies the conditions for a matrix norm.
38. Prove that $\|A\|_2$ satisfies the conditions for a matrix norm.
39. Prove that $\|A^T\|_\infty = \|A\|_1$ and hence satisfies the conditions for a matrix norm.
40. Determine the 1, 2, and maximum norms (i.e. norms (a), (b), and (c) of Example 9) of the following matrices:

$$\begin{bmatrix} 1 & 3 \\ 2 & 6 \end{bmatrix} \quad \begin{bmatrix} 2 & 0 \\ 1 & 2 \end{bmatrix}$$

(Reference Isaacson and Keller [24]).

We finally state the following lemmas. A statement of their proof may be found in Isaacson and Keller [24].

Lemma 1 *Every vector norm $\|\mathbf{x}\|$ is a continuous function of the components of $\mathbf{x}$.*

Lemma 2 *Every matrix norm $\|A\|$ is a continuous function of the elements of the matrix A.*

From our discussion we recall

$$\|B\mathbf{y}\| \le \|B\| \, \|\mathbf{y}\| \tag{36}$$

Also, since $\mathbf{y} = B^{-1}B\mathbf{y}$ we have

$$\|\mathbf{y}\| \le \|B^{-1}\| \cdot \|B\mathbf{y}\|$$

so that

$$\frac{\|\mathbf{y}\|}{\|B^{-1}\|} \le \|B\mathbf{y}\| \tag{37}$$

Combining Eq. (37) and (36), we obtain

$$\frac{\|\mathbf{y}\|}{\|B^{-1}\|} \le \|B\mathbf{y}\| \le \|B\| \cdot \|\mathbf{y}\| \tag{38}$$

If we use the inequalities in Eq. (36) with $B \equiv A^{-1}$, and in turn $\mathbf{y} \equiv \mathbf{r}$, $\mathbf{y} \equiv \mathbf{b}$, we obtain

$$\frac{\|\mathbf{r}\|}{\|A\|} \le \|A^{-1}\mathbf{r}\| \le \|A^{-1}\|\|\mathbf{r}\| \tag{39}$$

and

$$\frac{\|\mathbf{b}\|}{\|A\|} \le \|A^{-1}\mathbf{b}\| \le \|A^{-1}\| \|\mathbf{b}\| \tag{40}$$

But $\mathbf{r} = -A(\mathbf{x}_T - \mathbf{x}_C) \equiv A\mathbf{e}$, so that Eq. (39) becomes

$$\frac{\|\mathbf{r}\|}{\|A\|} \le \|\mathbf{e}\| \le \|A^{-1}\| \|\mathbf{r}\| \tag{41}$$

and $A^{-1}\mathbf{b} \equiv \mathbf{x}_T$, so that Eq. (40) may be written as

$$\frac{\|\mathbf{b}\|}{\|A\|} \le \|\mathbf{x}_T\| \le \|A^{-1}\| \|\mathbf{b}\| \tag{42}$$

Since all quantities in the inequalities (41) and (42) are positive, we may combine (41) and (42) in an obvious manner, so that

$$\frac{\|\mathbf{r}\|}{\|A\| \|A^{-1}\| \|\mathbf{b}\|} \le \frac{\|\mathbf{e}\|}{\|\mathbf{x}_T\|} \le \frac{\|A\| \|A^{-1}\| \|\mathbf{r}\|}{\|\mathbf{b}\|} \tag{43}$$

Here, $\|\mathbf{e}\|/\|\mathbf{x}_T\|$ is the relative error and $\|\mathbf{r}\|/\|\mathbf{b}\|$ is the relative (to the right-hand side vector $\mathbf{b}$) residual. Thus the reliance we can place on $\mathbf{r}$ as a measure of how good is the computed solution $\mathbf{x}_C$ is clearly dependent upon the quantity $\|A\| \|A^{-1}\|$ occurring in both the inequalities contained in Eq. (43). Because of the crucial nature of this quantity we give it a special name; the *condition number* of the matrix A, represented as cond (A). In Eq. (43) it is clear that the *larger* is cond (A) then the less reliable becomes the residual vector $\mathbf{r}$ as a guide to how good is our solution $\mathbf{x}_C$ (i.e. how small is our $\mathbf{e}$ relative to $\mathbf{x}_T$). However, for cond (A) close to 1 the bound on the relative error in Eq. (43) becomes sharp and hence in this case the size of $\mathbf{r}$ is a reliable guide to the excellence of the solution $\mathbf{x}_C$.

The condition number plays a significant role when we consider the effects of rounding error on the coefficient matrix A. Let us assume that perturbations, caused by round-off error, occur in the element of A, so that

$$A_C = A + A_E$$

where A_C is the "computed" matrix and A_E is the matrix of perturbations.

If we assume the system $A_C\mathbf{x} = \mathbf{b}$ has a solution $\mathbf{x}_C$, then

$$\mathbf{x}_T = A^{-1}\mathbf{b} = A^{-1}A_C\mathbf{x}_C = A^{-1}[A + A_E]\mathbf{x}_C$$

$$\therefore \quad \mathbf{x}_T = \mathbf{x}_C + A^{-1}A_E\mathbf{x}_C$$

Hence

$$\mathbf{x}_T - \mathbf{x}_C = A^{-1}A_E\mathbf{x}_C$$

from which

$$\|\mathbf{x}_T - \mathbf{x}_C\| \le \|A^{-1}\| \cdot \|A_E\| \|\mathbf{x}_C\| = \text{cond}(A) \cdot \frac{\|A_E\|}{\|A\|} \|\mathbf{x}_C\|$$

Thus

$$\frac{\|\mathbf{x}_T - \mathbf{x}_C\|}{\|\mathbf{x}_C\|} \le \text{cond}(A) \cdot \frac{\|A_E\|}{\|A\|}$$

Thus the computed solution can vary from the theoretical solution, as a result of round-off errors occurring in A's elements, by an amount directly proportional to the condition number of A. Thus, again, the larger is the condition number the worse the computed solution can be. That is, the larger the condition number, the more ill-conditioned the system becomes.

Thus, since $\mathbf{r}$ cannot be used to signal the correctness of the computer solution, we would require a numerical value for the condition number. This involves A^{-1}, which in turn requires us to solve

$$AX = I$$

using the $(P)LU$ decomposition of A, and of course this is the very computation we are trying to monitor. Hence, determining exactly the condition number is not a practical proposition. However, there are means by which an estimate (usually reliable) of the condition number are possible. The practical details may be found in Forsythe, Malcolm, and Moler [12]. The estimate of this entity does one thing only: it puts the user on his guard! If the condition number is large the numerical results should be treated with caution. If the condition number is small (close to 1) then the user may safely predict that his numerical results are reliable. In addition to the condition number being a sign of ill conditioning, there are also other noticeable signs when an ill-conditioned system is being solved.

Example 10   Consider solving

$$0.16x_1 + 0.21x_2 + 0.30x_3 = 1.569$$

$$0.32x_1 + 0.40x_2 + 0.75x_3 = 3.320$$

$$0.28x_1 + 0.37x_2 + 0.50x_3 = 2.713$$

Interchanging the first and second row and eliminating x_1 from the last two equations, we obtain

$$0.32x_1 + 0.40x_2 + 0.75x_3 = 3.320$$

$$0.01x_2 - 0.075x_3 = -0.091$$

$$0.02x_2 - 0.15625x_3 = -0.192$$

Interchanging the latter two equations and eliminating x_2, we finally obtain

$$0.32x_1 + 0.40x_2 + 0.75x_3 = 3.320$$

$$0.02x_2 - 0.15625x_3 = -0.192$$

$$-0.00625x_3 = -0.01$$

During the forward elimination it is seen that the pivotal elements lose significant figures as the elimination proceeds. The original coefficients were given to two decimal places. In the reduced system, the first pivot 0.32 has (still) two significant figures, 0.02 has just one, and 0.00625 has none! This loss of significant figures is an indication of ill conditioning.

If we suspect (and even if we do not!) that a system is ill conditioned, the computed solution from the Gaussian elimination algorithm may be improved in the following way.

Assume we obtain the $(P)LU$ decomposition of the matrix A and computed a solution $\mathbf{x}^{(1)}$, say. In general, this will differ from $\mathbf{x}_T$ as indicated above. Compute in *double precision* the residual

$$\mathbf{r}^{(1)} = A\mathbf{x}^{(1)} - \mathbf{b} \tag{44}$$

Let

$$\mathbf{x}_T - \mathbf{x}^{(1)} = \boldsymbol{\varepsilon}^{(1)}$$

Then

$$A\boldsymbol{\varepsilon}^{(1)} = A\mathbf{x}^T - A\mathbf{x}^{(1)} = \mathbf{b} - A\mathbf{x}^{(1)} = -\mathbf{r}^{(1)} \tag{45}$$

That is, the "error" vector $\boldsymbol{\varepsilon}^{(1)}$ is itself the solution of a system of linear equations with the common coefficient matrix A but with a new right-hand side $-\mathbf{r}^{(1)}$. Using the $(P)LU$ decomposition of A the system (45) can be solved, with little extra work (how much?) and hence obtain

$$\mathbf{x}^{(2)} = \mathbf{x}^{(1)} + \boldsymbol{\varepsilon}^{(1)}$$

Compute

$$\mathbf{r}^{(2)} = A\mathbf{x}^{(2)} - \mathbf{b}$$

in *double precision*. If we define

$$\boldsymbol{\varepsilon}^{(2)} = \mathbf{x}_T - \mathbf{x}^{(2)}$$

then

$$A\boldsymbol{\varepsilon}^{(2)} = -\mathbf{r}^{(2)} \tag{46}$$

can be solved for $\boldsymbol{\varepsilon}^{(2)}$, again with little extra computation. This procedure can be repeated until

$$\frac{\|\mathbf{x}^{(i)} - \mathbf{x}^{(i-1)}\|}{\|\mathbf{x}^{(i)}\|} \leq \text{tol} \tag{47}$$

where tol is "sufficiently small," i.e. either 2^{-p} in p-precision arithmetic or a larger number if this degree of accuracy is not required.

It must be emphasized that the computation of the residual at *each stage* must be carried out in (at least) double precision if any improvement in the sequence of vectors $\mathbf{x}^{(i)}$ is to be achieved. This additional iterative procedure is called *iterative improvement.* It should be part of any worthwhile linear equation solver. If one or two iterations are required to satisfy (47) with tol $= 2^{-p}$, then it can safely be assumed that the system is well conditioned. If a large number of iterations is necessary for convergence then this is another signal that the user is dealing with a badly conditioned system. The precise rate at which the iterates $\mathbf{x}^{(i)} \rightarrow \mathbf{x}_T$ depends intimately on cond (A). If we are using a floating point number system with base β and precision p (see Chapter 2), then it can be shown (see Forsythe and Moler [13]) that

$$\text{cond}\,(A) \simeq \beta^p \|\boldsymbol{\varepsilon}^{(1)}\| / \|\mathbf{x}^{(1)}\| \tag{48}$$

so that from the first two vectors $\mathbf{x}^{(1)}$ and $\boldsymbol{\varepsilon}^{(1)}$ we may obtain an *estimate* of cond (A).

If cond $(A) = \beta^q$, say, then

$$\frac{\|\boldsymbol{\varepsilon}^{(1)}\|}{\|\mathbf{x}^{(1)}\|} \simeq \beta^{q-p}$$

If $q - p \geq 0$ then no figures of $\mathbf{x}^{(1)}$ are correct—the system is so ill conditioned (q is so large in relation to p) that iterative improvement will not help. To obtain an accurate solution the whole computation, representation and Gaussian elimination have to be calculated in double precision. This then implies using double-double precision for the calculation of the residual. The above analysis can be carried out again until the final degree of precision p is achieved—for very badly conditioned matrices this will imply that multiple precision arithmetic should be available.

If $q - p < 0$ then the number of iterations needed for the convergence of the sequence of iterates generated by the iterative improvement is given by

$$\frac{p}{\log_\beta (\|\mathbf{x}^{(1)}\| / \|\boldsymbol{\varepsilon}^{(1)}\|)}$$

(see Forsythe and Moler [13] for details and also a proof of the convergence of the method of iterative improvement).

Example 11 Consider three springs with stiffness coefficients α, β, α (>0) which support two points B and C and which are secured at two fixed points A and D (see Figure 1).

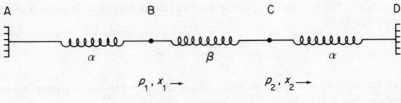

Figure 1

Let known forces p_1 and p_2 be applied at B and C respectively, causing displacements x_1 and x_2 respectively, where these displacements are unknown. To obtain the values of x_1 and x_2, the equation obtained by applying the principle of equilibrium† at these two points B and C must be solved. These equations can be written as

$$p_1 = (\alpha + \beta)x_1 - \beta x_2$$

$$p_2 = -\beta x_1 + (\alpha + \beta)x_2$$

Thus in vector notation we require to solve for $\mathbf{x}$, given A and $\mathbf{p}$, where

$$\mathbf{x} = (x_1, x_2)^{\mathrm{T}} \quad \mathbf{p} = (p_1, p_2)^{\mathrm{T}} \quad \text{and} \quad A = \begin{bmatrix} \alpha + \beta & -\beta \\ -\beta & \alpha + \beta \end{bmatrix}$$

It is easily found that

$$A^{-1} = \begin{bmatrix} \dfrac{\alpha + \beta}{\alpha(\alpha + 2\beta)} & \dfrac{\beta}{\alpha(\alpha + 2\beta)} \\ \dfrac{\beta}{\alpha(\alpha + 2\beta)} & \dfrac{\alpha + \beta}{\alpha(\alpha + 2\beta)} \end{bmatrix}$$

Therefore using the maximum norm (or the 1-norm (why?)),

$$\text{cond}(A) = \frac{\alpha + 2\beta}{\alpha} = 1 + 2\frac{\beta}{\alpha}$$

Thus, as we would expect, the conditioning of the system is dependent upon the parameters α and β. In particular, for β/α small, the system is well conditioned, e.g. for $\beta = 0.01\alpha$,

$$A = \frac{\alpha}{100} \begin{bmatrix} 101 & -1 \\ -1 & 101 \end{bmatrix}$$

is well conditioned. For $\beta = 100\alpha$, then

$$A = \alpha \begin{bmatrix} 101 & -100 \\ -100 & 101 \end{bmatrix}$$

and the system is not so well conditioned since $\text{cond}(A) = 1 + 200 = 201$.

---

† The principle of equilibrium states that, for equilibrium to exist at a point, all the forces acting at the point, must have algebraic sums of their components in any two direction at right angles which vanish. The force which is caused by extending a spring of stiffness α through a distance x is αx in the direction opposite to the direction of x positive.

Exercises

41. In Example 11 we considered the conditioning in terms of the maximum norm. If, instead, we were to consider the 2-norm, what is the crucial relation between β and α for ill conditioning?

42. Consider the problem depicted in Figure 1 where now the stiffness coefficient of the spring nearest to D is replaced by γ. For the unknowns x_1 and x_2 what are the constraints on the stiffness coefficients such that the resulting system of equations is well conditioned with respect to the maximum norm?

43. Generalize the model for three springs in Figure 1 to four springs supporting three points B, C, D (say), and secured at A and E. Let the stiffness coefficients of the springs be α, β, γ, and δ. What form has the matrix arising from this model? Find values of α, β, γ, and δ which give rise to an ill-conditioned system of equations with respect to the maximum norm.

44. Develop a computer program to solve a given system of linear equations by the method of iterative improvement using, as a basis, the decomposition method based on the algorithm on page 69. The maximum number of iterations which should be carried out before abandoning the run (with an appropriate message) should be read as data. The solution is deemed to be accurate enough when expression (47) using the maximum norm, is less than a quantity EPS, to be read as data.

45. Use your program of Exercise 44 to solve the system of equations derived in Exercise 43 for the following values of the parameters occurring in Table 1.

Table 1

p_1	1	2	1.5	1.6
p_2	1.1	1	2.3	2.3
p_3	2	1.2	4.5	3.4
α	0.5	1	0.1	0.01
β	1	1	0.1	1.0
γ	1	1	0.1	1.0
δ	1	1	0.1	0.01

The iterations should continue until expression (47) is less than 10^{-4} or the number of iterations exceeds six iterations.

46. Use your program of Exercise 44 to obtain correct to five decimal places (if possible!) the solution to Example 10. How many iterations are required?

47. Which of the following matrices give rise to an ill-conditioned system?

$$\text{(i)} \quad A = \begin{bmatrix} 1 & 2 & -1 \\ 3 & 4 & 0 \\ 1 & 1 & 0 \end{bmatrix} \quad \mathbf{b} = \begin{bmatrix} 1 \\ 1 \\ 1 \end{bmatrix}$$

$$\text{(ii)} \quad A = \begin{bmatrix} 1 & 2 & 3 \\ 4 & 5 & 6 \\ 7 & 8 & 8 \end{bmatrix} \quad \mathbf{b} = \begin{bmatrix} 0 \\ 2 \\ 1 \end{bmatrix}$$

$$\text{(iii)} \quad A = \begin{bmatrix} 1 & 0 & 0 \\ 0 & 1 & 0 \\ 1 & 1 & 1 \end{bmatrix} \quad \mathbf{b} = \begin{bmatrix} 0 \\ 2 \\ 1 \end{bmatrix}$$

88

$$\text{(iv)} \quad A = \begin{bmatrix} 1 & 1/2 & 0 \\ 1/2 & 1/3 & 1 \\ -1 & -1 & -1 \end{bmatrix} \quad b = \begin{bmatrix} 1/2 \\ 1/4 \\ 3/4 \end{bmatrix}$$

48. For the systems in Exercise 47 which are ill conditioned, use your program of Exercise 44 to solve the systems. How many iterations do you require?

8 Methods for Sparse Matrices

In this section we will describe briefly algorithms which are generally applicable to systems of linear equations whose coefficient matrix is sparse.

Example 12   In the numerical solution of partial differential equations (see Mitchell and Griffiths [32]) a method used extensively is that of finite differences where the partial derivatives occurring in the equation are replaced by *divided differences* (see Chapter 4). To apply the method, the *region* in which the solution to the partial differential equation is required, is discretized using a uniform grid. Let us consider Laplace's equation on the unit square, namely,

$$\frac{\partial^2 u}{\partial x^2} + \frac{\partial^2 u}{\partial y^2} = 0 \qquad 0 < x, y < 1 \tag{49}$$

We superimpose a uniform grid of mesh size h on the unit square as depicted in Figure 2. We denote a point of intersection of the grid lines, called a grid point, by (ih, jh) or (i, j). We denote the solution at the grid point by $u(ih, jh) = u_{ij}$. Define the x-central difference operator δ_x^2 by

$$\delta_x^2 u_{ij} \equiv u_{i+1j} - 2u_{ij} + u_{i-1j}$$

and the y-central difference operator δ_y^2 by

$$\delta_y^2 u_{ij} = u_{ij+1} - 2u_{ij} + u_{ij-1}$$

Then an approximation to the solution of (49) is obtained by solving

$$(\delta_x^2 + \delta_y^2)u_{ij} = 0 \tag{50}$$

where i and j take on all possible values corresponding to the grid points in the interior of the unit square.

To guarantee a unique solution to (49) we must also impose boundary conditions on the boundary of the unit square. In particular, let us assume

$$u(x, y) = x + y \qquad 0 \le x \le 1 \qquad y = 0, 1 \text{ and } 0 \le y \le 1 \qquad x = 0, 1 \tag{51}$$

If we assume $h = 1/4$, then applying Eq. (50) subject to conditions over the resulting grid gives rise to the *sparse* system of linear equations.

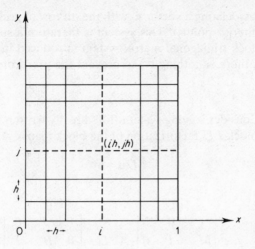

Figure 2

$$
\begin{bmatrix}
-4 & 1 & 0 & 1 & 0 & 0 & 0 & 0 & 0 \\
1 & -4 & 1 & 0 & 1 & 0 & 0 & 0 & 0 \\
0 & 1 & -4 & 0 & 0 & 1 & 0 & 0 & 0 \\
1 & 0 & 0 & -4 & 1 & 0 & 1 & 0 & 0 \\
0 & 1 & 0 & 1 & -4 & 1 & 0 & 1 & 0 \\
0 & 0 & 1 & 0 & 1 & -4 & 0 & 0 & 1 \\
0 & 0 & 0 & 1 & 0 & 0 & -4 & 1 & 0 \\
0 & 0 & 0 & 0 & 1 & 0 & 1 & -4 & 1 \\
0 & 0 & 0 & 0 & 0 & 1 & 0 & 1 & -4
\end{bmatrix}
\begin{bmatrix}
u_{11} \\ u_{21} \\ u_{31} \\ u_{12} \\ u_{22} \\ u_{32} \\ u_{13} \\ u_{23} \\ u_{33}
\end{bmatrix}
=
\begin{bmatrix}
-0.5 \\ -0.5 \\ -2.0 \\ -0.5 \\ 0 \\ -0.75 \\ -2 \\ -0.75 \\ -3.5
\end{bmatrix}
\tag{52}
$$

Let us once again write our system of equations as

$$Ax = b \tag{2}$$

where now A is a sparse matrix. We have partitioned the 9×9 matrix A into 3×3 block matrices so that we could write Eq. (52) as

$$
\begin{bmatrix}
B & I & 0 \\
I & B & I \\
0 & I & B
\end{bmatrix}
\begin{bmatrix}
\mathbf{u}_1 \\ \mathbf{u}_2 \\ \mathbf{u}_3
\end{bmatrix}
=
\begin{bmatrix}
\mathbf{b}_1 \\ \mathbf{b}_2 \\ \mathbf{b}_3
\end{bmatrix}
\tag{53}
$$

where

$$
B = \begin{bmatrix}
-4 & 1 & 0 \\
1 & -4 & 1 \\
0 & 1 & -4
\end{bmatrix}
$$

and I is the 3×3 unit matrix. The vector of unknowns $\mathbf{u}$ has been partitioned in a corresponding fashion and the reader will note that the partitioning corresponds to associating a vector $\mathbf{u}_i$ with the ith row of grid points in Figure 2 (with $N = 3$ interior points). This system is therefore a special case of the more general block tridiagonal matrix system introduced in Exercise 32. As indicated there, there is nothing (?) to prevent the user from using Gaussian elimination.

Example 13 Consider solving system (52) directly written in the form (53). We obtain the block LU factorization of the block matrix A, namely

$$LU\mathbf{u} = \mathbf{b}$$

where

$$L = \begin{bmatrix} I & 0 & 0 \\ M_2 & I & 0 \\ 0 & M_3 & I \end{bmatrix} \qquad U = \begin{bmatrix} B & I & 0 \\ 0 & B_1 & I \\ 0 & 0 & B_2 \end{bmatrix}$$

where

$$M_2 = B^{-1} \qquad B_1 = B - M_2$$
$$M_3 = B_1^{-1} \qquad B_2 = B - M_3$$

where, of course, M_2 and M_3 are obtained by *solving*, respectively,

$$BM_2 = I \tag{54}$$

and

$$B_1 M_3 = I \tag{55}$$

Let us carry out the first few computations by hand. In Eq. (54), to find M_2 we obtain the LU factorization of B, namely

$$B = L_B U_B \quad \text{(say)}$$

where

$$L_B = \begin{bmatrix} 1 & 0 & 0 \\ -1/4 & 1 & 0 \\ 0 & -4/15 & 1 \end{bmatrix} \qquad U_B = \begin{bmatrix} -4 & 1 & 0 \\ 0 & -15/4 & 1 \\ 0 & 0 & -56/15 \end{bmatrix}$$

Hence in solving $L_B U_B M_2 = I$ we actually solve $L_B U_B \mathbf{m}_{2i} = \mathbf{e}_i$, $i = 1, 2, 3$, where $\mathbf{e}_i$ is the ith column of I from which we find

$$M_2 = \begin{bmatrix} -15/56 & -4/56 & -1/56 \\ -4/56 & -16/56 & -4/56 \\ -1/56 & -4/56 & -15/56 \end{bmatrix}$$

i.e. the matrix M_2 ($=B^{-1}$) is full, but still symmetric. Then

$$B_1 = B - M_2$$

$$= \begin{bmatrix} -209/56 & 60/56 & 1/56 \\ 60/56 & -208/56 & 60/56 \\ 1/56 & 60/56 & -209/56 \end{bmatrix}$$

i.e. B_1 is now full (but still symmetric). Then we solve $B_1 M_3 = I$ as above and hence finally obtain the required block LU factorization of A. We have carried out the first few steps to emphasize the fact that as the elimination proceeds the blocks of the original matrix A which were sparse have become filled in, i.e., the matrix factors become more dense than the original blocks of A. Hence the storage requirements essentially become

—so what was a sparse matrix has become a denser matrix.

For the more general situation (with large N) we would have had in Example 13

$$\begin{bmatrix} B & I & & & \\ I & B & I & & \\ & I & B & I & \\ & & \ddots & \ddots & I \\ & & & I & B \end{bmatrix} \quad \text{becoming} \quad \begin{bmatrix} B & I & & & \\ \blacksquare & \blacksquare & I & & \\ & \blacksquare & \blacksquare & I & \\ & & \ddots & \ddots & I \\ & & & \blacksquare & \blacksquare \end{bmatrix}$$

i.e. the matrix becomes full in between the lower outermost diagonals and the extremities of the B. This increases substantially the storage requirements and for examples which arise from partial differential equations we are considering values of N of the order of 100's or 1000's, so that we are dealing with matrices of order $10^4 \times 10^4$ or $10^6 \times 10^6$! To increase the storage in a manner outlined can quickly prove excessive even for the largest machines. This type of problem has been resolved in one of two ways:

(1) iterative methods;
(2) sparse matrix methods.

The first of these two methods is more classical and, algorithmically, is simple to explain. The second of these approaches is current. The sparse matrix methods use direct methods of solution but try to organize the ordering of the variables and equations in such a way so as to minimize (or nearly so) the amount of fill which can occur, as shown above. The following simple example will indicate the type of improvement that can be obtained—for fuller details the reader is directed to Tewarson [42], Willoughby [51], and George and Liu [16].

Example 14 Consider a grid of points shown in Figure 3. Assume we solve Laplace's equation (49) on this region using approximation (50). Using the

35	36	37	38	39	40	41	42	43	44	45	46	47	48	49	50	51
18	19	20	21	22	23	24	25	26	27	28	29	30	31	32	33	34
1	2	3	4	5	6	7	8	9	10	11	12	13	14	15	16	17

Figure 3

grid points numbered in rows as shown, the system of equations to be solved is

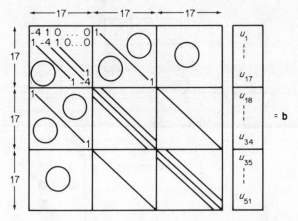

where the 51-vector **b** contains contributions from given values on the boundary of the rectangle. After *LU* decomposition the corresponding blocks look like

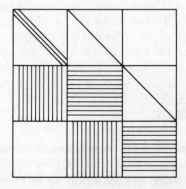

Instead, consider numbering the grid points according to columns, as shown in Figure 4.

3	6	9	12	15	18	21	24	27	30	33	36	39	42	45	48	51	
2	5	8	11	14	17	20	23	26	29	32	35	38	41	44	47	50	
1	4	7	10	13	16	19	22	25	28	31	34	37	40	43	46	49	

Figure 4

Writing the system of linear equations associated with the unknowns $u_1, u_2, \ldots, u_{51}$, it is found that we now have the matrix system

$$\begin{bmatrix} B & I & & & \\ I & B & I & & \\ & \ddots & \ddots & \ddots & \\ & & & & I \\ & & & I & B \end{bmatrix} \begin{bmatrix} u_1 \\ u_2 \\ \vdots \\ u_{51} \end{bmatrix} = \mathbf{b}$$

where

$$B = \begin{bmatrix} -4 & 1 & 0 \\ 1 & -4 & 1 \\ 0 & 1 & -4 \end{bmatrix} \qquad I = \begin{bmatrix} 1 & 0 & 0 \\ 0 & 1 & 0 \\ 0 & 0 & 1 \end{bmatrix}$$

and $\mathbf{b}$ is the associated 51-vector of right-hand sides. After LU factorization the corresponding blocks now have the form

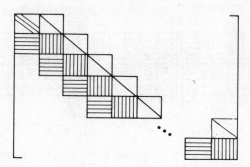

The LU decomposition here has filled all the 3×3 unit matrices below the main diagonal and, essentially, all the main diagonal blocks. A comparison of the "fill" in these two orderings indicates that when numbering by rows, for the present example an extra *1020* nonzeros are introduced whereas when numbering by columns an extra *128* nonzeros are introduced!

Exercises

49. Compare the number of operations needed to solve the system of equations in Example 14 when
    (a) the unknowns are ordered by rows
    (b) the unknowns are ordered by columns

50. For the problem of Example 14 assume we decide to number the unknowns in the following way (see Figure 5). The philosophy behind this numbering system is to introduce a "separator" which divides the given region into smaller regions; within these smaller regions there is again a choice of ordering the points. Assume within each of the two subareas the unknowns are numbered by rows, as shown. By repeating the steps described in Example 14 determine the form of the block factors LU and hence the fill that takes place for this ordering.

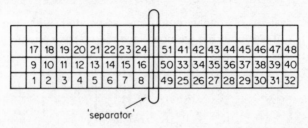

Figure 5

51. Repeat Exercise 50 numbering the unknowns by columns in the separate subareas.
52. Consider a generalization of Exercise 50 where two separators are introduced as shown in Figure 6. Assume the unknowns are numbered by rows within each of the three subareas. What is the fill that takes place for this ordering of the unknowns? What is the situation if the unknowns within each subarea are numbered by columns? Can you extrapolate this and determine, for this configuration; what the ordering of points should be to produce minimum fill?

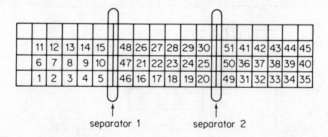

Figure 6

The above examples have been used to give an indication of possible savings by reordering the unknowns in a system of linear equations. The field of direct methods for sparse linear systems is currently an area of intense research and a full account is both beyond the scope of this text and to any degree not possible at this time. For a good coverage of the techniques the reader is referred to George and Liu [16].

An alternative to the direct methods is the class of methods known as "iterative methods for sparse linear equations."

Consider once again our sparse matrix system (2). We want to set up an *iteration* procedure by which we may start with an initial estimate (guess) $\mathbf{x}_{(0)}$ of the solution $\mathbf{x}$ and produce a sequence of iterates $\{\mathbf{x}_{(n)}\}$ by an "appropriate" method whose general form will be

$$\mathbf{x}_{(n+1)} = M\mathbf{x}_{(n)} + \mathbf{k} \tag{56}$$

where M is an $N \times N$ matrix and $\mathbf{k}$ is an N-vector.

Example 15 If we write matrix A as

$$A = D + L + U \tag{57}$$

where D is a diagonal matrix and L and U are lower and upper triangular matrices respectively, with zero elements on the diagonals, then Eq. (2) may be written as

$$\mathbf{x} = -D^{-1}(L+U)\mathbf{x} + D^{-1}\mathbf{b} \tag{58}$$

or

$$\mathbf{x} = -(D+L)^{-1}U\mathbf{x} + (D+L)^{-1}\mathbf{b} \tag{59}$$

for example. In terms of Eq. (56), Eq. (58) has

$$M \equiv -D^{-1}(L+U) \qquad \text{and} \qquad \mathbf{k} \equiv D^{-1}\mathbf{b} \tag{60}$$

and Eq. (59) has

$$M \equiv -(D+L)^{-1}U \qquad \text{and} \qquad \mathbf{k} \equiv (D+L)^{-1}\mathbf{b} \tag{61}$$

Writing A as indicated in Eq. (56), then with M defined by Eq. (60) we have the *point Jacobi* method:

$$\mathbf{x}_{(n+1)} = -D^{-1}(L+U)\mathbf{x}_{(n)} + D^{-1}\mathbf{b} \qquad n = 0, 1, \ldots \tag{62}$$

This method clearly satisfies our two requirements in that we *have* obtained the iterative method explicitly and no complicated matrix inversion is required since D, by definition, is diagonal and so the inverse D^{-1} is also diagonal with the reciprocals of the elements of D along its diagonal; i.e. if

$$D = \begin{bmatrix} d_{11} & & & 0 \\ & d_{22} & & \\ & & \ddots & \\ 0 & & & d_{NN} \end{bmatrix}$$

then

$$D^{-1} = \begin{bmatrix} 1/d_{11} & & & 0 \\ & 1/d_{22} & & \\ & & \ddots & \\ 0 & & & 1/d_{NN} \end{bmatrix}$$

We have assumed here that none of the diagonal elements d_{ii} are zero. Clearly if they are D^{-1} is not defined. In this case when A does have zero elements on the main diagonal, we interchange rows so that the resulting system has nonzero main diagonal elements and the inverse of the diagonal so defined is well defined.

The algorithm proceeds by taking an initial guess to the solution vector $\mathbf{x}$ and determining a new approximation to the solution by evaluating the right-hand side of Eq. (62). The point to note is that the successive components of $\mathbf{x}_{(1)}$ are always defined in terms of the vector $\mathbf{x}_{(0)}$. This obviously seems a waste of effort when we consider that $x_{1(1)}$ (the first element of $\mathbf{x}_{(1)}$) will be calculated at the first stage of the first iteration and once we have calculated $x_{1(1)}$ we can then use it to calculate $x_{2(1)}$ by replacing the element $x_{1(0)}$ by $x_{1(1)}$. Similarly, when $x_{2(1)}$ has been calculated we update $x_{2(0)}$ by overwriting it with $x_{2(1)}$, and so on. This modification to the Jacobi method where the most recent information is used in the next stage of the iteration is called the *point Gauss–Seidel* method. The method in fact arises out of writing M as given by Eq. (59), namely

$$\mathbf{x}_{(n+1)} = -(D+L)^{-1}U\mathbf{x}_{(n)} + (D+L)^{-1}\mathbf{b} \qquad n = 0, 1, 2, \ldots \qquad (63)$$

That this algorithm is equivalent to that just described and does not require the inversion of a complicated matrix (as would appear at first sight) is easily shown as follows.

Let the matrix A have elements a_{ij}, $i, j = 1, 2, \ldots, n$, then

$$D = \begin{bmatrix} a_{11} & & & 0 \\ & a_{22} & & \\ & & \ddots & \\ 0 & & & a_{NN} \end{bmatrix} \qquad L = \begin{bmatrix} 0 & & & 0 \\ a_{21} & & \ddots & \\ \vdots & & & \\ a_{N1} & \cdots & a_{NN-1} & 0 \end{bmatrix}$$

and

$$U = \begin{bmatrix} 0 & a_{12} & \cdots & a_{1N} \\ & & & \vdots \\ & \ddots & & a_{N-1N} \\ \mathbf{0} & & & 0 \end{bmatrix}$$

Thus considering the first components of Eq. (63) we obtain for $n = 0$,

$$a_{11}x_{1(1)} = -\sum_{j=2}^{N} a_{ij}x_{j(0)} + b_1$$

or

$$x_{1(1)} = -\frac{1}{a_{11}}\left[\sum_{j=2}^{N} a_{ij}x_{j(0)} - b_1\right]$$

Then

$$a_{22}x_{2(1)} + a_{21}x_{1(1)} = -\sum_{j=3}^{N} a_{2j}x_{j(0)} + b_2$$

or

$$x_{2(1)} = -\frac{1}{a_{22}}\left[a_{21}x_{1(1)} + \sum_{j=3}^{N} a_{2j}x_{j(0)} - b_2\right]$$

Continuing this process, we find that

$$x_{i(1)} = -\frac{1}{a_{ii}}\left[\sum_{j=1}^{i-1} a_{ij}x_{j(1)} + \sum_{j=i+1}^{N} a_{ij}x_{j(0)} - b_i\right]$$

and at the $(1+n)$th iteration,

$$x_{i(n+1)} = -\frac{1}{a_{ii}}\left[\sum_{j=1}^{i-1} a_{ij}x_{j(n+1)} + \sum_{j=i+1}^{N} a_{ij}x_{j(n)} - b_i\right] \qquad i = 1, 2, \ldots, N$$

Example 16 Formulate the point Jacobi and point Gauss–Seidel methods for the system of linear equations

$$20x_1 + 2x_2 - x_3 = 25$$
$$2x_1 + 13x_2 - 2x_3 = 30$$
$$x_1 + x_2 + x_3 = 2$$

which has the solution $x_1 = 1$, $x_2 = 2$, $x_3 = -1$.

Using the definition of the point Jacobi method, Eq. (62), we obtain

$$x_{1(1)} = \tfrac{1}{20}(-2x_{2(0)} + x_{3(0)} + 25)$$
$$x_{2(1)} = \tfrac{1}{13}(-2x_{1(0)} + 2x_{3(0)} + 30)$$
$$x_{3(1)} = (-x_{1(0)} - x_{2(0)} + 2)$$

At the $(1+n)$th iteration this becomes

$$x_{1(n+1)} = \tfrac{1}{20}(-2x_{2(n)} + x_{3(n)} + 25)$$
$$x_{2(n+1)} = \tfrac{1}{13}(-2x_{1(n)} + 2x_{3(n)} + 30)$$
$$x_{3(n+1)} = (-x_{1(n)} - x_{2(n)} + 2)$$

where $x_{1(0)}$, $x_{2(0)}$, and $x_{3(0)}$ are initial guesses.

In contrast, the point Gauss–Seidel method, Eq. (63), may be written for the first iteration

$$x_{1(1)} = \tfrac{1}{20}(-2x_{2(0)} + x_{3(0)} + 25) \quad = \frac{25}{20} = 1.25$$
$$x_{2(1)} = \tfrac{1}{13}(-2x_{1(1)} + 2x_{3(0)} + 30)$$
$$x_{3(1)} = (-x_{1(1)} - x_{2(1)} + 2)$$

$(n+1)$

and at the $(1+n)$th iteration

$$x_{1(n+1)} = \tfrac{1}{20}(-2x_{2(n)} + x_{3(n)} + 25)$$

$$x_{2(n+1)} = \tfrac{1}{13}(-2x_{1(n+1)} + 2x_{3(n)} + 30)$$

$$x_{3(n+1)} = (-x_{1(n+1)} - x_{2(n+1)} + 2)$$

where $x_{2(0)}$ and $x_{3(0)}$ are initial guesses.

Starting with the initial guesses $x_{1(0)} = x_{2(0)} = x_{3(0)} = 0$, the two algorithms produce the results given in Tables 2 and 3.

Table 2 Point Jacobi results

Iteration $(n+1)=$	x_1	x_2	x_3
1	0	0	0
2	1.25	2.307 692	2
3	1.119 231	2.423 077	−1.557 692
4	0.929 807 7	1.895 858	−1.542 308
5	0.983 298 8	1.927 367	−0.825 665 7
6	1.015 98	2.029 39	−0.910 665 7
7	1.001 528	2.011 285	−1.045 37
8	0.996 603	1.992 785	−1.012 813
9	1.000 081	1.998 551	−0.989 387 9
10	1.000 675	2.001 62	−0.998 632 2
11	0.999 906 4	2.000 106	−1.002 296
12	0.999 874 6	1.999 661	−1.000 013
13	1.000 033	2.000 017	−0.999 535 8
14	1.000 021	2.000 066	−1.000 051

Table 3 Point Gauss–Seidel results

Iteration $(n+1)=$	x_1	x_2	x_3
1	1.25	2.115 385	−1.365 385
2	0.970 192 3	1.948 373	−0.918 565 1
3	1.009 234	2.011 108	−1.020 342
4	0.997 872 1	1.997 198	−0.995 069 9
5	1.000 527	2.000 677	−1.001 204
6	0.999 872 1	1.999 834	−0.999 706 5
7	1.000 031	2.000 04	−1.000 072
8	0.999 992 4	1.999 99	−0.999 982 5
9	1.000 002	2.000 002	−1.000 004
10	0.999 999 6	1.999 999	−0.999 998 9
11	1	2	−1

The faster convergence of the Gauss–Seidel method is evident from the tables. This is often the case although examples can be constructed where the point

Jacobi method produces a convergent sequence whereas the point Gauss–Seidel method does not. The convergence properties of the algorithms will occupy our attention for the remainder of this section.

Exercises

53. Write a computer program for the point Jacobi method. The program should be written for an $N \times N$ matrix A, N-vector B, and solution vector X for given N. The coefficients of A and B should be read as data and initial guesses for X also, input. The iterations should be allowed to continue until the maximum absolute relative difference of the elements of successive iterates is less than a parameter TOL (read as data) or the number of iterations exceeds MAX (also read as data).

54. Write a program for the point Gauss–Seidel method in a similar manner to that outlined in Exercise 53.

55. Calculate the solution of the system of equations given in Example 12 for TOL = 0.001 and MAX = 20. Compare the number of iterations required by the two methods of Exercise 53 and 54.

56. Using the number of iterations used in Exercise 55, calculate the number of operations required for solving the problem using either of the point Jacobi method or the point Gauss–Seidel method. How many operations would be necessary to solve the problem directly using the Cholesky method? How do the storage requirements compare between the three methods?

If an iterative method is to be of practical use, conditions under which the sequence of iterates produced by the algorithm converges to the required solution must be determined.

Consider Eq. (56),

$$\begin{aligned}
\mathbf{x}_{(n+1)} &= M x_{(n)} + \mathbf{k} \\
&= M[M\mathbf{x}_{(n-1)} + \mathbf{k}] + \mathbf{k} \\
&\vdots \\
&= M^{n+1}\mathbf{x}_{(0)} + (I + M + \cdots + M^n)\mathbf{k}
\end{aligned}$$

If $\mathbf{x}_{(n)} = \mathbf{x}_{(n+1)} = \mathbf{x}$, then $(I - M)\mathbf{x} = \mathbf{k}$, so that this is equivalent to having written the original system of equations $A\mathbf{x} = \mathbf{b}$ in the form

$$(I - M)\mathbf{x} = \mathbf{k}$$

Hence, assuming the existence of $(I - M)^{-1}$, the solution we seek is

$$\mathbf{x} = (I - M)^{-1}\mathbf{k}$$

In order that the sequence generated by Eq. (56) converges to $\mathbf{x}$ as n increases, we clearly require

$$\lim_{n \to \infty} \|\mathbf{x}_{(n+1)} - \mathbf{x}\| = \lim_{n \to \infty} \|M^{n+1}\mathbf{x}_{(0)} + (I + M + \cdots + M^n)\mathbf{k} - (I - M)^{-1}\mathbf{k}\| = 0$$

for arbitrary starting value $\mathbf{x}_{(0)}$. It is clearly necessary in this case (for such $\mathbf{x}_{(0)}$) that $\lim_{n \to \infty} M^{n+1} = 0$. However, this also proves to be sufficient for the convergence of $\{\mathbf{x}_{(n)}\}$ since, if $(I - M)^{-1}$ exists we obtain

$$(I + M + \cdots + M^n)\mathbf{k} = (I - M)^{-1}(I - M^{n+1})\mathbf{k}$$

and

$$\lim_{n \to \infty} (I - M)^{-1}(I - M^{n+1})\mathbf{k} = (I - M)^{-1}\mathbf{k}$$

since by assumption

$$\lim_{n \to \infty} M^{n+1} = \mathbf{0}$$

Thus, if $(I - M)^{-1}$ exists the necessary and sufficient condition that the sequence generated by Eq. (56) converges to the solution $\mathbf{x}$ of Eq. (2) is that $\lim_{n \to \infty} M^{n+1} = \mathbf{0}$.

We hence require conditions on M which will ensure that this is the case. First we introduce the following definitions.

Definition The *spectral radius* of an $(N \times N)$ matrix M is defined to be the maximum modulus eigenvalue of M, denoted by $\rho(M)$. I.e.

$$\rho(M) \equiv \max_i |\lambda_i|$$

where λ_i, $i = 1, 2, \ldots, N$, are the eigenvalues of M.

Definition A matrix M is said to be *convergent* if $\lim_{n \to \infty} M^n = \mathbf{0}$.

Using these definitions we then obtain:

Theorem 2 *A necessary and sufficient condition that M be convergent is that* $\rho(M) < 1$.

Then:

Theorem 3 *If M is convergent* $(I - M)$ *is nonsingular.*

Proof If we denote the eigenvalues of $I - M$ by μ_i then the eigenvalues are given by

$$\det((I - M) - \mu I) = 0$$

If one of the μ_i is zero then

$$\det((I - M) - \mu I) = \det(I - M) = 0$$

and the matrix $(I - M)$ is singular. But

$$\det((I - M) - \mu I) = \det((1 - \mu)I - M)$$

Hence the eigenvalues of $I - M$ are related to the eigenvalues λ of M by $\lambda = 1 - \mu$. I.e.

$$\mu = 1 - \lambda$$

Thus if we ensure all the eigenvalues λ of M are less than 1 in modulus we will obtain a nonsingular matrix $I - M$, and the theorem is proved.

It is most often difficult to obtain $\rho(M)$ in practice. Consequently, we often accept *sufficient conditions* under which the sequence of iterations defined by Eq. (56) will converge. Such a sufficient condition follows from the following observation. If λ is an eigenvalue of M then for any eigenvector $\mathbf{x}$,

$$M\mathbf{x} = \lambda\mathbf{x}$$

Hence

$$\|\lambda\mathbf{x}\| = |\lambda| \, \|\mathbf{x}\| = \|M\mathbf{x}\| \le \|M\| \, \|\mathbf{x}\|$$

Hence

$$|\lambda| \le \|M\|$$

in any norm. In particular,

$$\max_{1 \le i \le N} |\lambda_i| \le \|M\|$$

or

$$\rho(M) \le \|M\|$$

Hence:

Lemma 3 *If $\|M\| < 1$, then $\rho(M) < 1$ and we have a convergent matrix by Theorem 2. That is, if we find any norm for which $\|M\| < 1$, the resulting sequence of iterates $\{\mathbf{x}\}_{(i)}$ will converge to the unique solution $\mathbf{x}$ of $A\mathbf{x} = \mathbf{b}$.*

Using the information we have gained on the conditions for M to be convergent, we would now like to be able to give general conditions on the properties of A which would guarantee the convergence of the point Jacobi and point Gauss–Seidel methods. Unfortunately this is a difficult task for general A. We therefore content ourselves with stated conditions for some matrices with common properties.

Definition† An $(N \times N)$ matrix $A = (a_{ij})$ is *row diagonally dominant* if

$$|a_{ij}| \ge \sum_{\substack{j=1 \\ j \ne i}}^{N} (a_{ij}) \qquad i = 1, 2, \ldots, n \tag{64}$$

† A similar definition holds for column diagonal dominance by interchanging i and j in the summation in Eq. (64).

with *strict* inequality for at least one value of *i*. The matrix A is *strictly* row diagonally dominant if the inequality in Eq. (64) is replaced by a strict inequality

Theorem 4 *If A is strictly diagonally dominant the point Jacobi method is convergent.*

Proof By definition,

$$M = \left\{ -\frac{a_{ij}}{a_{ii}} \right\} \qquad i, j = 1, 2, \ldots, N$$

where $a_{ii} \neq 0$.† But if A is strictly diagonally dominant,

$$|a_{ii}| > |a_{ij}| \qquad \forall i, j$$

Hence

$$\|M\|_\infty = \max_{1 \leq i \leq N} \sum_{\substack{j=1 \\ j \neq i}}^{N} \left| \frac{a_{ij}}{a_{ii}} \right| < 1$$

and by Lemma 3, the sequence of Jacobi iterates converges.

Theorem 5 *The point Gauss–Seidel method is convergent if the matrix is strictly row diagonally dominant.*

Theorem 6 *The point Gauss–Seidel method is convergent if the matrix A is positive definite.*

Theorem 7 *The point Jacobi and Gauss–Seidel methods are convergent if A is strictly column diagonally dominant.*

Theorem 8 *The point Jacobi method is convergent if the matrix A is irreducible‡ and diagonally (row or column) dominant.*

We note, however, that the theorems are not necessary conditions for convergence (consider the system of equations of Example 16 in the light of Theorems 4, 5, and 8).

---

† If for the original A a diagonal element exists which is zero then the point Jacobi method requires (as does the point Gauss–Seidel method) that the given system of equations be reorganized by interchange or rows) so that the $a_{ii} \neq 0$, $i = 1, 2, \ldots, N$.

‡ A matrix is irreducible if the solution to the original system of equations cannot be reduced to the solution of two or more systems of lower order.

CHAPTER 4

Interpolation

1 Introduction

In Chapter 1, we met several physical problems which exemplified the different rôles which numerical methods can play. As we saw, one of these rôles is that of *approximation*. For example, assume a metal rod has been heated for a sufficiently long period of time so that the temperature at a given point in the rod remains constant over successive time intervals. Assume it is required to approximate (measure) the temperature at any point x, measured from one end of the rod by an easily calculable formula; a quadratic polynomial in x perhaps, namely

$$y = a_2 x^2 + a_1 x + a_0 \tag{1}$$

where a_2, a_1, and a_0 are known constants and y is the measured or approximated temperature. Thus, we can determine the temperature in the rod approximately by substituting that value of x which represents the required point.

This simple example serves to illustrate several important points. First, we have stated that a_2, a_1, and a_0 are *known* constants. The question arises "how should we choose these constants?" For example, a quadratic of the form (1) is obviously $2x^2 + 3x + 1$; so is $3x^2 + 2x + 1$, each giving a different value for the required temperature.†

Second, when specifying a function to approximate a certain physical phenomenon, it is clearly necessary to specify in what region the approximation is valid. For example, if the length of the rod is defined by $0 \le x \le 100$ cm, it is fallacious to use the approximating function at either $x = 101$ or $x = -3$ cm!

Third, why use a quadratic polynomial as the approximating function y? Would a cubic polynomial give better results or would a polynomial of degree

---

† An answer to this question is, perhaps, "Choose a_2, a_1, and a_0 to give the best approximation to the required physical phenomenon". However, this evades to a certain extent the question of choosing a_2, a_1, and a_0, since what do we mean by *best*?

one give just as good results as the quadratic? In fact, why choose polynomials in x at all as our approximating functions? Would some other class of functions do as well or better than a comparable polynomial in x? For example, combinations of trigonometric functions such as $\sin(x)$, $\cos(x)$, or rational functions such as

$$\frac{1}{a+bx}, \quad \frac{cx+d}{ax+b}$$

In this and the two following chapters, we will endeavor to answer these questions. The division of the material into the three chapters comes about according to the three ways in which we answer the first of our questions, namely how do we choose the parameters $a_0, a_1, \ldots$? The present chapter deals with a form of approximation known as *exact approximation*.

Consider the approximation of a function $f(x)$ whose values $f(x_i)$, which we represent notationally by f_i, are given at $n+1$ *distinct* tabular points x_i, $i = 0, 1, \ldots, n$, not necessarily in ascending order of magnitude.[†] Determine an approximating function $y(x)$ which takes values f_i at the tabular points. We then say $y(x)$ is the *exact approximant*. That is, exact approximation occurs when

$$y_i = f_i \qquad i = 0, 1, \ldots, n \tag{2}$$

This problem may be represented diagrammatically by Figure 1. Here a discrete set of point values of a function $f(x)$ at the tabular points x_i, $i = 0, 1, \ldots, n$, is represented by the heavy dots. If this set of points is all that is given, then in general, it is impossible to determine what function these values represent. Hence, to determine values intermediate to the tabular points it

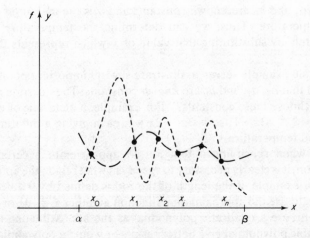

Figure 1

[†] That is, x_i need not necessarily be greater than x_{i-1} and x_i need not be less than x_{j+3}, etc.

is necessary to approximate the function f by some other function. Two such functions which both pass through the dots and hence satisfy Eq. (2) are shown in the figure; clearly they are not the same and have considerably different behaviors. Who is to say which is correct or even better?

If the tabular points x_i, $i = 0, 1, \ldots, n$, lie in an interval $[\alpha, \beta]$ (like that depicted in Figure 1) and the value of x for which it is required to evaluate $f(x)$ is also in $[\alpha, \beta]$, then the approximation problem is called *interpolation*. If in contrast, x does not lie in $[\alpha, \beta]$, the problem is one of *extrapolation*. In the former case $y(x)$ is called an interpolating function whereas in the latter case $y(x)$ is called an extrapolating function. In this chapter we will be concerned in the main with the problem of interpolation. However, we will deal briefly with extrapolation in section 11.

Example 1 A common example of the need for interpolation arises in the case of requiring values of a function at points other than those given in a table. For example, if we are given the values of $\sin(x)$ in Table 1 at the tabular points indicated, we might require the value of $\sin(x)$ at $x = 1.375$ radians. To determine this values of $\sin(x)$ we can approximate the value of $\sin(x)$ between $x = 1.35$ and 1.40 by an interpolating function $y(x)$.

Table 1

x	1.100	1.1500	1.2000	1.2500	1.3000	1.3500	1.4000	1.4500	1.5000
$\sin(x)$	0.8912	0.9128	0.9320	0.9490	0.9636	0.9757	0.9854	0.9927	0.9975

If we assume $\sin(x)$ behaves like a straight line between $x = 1.35$ and 1.40, then we can interpolate $\sin(x)$ by the straight line

$$y = a_1 x + a_0$$

where the parameters a_1 and a_0 are to be determined such that

$$y(x = 1.35) = a_1 \times 1.35 + a_0 = \sin(1.35) = 0.9757 \tag{3}$$

and

$$y(x = 1.40) = a_1 \times 1.40 + a_0 = \sin(1.40) = 0.9854 \tag{4}$$

where we have used the values of $\sin(x)$ from Table 1. From Eqs. (3) and (4) we can determine the constants a_1 and a_0, namely $a_1 = 0.1940$ and $a_0 = 0.7138$, so that the linear interpolating function is

$$y(x) = 0.1940x + 0.7138$$

If we assume instead that $\sin(x)$ behaves like a quadratic polynomial in a small interval about the point where interpolation is required, then we can interpolate $\sin(x)$ by

$$y(x) = a_2 x^2 + a_1 x + a_0$$

106

In this case the constants a_2, a_1, and a_0 can be determined by equating $y(x)$ at $x = 1.30$, 1.35, and 1.40, or at $x = 1.35$, 1.40, and 1.45, depending on which interval we assume the quadratic behavior of sin (x). For the former case we obtain the three equations

$$y(x = 1.3) = a_2(1.3)^2 + a_1(1.3) + a_0 = \sin(1.3) = 0.9636$$

$$y(x = 1.35) = a_2(1.35)^2 + a_1(1.35) + a_0 = \sin(1.35) = 0.9757$$

$$y(x = 1.40) = a_2(1.4)^2 + a_1(1.4) + a_0 = \sin(1.4) = 0.9854$$

to solve for a_2, a_1, and a_0. Solving this system of equations we find that $y(x) = -0.48x^2 + 1.514x - 0.1934$. (The quadratic polynomial based on $x = 1.35$, 1.40, 1.45 is left as an exercise for the reader—see Exercise 2.)

The procedure outlined in Example 1 can be repeated for polynomials of higher degree until we are satisfied that the interpolating function has the behavior sufficiently close to sin (x) or the number of data points runs out!

Having obtained a criterion for choosing the parameters in the interpolating function, it remains to discuss the nature of the functions which these parameters multiply. That is, in the case of two available parameters should the interpolating function be of the form

$$a_1x + a_0, \qquad \frac{1}{a_1x + a_0} \qquad \text{or} \qquad a_1 \cos(x) + a_0 \sin(x)?$$

In all cases two tabular points and their corresponding data points will allow the determination of the available parameters (provided that $f(x_i) \neq 0$ for the rational interpolant—why?).

Whereas there is no a priori way of determining mathematically the answer to this question for all cases, the question can be best answered from the considerations of computer time. For practical problems, numerical methods today are heavily dependent upon computers. Thus the operations involved in an algorithm should be chosen, when all other considerations are equal, on the basis of which one requires least computer time. Since polynomials require less time to calculate than combinations of trigonometric functions (each one of which will require the call of a subroutine), rational functions (which require divisions as well as multiplication), etc., it can be seen that polynomial interpolation requires less computer time. For this reason, interpolation using polynomials is usually *discussed*† most fully.

Exercises

1. For the values in Table 1 derive the quadratic polynomial which interpolates sin (x) using the tabular points $x = 1.25$, 1.30, and 1.35. Hence interpolate sin (1.325), sin (1.275), and sin (1.375).

† This sentence has been carefully worded to avoid misleading the reader into thinking that interpolation is a *process* which is often used practically. In fact, very little practical interpolation is carried out today by numerical analysts since required function values are generated directly by the computer by other means. However, interpolation forms the core of a great deal of the theory of numerical analysis and it is for this reason the interpolation is discussed in detail.

2. Derive the quadratic polynomial which interpolates the tabular points 1.35, 1.40, 1.45 in Table 1. Hence interpolate $\sin (1.375)$. How does the present value compare with the value obtained in Exercise 1? Can you explain the difference? Can you make a general deduction from these cases for the positioning of tabular points relative to the required point of interpolation ($x = 1.375$ above)?

3. Derive an interpolating function of the form $y(x) = 1/(a_1 x + a_0)$ which interpolates $\sin (x)$ for $x = 1.375$. What form does the interpolating function take when three data points are used?

4. Using interpolating formulas based on: first, powers of x (linear and quadratic polynomials as in Exercises 1 and 2); and second, on rational functions involving a denominator which is linear and then one which is a quadratic, interpolate for $\sin (x)$ using values in Table 1. Compare the results obtained using the different formulas.

2 Polynomial Interpolation

For reasons outlined in the previous section, we will restrict our attention in the main to interpolating functions which are polynomials in x, i.e. our interpolating function $y(x)$ can be written as

$$y(x) = \sum_{i=0}^{n} a_i x^i$$

for some given value of n. If n is 1, then we are using linear interpolation since y is then a straight line. If $n = 2$, we are using quadratic interpolation; $n = 3$, cubic interpolation; and so on. Before considering the construction of such polynomial interpolants of a specified degree n, we require to consider the question of existence and uniqueness of such interpolants. That is, we require to determine whether for $n + 1$ tabular points x_i, $i = 0, \ldots, n$, and their corresponding data points f_i, there exists an interpolating polynomial, and if so when is it unique? First let us consider the *existence* of such an interpolating polynomial.

Consider the case where $n = 1$, i.e. there are just two data points f_1 and f_0 at x_1 and x_0 respectively. By definition, the interpolating polynomial of degree 1 is the straight line $y \equiv a_1 x + a_0$ which satisfies the relations

$$y(x_1) \equiv a_1 x_1 + a_0 = f_1 \tag{5}$$

$$y(x_0) \equiv a_1 x_0 + a_0 = f_0 \tag{6}$$

Thus we see that by subtracting Eq. (6) from Eq. (5) we obtain

$$a_1(x_1 - x_0) = f_1 - f_0 \quad \text{or} \quad a_1 = \frac{f_1 - f_0}{x_1 - x_0}$$

Substituting the expression for a_1 into either of Eqs. (5) and (6) will give the result

$$a_0 = \frac{f_0 x_1 - f_1 x_0}{x_1 - x_0}$$

so that the linear interpolant is given by

$$y(x) = \frac{(f_1 - f_0)x}{x_1 - x_0} + \frac{f_0 x_1 - f_1 x_0}{x_1 - x_0} \tag{7}$$

Exercises

5. If a polynomial $a_2 x^2 + a_1 x + a_0$ were used to interpolate the two data points f_0 and f_1, what precise form would the interpolant take? Would the form indicate that there was just one interpolant or many? Why?
6. If in contrast to the interpolant of Exercise 5, an interpolant of the form $y = a_0$ were proposed, what would you conclude in general about interpolating the function $f(x)$ given f_0 and f_1 at x_0 and x_1 respectively? When would such an interpolation be reasonable?
7. What in general is the lowest degree of the polynomial which interpolates the function $f(x)$, given the values f_0, f_1, and f_2 at x_0, x_1, x_2 respectively? What condition would be implied if a polynomial of lower degree interpolated these points?
8. Determine explicitly the interpolating polynomial for Exercise 7.
9. Given $n + 1$ data points f_i, $i = 0, 1, \ldots, n$, corresponding to distinct tabular points x_i, $i = 0, 1, \ldots, n$, what in general is the lowest degree of the interpolating polynomial?

Consider now the case where values of the function $f(x)$ are given at the $n + 1$ distinct tabular points $x_0, x_1, \ldots, x_n$. Let us construct the simplest polynomial which satisfies Eq. (2) for $i = 0, 1, \ldots, n$. From Exercises 5 to 9 it will be clear that for $n + 1$ data points f_i, $i = 0, 1, \ldots, n$ a polynomial of degree n will be required, in general, to satisfy Eq. (2). Let this polynomial be represented by

$$y(x) = a_n x^n + a_{n-1} x^{n-1} + \cdots + a_1 x + a_0 \tag{8}$$

We therefore require to determine the parameters a_i, $i = 0, 1, \ldots, n$, such that y defined by Eq. (8) satisfies Eq. (2). That is,

$$y(x_i) = a_n x_i^n + a_{n-1} x_i^{n-1} + \cdots + a_1 x_i + a_0 = f(x_i) \qquad i = 0, 1, \ldots, n$$

This, when written out in full, is equivalent to the system of linear equations given by

$$\begin{aligned}
a_n x_0^n + a_{n-1} x_0^{n-1} + \cdots + a_1 x_0 + a_0 &= f(x_0) \\
a_n x_1^n + a_{n-1} x_1^{n-1} + \cdots + a_1 x_1 + a_0 &= f(x_1) \\
&\vdots \\
a_n x_n^n + a_{n-1} x_n^{n-1} + \cdots + a_1 x_n + a_0 &= f(x_n)
\end{aligned} \tag{9}$$

where, of course, the right-hand sides of (9) are known. To determine the a_i, $i = 0, 1, \ldots, n$, we require to solve the system of equations (9). The condition which guarantees a *unique* solution for the a_i, $i = 0, 1, \ldots, n$, is that the

Vandermonde determinant

$$\begin{vmatrix} 1 & x_0 & x_0^2 & \cdots & x_0^n \\ 1 & x_1 & x_1^2 & \cdots & x_1^n \\ \vdots & & & & \vdots \\ 1 & x_n & x_n^2 & \cdots & x_n^n \end{vmatrix}$$

is nonzero. This is guaranteed if the tabular points x_i are distinct for $i = 0, 1, \ldots, n$.

Consequently we have shown that the required polynomial interpolant exists and is unique. To solve the set of equations (9), particularly for large n, poses a difficult problem. In fact, it can be shown that there is no need to solve Eq. (9) for the a_i. This comes about from the considerations of the following section.

3 Lagrangian Interpolation

We require an interpolating polynomial of degree n which passes through the $n+1$ points (x_0, f_0), $(x_1, f_1), \ldots, (x_n, f_n)$. Therefore construct functions which satisfy the relations

$$L_j(x_i) = \begin{cases} 1 & j = i, j = 0, 1, \ldots, n \\ 0 & j \neq i, i = 0, 1, \ldots, n \end{cases} \tag{10}$$

The interpolating polynomial defined by

$$y(x) = \sum_{j=0}^{n} L_j(x) f_j \tag{11}$$

clearly satisfies Eq. (2). It therefore remains to determine the precise analytical form of the $L_j(x)$. To do this, consider a particular function $L_J(x)$. By Eq. (10) we require that

$$L_J(x_i) = 0 \quad \text{for } i = 0, 1, \ldots, n, i \neq J$$

This therefore suggests that $L_J(x)$ has factors of the form

$$(x - x_0), (x - x_1), \ldots, (x - x_{J-1}), (x - x_{J+1}), \ldots, (x - x_n) \tag{12}$$

Now since $y(x)$ is to be a polynomial of degree n or less, then since in Eq. (11) we have a linear combination of the $L_j(x)$, the $L_j(x)$ must in turn by polynomials of degree n or less. Counting the factors in Eq. (12), we see that there are n such factors. Thus define

$$\pi_{n+1}(x) = \prod_{i=0}^{n} (x - x_i) \qquad \text{the polynomial of degree } n+1 \text{ in } x$$

and

$$\pi_{nJ}(x) = (x - x_0)(x - x_1) \cdots (x - x_{J-1})(x - x_{J+1}) \cdots (x - x_n) \tag{13}$$

$$= \prod_{\substack{i=0 \\ i \neq J}}^{n} (x - x_i) \qquad \text{the product of factors with } (x - x_J) \text{ omitted}$$

Then $\pi_{nJ}(x)$ is a polynomial of degree n and satisfies Eq. (10) for $j = J$, $i \neq J$. However, when $i = J$, $\pi_{nJ}(x)$ is not unity. To ensure that it is, divide $\pi_{nJ}(x)$ by $\pi_{nJ}(x_J) = (x_J - x_0)(x_J - x_1) \cdots (x_J - x_{J-1}) \cdots (x_J - x_n)$ to give

$$L_J(x) = \frac{\pi_{nJ}(x)}{\pi_{nJ}(x_J)} \tag{14}$$

Then this function satisfies Eq. (10) exactly for $j = J$, $i \neq J$. Generalizing this result for all $j = 0, 1, \ldots, n$, we see that the required form of $L_j(x)$ is

$$L_j(x) = \frac{\pi_{nj}(x)}{\pi_{nj}(x_j)} \qquad j = 0, 1, \ldots, n \tag{15}$$

The interpolating polynomial Eq. (11) with the $L_j(x)$ defined by Eq. (15) is called the *Lagrangian* interpolating polynomial.

Example 2 For the case $n = 1$, we have two data points f_0 and f_1. The Lagrangian interpolating polynomial gives the usual *linear* interpolating formula (7),

$$y(x) = \frac{(x - x_1)}{(x_0 - x_1)} f_0 + \frac{(x - x_0)}{(x_1 - x_0)} f_1$$

in reorganized form.

Example 3 By substituting $n = 2$ into *Eq.* (15), the second-degree Lagrangian interpolating polynomial can be obtained, namely

$$y(x) = \frac{(x - x_1)(x - x_2)}{(x_0 - x_1)(x_0 - x_2)} f_0 + \frac{(x - x_0)(x - x_2)}{(x_1 - x_0)(x_1 - x_2)} f_1 + \frac{(x - x_0)(x - x_1)}{(x_2 - x_0)(x_2 - x_1)} f_2$$

which is exactly the interpolating polynomial of Exercise 8 in reorganized form.

Example 4 The derivation of the Lagrangian interpolating polynomial of degree n given above duplicates to some extent our work in proving the existence of an interpolating polynomial. Since it is the principle of this book to ask as little as possible prerequisite knowledge of the reader and since we merely *stated* that existence and uniqueness is guaranteed when the Vandermonde determinant is nonzero (we did not prove it), we can reassure the reader by a very simple proof of the uniqueness of the interpolating polynomial that the Lagrangian interpolating polynomial is nothing more than a reorganization of Eq. (8). This proof, however, requires the following lemma.

Lemma 1 *A polynomial of degree n has more than n zeros if and only if it is identically zero.*

(The interested reader is referred to Theorem 4.2 of [47] for a proof of this lemma.)

In other words, if we find that a supposed nth-degree polynomial is zero for more than n values, then the polynomial must be zero everywhere.

Using this result we can prove the uniqueness of the interpolating polynomial of degree n or less which passes through the points (x_0, f_0), $(x_1, f_1), \ldots, (x_n, f_n)$.

Proof Assume there are two interpolating functions $P(x)$ and $Q(x)$ which satisfy

$$P(x_i) = Q(x_i) = f_i \qquad i = 0, 1, \ldots, n$$

such that P and Q are *not* identical. We can form their difference and represent it by

$$D(x) = P(x) - Q(x)$$

Since P and Q are both of degree n or less, $D(x)$ must also be of degree less than or equal to n (why?). But by assumption both polynomials interpolate $f_i, i = 0, 1, \ldots, n$. Hence,

$$D(x_i) = P(x_i) - Q(x_i) = 0 \qquad i = 0, 1, \ldots, n$$

which indicates that $D(x)$ has at least $n + 1$ zeros. By Lemma 1, since $D(x)$ is at most degree n, $D(x)$ must be identically zero. That is, $P(x)$ and $Q(x)$ are identical and our assumption of more than one interpolating polynomial was false. Hence the interpolating polynomial is unique.

Example 5 In the development of the Lagrangian interpolating polynomial, nothing has been stated about the interpolating points other than that they are distinct. The question naturally arises: is there any advantage in taking the interpolating points equally spaced when their position is our choice? Clearly the interpolating formula is valid whether the points $x_i, i = 0, 1, \ldots n$, are equally spaced or not. However, if the points *are* equally spaced the Lagrangian interpolating polynomial takes on a simplified form. To see this, let us assume that the distance between the interpolating points is h, that is, $x_i - x_{i-1} = h, i = 1, 2, \ldots, n$, and that, for convenience, we have $2r + 1$ interpolating points. Thus we assume x_r is the mid-point of the interval. With this notation we also let x be represented as a fractional multiple of h, so

$$x = x_r + mh$$

If we substitute this notation into Eq. (11) we obtain

$$y(x_r + mh) = \sum_{j=0}^{2r} L_j(x_r + mh) f(x_j)$$

or

$$y(mh) = \sum_{j=0}^{2r} L_j(mh) f(x_j)$$

where

$$L_j(mh) = \frac{\pi_{nj}(mh)}{\pi_{nj}(jh)} = \prod_{\substack{i=0 \\ i \neq j}}^{2r} \frac{(x_r + mh - x_r - ih)}{(x_r + jh - x_r - ih)} \qquad j = 0, 1, \ldots, 2r$$

Thus

$$L_j(mh) = \prod_{\substack{i=0 \\ i \neq j}}^{2r} \frac{(m-i)h}{(j-i)h} = \prod_{\substack{i=0 \\ i \neq j}}^{2r} \frac{(m-i)}{(j-i)} \qquad j = 0, 1, \ldots, 2r$$

Consequently, in this case the Lagrangian interpolating polynomial reduces to a simplified form which *does not* depend on the distance h, as long as this distance is constant for each successive interpolating pair (x_i, x_{i+1}).

This form of the interpolating polynomial is often called the *normalized* Lagrangian interpolating polynomial.

In the present chapter, we will discuss several forms of the interpolating polynomials of degree n. In the light of the preceding discussions on the uniqueness of the interpolating polynomials, it should be borne in mind that these different formulations are reorganizations of the *same* basic Lagrangian interpolating polynomial. This, of course, does not mean to say that one formulation is as good as another in certain situations. (As we have already seen, the case of equally spaced tabular points yields a simplified form of the Lagrangian interpolating polynomial; the answers one obtains using either formulation, ignoring rounding-error differences, are the same.)

Exercises

10. *Derive* the Lagrangian interpolating polynomial which is exact for functions $f(x)$ which are polynomials of degree up to and including those of degree 3. Show that this is equivalent to the interpolating polynomial represented by Eq. (11) with $n = 3$.
11. What form does the normalized Lagrange interpolating formula take for:
    (a) $m = 0.2, r = 3$
    (b) $m = 3.2, r = 1$
    (c) $m = 0.6, r = 1$
    (d) $m = -1.8, r = 2$?
12. Use the normalized Lagrangian interpolating polynomial of degree 4 using the points in the table below to evaluate approximately $f(1.25)$ and $f(1.15)$.

x	1.0	1.1	1.2	1.3	1.4
$f(x)$	0.3679	0.3329	0.3012	0.2725	0.2466

13. What modification to the normalized Lagrangian interpolating formula is required if the number of interpolating points is *even*? Write a computer program to calculate the Lagrangian interpolating polynomial for a general positive integer n and function values $f_i, i = 0, 1, \ldots, n$.

14. Use the program of Exercise 13 to calculate sin (0.6), sin (1.2), sin (2.0), and sin (2.6) using values of sin (x) for $x = 0.0$, 0.8, 1.6, 2.4, *and* 3.0 contained in Table 2.

Table 2

x	0.0000	0.4000	0.800	1.0000	1.4000	1.6000
sin (x)	0.0000	0.3894	0.7174	0.8415	0.9854	0.9996
x	1.8000	2.2000	2.4000	2.8000	3.0000	
sin (x)	0.9738	0.8085	0.6755	0.3350	0.1411	

Similarly, calculate exp (0.6), exp (1.2), exp (2.0), and exp (2.6) using the values of exp (x) in Table 3.

Table 3

x	0.0000	0.8000	1.6000	2.4000	3.0000
exp (x)	1.0000	2.2255	4.9530	11.0232	20.0855

15. If your computer center possesses a graph plotter:
    (a) Plot, on one graph, curves of sin (x) and the Lagrangian formula based on the values of sin (x) given in Table 2 for $x = 0.0$, 0.8, 1.6, 2.4, and 3.0, for
    (i) $0 \le x \le 3.1$
    (ii) $-3.1 \le x \le 6.1$
    (b) Plot on one graph, curves of exp (x) and the Lagrangian formula based on the five points in Table 3, for
    (i) $0 \le x \le 3.1$
    (ii) $-3.1 \le x \le 6.1$
    By reading into your computer program a value of n equal to 10 for the interpolation of sin (x) and using the remaining values of Table 2, compute the Lagrangian interpolating polynomial and plot its value for ranges (i) and (ii) above. In contrast to part (a) of this question, what are your conclusions about this latest run?

16. In Exercise 15 the Lagrangian interpolating polynomial of degree 10 gave certain behavior when used to represent sin (x) over the interval $-3.1 \le x \le 6.1$. This behavior was in contrast to using the interpolating polynomial of degree 4. What behavior is exhibited by the interpolating polynomial of degree 4 using the points of Table 4?

Table 4

x	−2.5000	−1.0000	0.50000	2.0000	3.5000
sin (x)	−0.5985	−0.8415	0.4794	0.9093	−0.3508

Does this interpolating polynomial give better (worse) results than those obtained from the polynomial of degree 10 for the range
(a) $0 \le x \le 3.1$
(b) $-3.1 \le x \le 6.1$?
Can you generalize your conclusions?

17. Conduct a similar set of experiments for exp (x); namely, obtain an interpolating polynomial of high degree (10) based on a subinterval of an interval in which interpolation is required† and compare the results obtained by using an interpolating polynomial of lower degree based on the complete interval.

Exercises 14, 15, 16, and 17 are clearly ones which are not meant to be solved by paper-and-pencil methods. If, however, such an attempt were tried, the reader would be impressed by the considerable amount of work involved when changing the degree of the Lagrangian interpolating polynomial. That is, for each degree polynomial *all* the coefficient functions $L_i(x)$ have to be recalculated without being able to use any of the information gained in calculating the $L_i(x)$ of lesser degree. We will consider this point later in section 7.

4   The Error Term in Lagrangian Interpolation

Exercises 14, 15, 16, and 17, in particular, will have inpressed upon the reader that, in general, an error exists at nontabular points when interpolating $f(x)$ by a function $y(x)$ for functions $f(x)$ which are *not* polynomials of degree n or less. To determine the form of this error term when interpolating $f(x)$ by the Lagrangian interpolating polynomial of degree n, we require *Rolle's theorem*, which states:

> If $f(x)$ is continuous on $\alpha \le x \le \beta$ and has a derivative $f^{(1)}(x)$ for $\alpha \le x \le \beta$, and if $f(\alpha) = f(\beta) = 0$, then there is at least one root ξ satisfying $\alpha \le \xi \le \beta$ such that $f^{(1)}(\xi) = 0$.

(A proof of this theorem is not needed but the interested reader may obtain a proof from Jones and Jordan [27]).

Assume $f(x)$ has at least $n + 1$ derivatives. Let the error term be represented by

$$E(x) = f(x) - y(x)$$

Consider the function F defined by

$$F(z) = f(z) - y(z) - (f(x) - y(x))(\pi_{n+1}(z)/\pi_{n+1}(x)) \tag{16}$$

By substituting $z = x_0, x_1, \ldots, x_n$ and x into Eq. (16), it may be verified that $F(z)$ has $n + 2$ zeros where, for the present, we assume x is *not* a tabular point where we have used Eq. (2).

Applying Rolle's theorem $n + 1$ times to $F(z)$, we find that

$$F^{(n+1)}(z) = f^{(n+1)}(z) - y^{(n+1)}(z) - (f(x) - y(x))[(n + 1)!/\pi_{n+1}(x)]$$

has at least one zero in the smallest interval containing $x_0, x_1, \ldots, x_n$, and x, where $n!$ is the usual notation for $n(n - 1)(n - 2) \cdots 3 \cdot 2 \cdot 1$.

---

† Strictly, this situation includes the problem of extrapolation; see sections 1 and 11.

Let such a root be ξ. Then

$$F^{(n+1)}(\xi) = f^{(n+1)}(\xi) - y^{(n+1)}(\xi) - (f(x) - y(x))[(n+1)!/\pi_{n+1}(x)] = 0$$

But $y^{(n+1)}(\xi) = 0$ since $y(x)$ is a polynomial of degree n or less. Hence

$$f^{(n+1)}(\xi) - (f(x) - y(x))[(n+1)!/\pi_{n+1}(x)] = 0$$

Thus using the definition of the error term we see that

$$E(x) = \frac{f^{(n+1)}(\xi)\pi_{n+1}(x)}{(n+1)!} \tag{17}$$

Whereas at first sight this might appear to give a great deal of information as to the amount by which the interpolating function $y(x)$ is in error, in general the $(1+n)$th derivative of $f(x)$ will not be known, and even if it were, ξ is not. Thus, $E(x)$ defined by Eq. (17) is of use in general only if we can *estimate* an upper bound to $f^{(n+1)}(\xi)$ for ξ contained in the smallest interval containing $x_0, x_1, \ldots, x_n$, and x.

From Eq. (17) we see that $E(x)$ is dependent upon the degree n of the interpolating polynomial. Thus, provided $f^{(n+1)}(\xi)$ is not a rapidly increasing function of n (see Exercise 23), the size of the error term will decrease with increasing n. That is, increasing the degree of the interpolating polynomial, with the above proviso, will often yield a more accurate interpolation. If $f(x)$ is a polynomial of degree n or less, then from Eq. (17) the error will be zero. That is, *the interpolating polynomial y of degree n reproduces exactly the function f of degree n or less.*

An alternative approach to the derivation of the error term is given in [6].

Exercises

18. If an interpolating function $y(x)$ of degree 2 is uniquely defined on an interval $[0, 1]$, what would the maximum error be if $y(x)$ were used to interpolate a function f which is
    (a) a polynomial of degree 3
    (b) a rational function, $1/(a_0 + a_1 x)$
    (c) $\sin(x)$?
    What would be the maximum error if $y(x)$ were degree 6 in each case?

19. It is required to interpolate x correct to three decimal places using an interpolating polynomial of degree 4 for the interval $(-1, 1)$, using tabular points spaced at intervals of 0.1 starting at -1. In what part of this interval can such accuracy be expected? In what part of the interval will three decimal places of accuracy be guaranteed if a fifth-degree interpolating polynomial is used instead?

20. It is required to interpolate $\log \cos(x)$ using linear interpolation on equally spaced interpolating points to obtain an absolute error less than 10^{-6} over the interval $[0, 1]$. What is the minimum number of interpolating points which will give the required accuracy?

21. What are the maximum error bounds for the normalized Lagrangian interpolation formulas of Exercise 11 where it is assumed that M_a, M_b, M_c, and M_d are the bounds on the necessary derivatives of the function f?

22. What are the error bounds in the normalized Lagrangian interpolating formulas used in Exercise 12, given that the tabulated function is $\exp(-x)$? Is this compatible with the computed errors?

23. What are the error bounds obtained when interpolating the function $f(x) = (2.025 - x)^{1/2}$ for x in the interval $1 \le x \le 2$, using:

   (i) $n = 1$ with $x_0 = 1$, $x_1 = 2$
   (ii) $n = 2$ with $x_0 = 1$, $x_1 = 1.5$, $x_2 = 2$
   (iii) $n = 3$ with $x_0 = 1$, $x_1 = 1.3$, $x_2 = 1.6$, $x_3 = 2.0$?

   On the basis of these error bounds alone, which of the three values of n would indicate the best degree of the interpolating polynomial?

5 Divided Differences

Given $n + 1$ tabular points (which are distinct) at which values of a function f are given, we have seen in section 3 that there exists a unique polynomial of degree at most n which interpolates these values. The reader could quite justifiably ask why we should bother considering any other interpolation formula based upon these $n + 1$ points when by reorganization they must be equivalent to the Lagrangian formula. The answer to this question, which necessitates our considering further reorganizations of the Lagrangian interpolating function, is that in certain circumstances the particular generality of the Lagrangian interpolating polynomials does not lend itself to convenient and efficient evaluation. A good example of this, which we will consider in more detail in the next section, occurs when we interpolate given data on a fairly large number of points. If the form of the data suggests that the function being interpolated were a straight line, using the general Lagrangian interpolating formula based upon all the points would clearly involve a considerable amount of work which would in the end possibly produce just a straight line. In other words, in certain circumstances it might be desirable to interpolate the given data with a straight line, evaluate the error bound, and if this is not sufficiently small introduce an extra data point and interpolate using a quadratic, evaluate the error term, and repeat the process until the error is sufficiently small. If we were to use the Lagrangian interpolating polynomials for this process, the introduction of an extra data point into the interpolation would necessitate a *complete* recalculation of the Lagrangian polynomials. If a method could be devised which did not require this extra work each time a new data point was added, then this interpolation formula would be superior in terms of efficiency to the Lagrangian interpolation.

In this section we want to consider an idea which will be used in, and suggests methods for, developing interpolating formulas which make use of particular properties of the data used and which hence will be more efficient and convenient than the general Lagrangian interpolating polynomials. The idea we require to consider here is that of *divided differences* of functions on unequally spaced intervals. In a sense we have already met this idea in our discussion of linear interpolation in section 3, namely the determination of the coefficients a_0 and a_1 in the linear interpolation formula involved the *first* divided difference of $f(x)$ relative to x_0 and x_1, i.e.

$$\frac{f(x_1) - f(x_0)}{x_1 - x_0} \tag{18}$$

Notationally we represent this by $f[x_0, x_1]$, where from Eq. (18) $f[x_0, x_1] = f[x_1, x_0]$ is symmetric in the arguments. It is clear that the first divided difference for a *linear* function f is independent of x_0 and x_1. Consequently, the assumption that another function g is *approximately* linear is tantamount to saying that the first divided difference is (approximately) independent of the tabular points appearing in it, namely

$$\frac{g(x) - g(x_0)}{(x - x_0)} \doteq \frac{g(x_1) - g(x_0)}{x_1 - x_0}$$

or

$$g(x) \doteq g(x_0) + (x - x_0)\left\{ \frac{g(x_1) - g(x_0)}{x_1 - x_0} \right\}$$

which is precisely formula (7) for a function g expressed in terms of observed values $g(x_1)$, $g(x_0)$ at $x = x_1$ and x_0 respectively.

If we introduce the concept of zeroth divided differences of f relative to x_0 as

$$f[x_0] = f(x_0)$$

then we can represent the linear interpolation completely in terms of zeroth and first divided differences as

$$g(x) \doteq g[x_0] + (x - x_0)g[x_0, x_1]$$

If we call the interpolation on the right-hand side $y_{01}(x)$, where 01 indicates the interpolation formula involving first divided differences[†] based on x_0 and x_1, we then have

$$g(x) \doteq y_{01}(x) = g[x_0] + (x - x_0)g[x_0, x_1]$$

Thus for two other points, x_2 and x_3, say, we also have

$$g(x) \doteq y_{23}(x) = g[x_2] + (x - x_2)g[x_2, x_3]$$

so that divided differences give us a convenient notation for writing down the particular formula.

The first divided difference $f[x_0, x_1]$ is called the *secant slope* of f. Unless f is linear, then clearly this slope will depend upon x_1 and x_0. If, however, f were a quadratic function, the secant slope $f[x_1, x]$ would itself be a linear function of x for fixed x_1. Thus in this case the divided difference of the two divided differences based on three points x_0, x_1, and x_2 will be independent

---

[†] Two subscripts indicate first divided differences, n subscripts indicate $(n-1)$st divided differences.

of the abscissas. We represent this *second* divided difference by

$$f[x_0, x_1, x_2] = \frac{f[x_1, x_2] - f[x_0, x_1]}{x_2 - x_0}$$

Consequently, using the fact that this is independent of the abscissas for quadratic f, we can approximate another function g which we assume is approximately quadratic in the interval $[x_0, x_2]$ by

$$g[x, x_0, x_1] \doteq g[x_0, x_1, x_2]$$

or

$$\frac{g[x_0, x_1] - g[x, x_0]}{x_1 - x} = g[x_0, x_1, x_2]$$

so

$$g[x, x_0] \doteq g[x_0, x_1] + (x - x_1)g[x_0, x_1, x_2]$$

and hence

$$\frac{g[x_0] - g[x]}{x_0 - x} \doteq g[x_0, x_1] + (x - x_1)g[x_0, x_1, x_2]$$

from which it follows that

$$g(x) \doteq g[x_0] + (x - x_0)g[x_0, x_1] + (x - x_0)(x - x_1)g[x_0, x_1, x_2]$$

If we now denote the interpolation formulas based on the points x_0, x_1, and x_2 by $y_{012}(x)$, we have

$$g(x) \doteq y_{012}(x) = g[x_0] + (x - x_0)g[x_0, x_1] + (x - x_0)(x - x_1)g[x_0, x_1, x_2] \quad (19)$$

so that interpolation of the function g based upon three points involves zeroth, first, and second divided differences of g. By taking the above idea to its logical conclusion we can define nth divided differences of a function of x based upon the points $x_0, x_1, \ldots, x_n$ by

$$f[x_0, x_1, \ldots, x_n] = \frac{f[x_1, \ldots, x_n] - f[x_0, \ldots, x_{n-1}]}{x_n - x_0}$$

From this we see that a divided difference of any order can be defined *recursively* in terms of divided differences of lower order. Also, the order of the arguments x_i appearing in the divided difference is of no consequence (see Exercise 24). That is,

$$f[x_0, x_1, \ldots, x_n] = f[x_n, x_{n-1}, \ldots, x_1, x_0]$$

More general properties of divided differences may be found in Hilderbrand [22].

Using a divided difference formula as the basis for an interpolating formula, we may obtain in a straightforward manner the form of the error term. For the linear interpolation, the error may be obtained as

$$E(x) = f(x) - y_{01}(x) = f[x_0] + (x - x_0)f[x_0, x] - f[x_0] - (x - x_0)f[x_0, x_1]$$
$$= (x - x_0)f[x_0, x] - (x - x_0)f[x_0, x_1]$$
$$= (x - x_0)(f[x_0, x] - f[x_0, x_1])$$
$$= (x - x_0)(x - x_1)f[x_0, x_1, x]$$

since

$$f[x_0, x] - f[x_0, x_1] = (x - x_1)f[x_0, x_1, x]$$

This is clearly only known exactly when $f[x_0, x_1, x]$ is known, which is tantamount to knowledge of $f(x)$ which we do not have. Thus, as in the case of Lagrangian interpolation, we use bounds on $E(x)$ to give an estimate of the accuracy achieved by the interpolation formula.

It is clear from the form of E that at $x = x_0$ and x_1 the error is zero and when $f[x_0, x_1, x]$ is zero the error is zero; this latter situation arises clearly when f is linear; the interpolating formula is exact (has no error) whenever interpolation on a linear function is effected.

Exercises

24. Prove by induction that the nth divided difference is a symmetric function of its arguments, i.e. the order of the arguments of the nth-order divided difference is irrelevant.
25. Determine the form of the error term in the interpolation formula involving third-order divided differences of a function f.
26. Develop a computer subprogram which will produce as an end result the divided difference of a function based upon two given input tabular points x_0 and x_1 and corresponding function values f_0 and f_1. Using this procedure generate nth-order differences for given $n + 1$ function values and tabular points.
27. For three function values f_0, f_1, and f_2 at tabular points x_0, x_1, and x_2, does it matter which first divided differences are used to form the second divided difference from which the interpolating formula (19) is derived? Why?

The conclusion on successfully attempting Exercise 26 is that for given function values at given abscissas, the first step of the procedure generates a total of n first divided differences for $n + 1$ function values. Likewise, generating second-order differences we obtain $n - 1$ differences in total. Proceeding to the nth-order difference we obtain a single value $f[x_0, x_1, x_2, \ldots, x_n]$. Thus at each successive stage we can represent the divided differences in tabular form. Such tables are called *difference tables*. For example, for four function values we can form the following difference table:

	First divided difference	Second divided difference	Third divided difference

x_0 $f(x_0)$
    $f[x_0, x_1]$
x_1 $f(x_1)$ $f[x_0, x_1, x_2]$
    $f[x_1, x_2]$ $f[x_0, x_1, x_2, x_3]$
x_2 $f(x_2)$ $f[x_1, x_2, x_3]$
    $f[x_2, x_3]$
x_3 $f(x_3)$

Before the innovation of electronic computers, such difference tables were extensively used in numerical analysis because they presented a simple and convenient method of calculating numbers which appear directly in the formulas being used.

Example 6 For the function values in Table 5 interpolate the function at $x = 1.25$ using the first three divided differences.

Table 5

	x_i	$f(x_i)$	Δ	Δ^2	Δ^3
$i = 0$	1.1	0.909 09			
			-0.7576		
1	1.2	0.833 33		0.5830	
			-0.6410		$-0.416\ 77$
2	1.3	0.769 23		0.4580	
			-0.5494		
3	1.4	0.714 29			

To interpolate $f(x = 1.25)$ we use the interpolation formula based on y_{0123}. That is, we assume that the third divided difference $f[x_0, x_1, x_2, x_3]$ is independent of the tabular points x_0, x_1, x_2, and x_3. Consequently, in the interval spanned by these tabular points we assume that any other third divided difference formed with x belonging to this interval is approximately equal to $f[x_0, x_1, x_2, x_3]$. In particular, $f[x, x_1, x_2, x_3] \doteq f[x_0, x_1, x_2, x_3]$.

Using the expanded form of the left-hand side we find that

$$f(x) \doteq y_{0123} = f[x_1] + (x - x_1)f[x_1, x_2] + (x - x_1)(x - x_2)f[x_1, x_2, x_3]$$
$$+ (x - x_1)(x - x_2)(x - x_3)f[x_0, x_1, x_2, x_3] \tag{20}$$

The divided differences required in this formula have been boxed-in in Table 5 for clarity. Substituting these values into Eq. (20) we obtain the value of $f(0.125)$ as 0.799 98, whereas the theoretical value is 0.8.

Exercises

28. In Example 6, we used a particular relationship between the third divided difference involving the argument x and that with all four tabular points x_0, x_1, x_2, and x_3. Because of this, the interpolating formula involved the boxed differences indicated in Table 5. Devise an interpolating formula which involves the divided differences y_0, y_{01}, y_{012}, and y_{013}. Does the answer obtained by your formula differ from the one obtained in Example 6? Why?
29. Write down the different interpolating formulas which are possible using divided differences based upon any three points given in Example 6.
30. Can you suggest a criterion which could be used to decide which one of the available interpolation formulas should be used?

The error term in our divided-difference interpolating formula has been incurred by substituting $y_{012\cdots n}(x)$ for the actual function $f(x)$. Our actual error is $f(x) - y_{012\cdots n}(x)$ based upon the $n + 1$ tabular points x_i, $i = 0, 1, \ldots, n$. To obtain the precise form of the error we use the expanded form of the $(n + 1)$th divided difference $f[x, x_0, x_1, \ldots, x_n]$, namely

$$f[x, x_0, x_1, \ldots, x_n] = \frac{1}{\prod_{i=0}^{n}(x - x_i)} f[x] - \frac{1}{\prod_{i=0}^{n}(x - x_i)} f[x_0]$$

$$- \frac{1}{\prod_{i=1}^{n}(x - x_i)} f[x_0, x_1] - \frac{1}{\prod_{i=2}^{n}(x - x_i)} f[x_0, x_1, x_2] \ldots$$

$$- \frac{1}{(x - x_n)} f[x_0, x_1, x_2, \ldots, x_n] \tag{21}$$

so that

$$f[x] = f(x) = \prod_{i=0}^{n}(x - x_i) f[x, x_0, x_1, \ldots, x_n] + \prod_{i=0}^{n-1}(x - x_i) f[x_0, x_1, \ldots, x_n]$$

$$+ \cdots + (x - x_0) f[x_0, x_1] + f[x_0]$$

Also, we have

$$y_{012\cdots n}(x) = \prod_{i=0}^{n-1}(x - x_i) f[x_0, x_1, \ldots, x_n] + \prod_{i=0}^{n-2}(x - x_i) f[x_0, x_1, \ldots, x_{n-1}]$$

$$+ \cdots + (x - x_0) f[x_0, x_1] + f[x_0]$$

Clearly then

$$E(x) = \prod_{i=0}^{n}(x - x_i) f[x, x_0, x_1, \ldots, x_n] \tag{22}$$

Exercises

31. If the $n+1$ differences in a divided difference table were zero, whilst the nth differences were not, what would you conclude about the function f from which the differences were formed?

32. When interpolating given data, one has the choice of (at least) two ways of determining what number of points should be used in defining the interpolating polynomial: *Either* one could obtain an estimate of the error for an nth-degree interpolating function using Eq. (22) or the alternative error formula (17), *or* one could adopt the following strategy.

    Calculate the value of the interpolant at the given point x using (say) p tabular points. Recalculate the interpolant using $p+1$ tabular points. Compare the values. Repeat the process of introducing an extra tabular point into the interpolant until successive answers remain unchanged to the prescribed number of decimal places or until all the tabular points are used. Program such an algorithm based upon $n+1$ tabular points to calculate the interpolant at $x = 1.375$ for the function in Table 5 (given that $f(x) = 1/x$), which is accurate to eight decimal places. Your program should indicate the number of points used in the interpolation and the difference between successive interpolates.

6 Iterated Interpolation

The problem of determining, a priori, the degree of the interpolating polynomial for a given set of data points can be resolved to a certain extent by following the procedure suggested in Exercise 32. That is, divided differences of successive orders can be formed and the corresponding interpolation evaluated using these divided differences. For example, consider the four function values in Table 5. In the fourth column of the table we have calculated the first divided differences. We can therefore use the interpolation formulas based upon just these first differences. Thus,

$$f(x) \doteq y_{ii+1}(x) = f(x_i) + (x - x_i)f[x_i, x_{i+1}] \qquad i = 0, 1, 2 \tag{23}$$

Hence three approximations to the required $f(x)$ can be obtained. By comparing these values, the user can ascertain whether sufficient accuracy has been achieved.

The second divided differences can be calculated from the values in column four of Table 5. In our case we have two such differences. We can therefore use an interpolation formula based upon the second divided difference, namely

$$f(x) \doteq f(x_i) + (x - x_i)f[x_i, x_{i+1}] + (x - x_i)(x - x_{i+1})f[x_i, x_{i+1}, x_{i+2}] \qquad i = 0, 1 \tag{24}$$

Again, these two approximations can be compared and the accuracy assessed. Finally, if it is deemed necessary a formula involving third divided differences can be used, namely

$$f(x) \doteq f(x_0) + (x - x_0)f[x_0, x_1] + (x - x_0)(x - x_1)f[x_0, x_1, x_2]$$
$$+ (x - x_0)(x - x_1)(x - x_2)f[x_0, x_1, x_2, x_3] \tag{25}$$

So, by calculating the differences of successive orders and using them in the available formulas, the accuracy can be monitored.

In the above example, when we proceed from the interpolation polynomial (24) to formula (25), the addition of the extra point into either of the polynomials in (24) merely requires the calculation of

$$(x - x_0)(x - x_1)(x - x_2)f[x_0, x_1, x_2, x_3]$$

and adding this *cubic polynomial* to the existing quadratic polynomial which forms the first part of (25). Consequently this method of adding a point to the interpolating data gives rise to a more efficient method of calculating the new interpolating polynomial compared with the recalculation of Lagrangian polynomials which would be required when effecting the same process.

Thus at the stage of adding one point x_k to the p tabular points $x_i, x_{i+1}, \ldots, x_{i+p}$, whose corresponding interpolating polynomial is $y_{i\cdots i+p}$, the calculation of the new interpolating polynomial merely requires the calculation of the polynomial

$$(x - x_k) \prod_{j=i}^{(i+p)} (x - x_j)$$

and the calculation of the divided difference $f[x_k, x_i, x_{i+1}, \ldots, x_{i+p}]$ and forming their product and adding the result to $y_{ii+1\cdots i+p}$.

A more efficient procedure of adding in a new point into the interpolation formula can be obtained by recalling the uniqueness of the interpolating function. Thus the approximation obtained by forming the pth-order (say) divided differences of two $(p-1)$th-order divided differences and then using this in the interpolation polynomial of degree p is exactly that obtained by linearly interpolating upon the two corresponding interpolation polynomials of degree $(p-1)$ based upon the above mentioned $(p-1)$th-order divided differences.

Thus, instead of forming a table of differences, we advocate forming the table of linear interpolations based upon the entries in the preceding column, as shown in Table 6.

Using this approach, divided differences are never actually calculated; only the interpolated values $y_{01\cdots}(x)$ are stored in successive elements of the column.

Example 7 Let us use the above approach to interpolate for $x = 1.25$ and interpolate until three decimal places agree or we run out of points, whichever is the sooner.

By linear interpolation on the function values for $x = 1.25$ we obtain the three values in column 3 of Table 7.

In column 4 we have obtained a quadratic interpolation using the linear interpolation of the values in column 3. Finally, in column 5 we have obtained a cubic interpolation using the linear interpolation of the values in column 4. Comparing the value in column 5 we see that it is identical to that obtained using Eq. (20).

Table 6

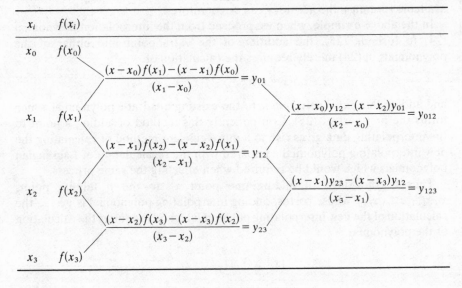

Table 7

x_i	$f(x_i)$	y_{ij}	y_{ijk}	y_{ijkl}
1.1	0.909 09			
		0.795 45		
1.2	0.833 33		0.799 82	
		0.801 28		0.799 98
1.3	0.769 23		0.800 14	
		0.796 70		
1.4	0.714 29			

Thus to three decimal places we could confidently predict as a result of the calculations giving rise to the values in column 4 that the required interpolate is 0.800. We have interpolated one stage further to stress to the reader the fact that our present algorithm is merely a reorganization of our previous method utilizing divided differences.

The above approach is perhaps the obvious way of forming the linear interpolations between respective entries in the table. This particular version of iterated linear interpolation is commonly called *Neville's method*. Another commonly used iterated linear interpolation technique uses a geometrical argument to indicate the choice of linear interpolations. This is *Aitken's method*. For $n+1$ data points at the tabular points x_i, $i = 0, 1, \ldots, n$, choose as a "pivot" point that point (x_i) which is furthest from the required interpolating point x.

Interpolate linearly between the pivot, x and successive remaining tabular points. Associate the interpolated value with the appropriate tabular point. When all the tabular values have been processed, drop the point from future consideration and then repeat the process of choosing a pivot point and interpolating linearly. Proceed until just two tabular points remain. Linear interpolation upon these two points will produce the required interpolate at x.

The algorithm for generating the respective interpolates may be described as follows.

For $j = 0, 1, \ldots, n - 1$, generate the interpolating polynomials by

$$p_{i0}(x) = f(x_i) \qquad i = 0, 1, \ldots, n$$

$$p_{ij+1}(x) = \frac{(x_i - x)p_{jj}(x) - (x_j - x)p_{ij}(x)}{x_i - x_j} \qquad i = j+1, j+2, \ldots, n \qquad (26)$$

In Example 6 this recursive formula would produce the tabular form shown in Table 8.

Table 8

j	0	1	2	3
x_0	p_{00}			
x_1	p_{10}	p_{11}		
x_2	p_{20}	p_{21}	p_{22}	
x_3	p_{30}	p_{31}	p_{32}	p_{33}

In the table we have assumed, for the moment, that the interpolated value x lies between x_2 and x_3. (This is no restriction since if it were not so, the ordering of the x_i could be changed so that the values of x in Table 8 were placed according to their distance from x.)

To assist the reader we have indicated in tabular form in Tables 9, 10, 11 the various stages involved in the Aitken process for the problem of Example 6. The circled entries in the tables represent the pivot points (the points furthest from $x = 1.25$).

Stage 1

Table 9

$j =$		0	1
$x_0 = \text{①.1}$		$p_{00} = 0.909\ 09$	
$x_1 = 1.2$		$p_{10} = 0.833\ 33$	$p_{11} = 0.795\ 450$
$x = 1.25 \rightarrow$			
$x_2 = 1.3$		$p_{20} = 0.769\ 23$	$p_{21} = 0.804\ 195$
$x_3 = 1.4$		$p_{30} = 0.714\ 23$	$p_{31} = 0.811\ 690$

Table 9 corresponds to the linear interpolation using Eq. (26) with $j = 0$, $i = 1, 2, 3$. Using these values, we proceed to stage 2 in Table 10 by reorganizing the points so that new x_1 becomes the furthest point from $x = 1.25$, the position occupied by x_3 in Table 9. The result is shown in Table 10.

Stage 2

Table 10

$j =$	0	1	2
$x_1 = \boxed{1.4}$	0.714 23	0.811 690	
$x_2 = 1.3$	0.769 23	0.804 195	$p_{22} = 0.800\ 447\ 5$
$x_3 = 1.2$	0.833 33	0.795 450	$p_{32} = 0.799\ 51$

$x \rightarrow$ (to the left of the rows)

Stage 3

Table 11

$j =$	0	1	2	3
$x_2 = \boxed{1.3}$	0.769 23	0.804 195	0.800 447 5	
$x_3 = 1.2$	0.833 33	0.795 450	0.799 51	0.799 98

$x \rightarrow$ (to the left of the rows)

In stage 3, Table 11, a choice of pivot exists. The result of interpolating linearly between values in the penultimate column produces, once again, 0.799 98 as the required estimate.

We have indicated the more popular versions of iterated linear interpolation; the enquiring reader will be able to produce many versions of such interpolation formulas based upon different ways of linear interpolation; these formulas will have various properties which can be exploited for particular properties of numerical data.

Exercises

33. What is the form of the error term for each entry in the methods of Aitken and Neville?
34. Program Neville's iterated interpolation formula for the data given in Table 12. Use your program to interpolate $f(x)$ at $x = 0.667$ so that successive interpolates in the columns of Neville's method agree to six decimal places or until no further data points are available, whichever is the sooner. As part of your program, output the fact that either the accuracy has been achieved before the last possible entry or that the answer provided is the last entry (in which case the accuracy cannot be guaranteed). In this latter situation, record the best agreement obtained by the elements of the penultimate column.

35. Repeat Exercise 34 for Aitken's method. At each stage compare the interpolated values at each stage with those obtained by the Neville process. Which of the two methods do you conclude is better suited for this particular problem? Why? Can you devise an example of interpolation where the converse is true? Are your conclusions the same for all orderings of the tabular points?

Table 12

x	$f(x)$
0.00	0.000 000
0.10	0.348 852
0.15	0.447 351
0.23	0.591 830
0.40	0.860 380
0.50	0.981 585
0.60	1.056 343
0.70	1.062 845
0.80	0.973 816
0.90	0.755 757
1.00	0.367 880
1.20	−1.125 159
1.40	−4.034 361
1.50	−6.246 288

7 Approximating a Polynomial of Degree n by One of Degree $n-1$

Consider the problem of interpolating a function f which is a polynomial of degree n. If we have at our disposal n values of f at distinct tabular points, how should we choose the location of these points such that the error (the absolute value of the difference between the function f and its interpolating function) is a minimum? In other words, how should we best interpolate a polynomial of degree n by one of degree $n-1$ when the position of the tabular points is at our disposal?

To make our discussion as general, yet as simple as possible, let us transform the general interval $[a, b]$, upon which our tabular points are usually given, to the interval $[-1, 1]$ by the transformation

$$\bar{x} = \frac{2x - (b+a)}{b-a} \tag{27}$$

That is, for any $x \in [a, b]$ we can transform our problem to one in a corresponding independent variable $\bar{x} \in [-1, 1]$. To reduce the notation to a minimum, however, we will omit the bar over the x and remember simply that we have transformed x lying in $[a, b]$ to a corresponding x lying in $[-1, 1]$.

Thus our problem is to find the location of the points $x_0, x_1, \ldots, x_{n-1} \in [-1, 1]$ so that

$$\max_{-1 \le x \le 1} |f(x) - y(x)|$$

is minimized. Without loss of generality let us assume the leading coefficient of the polynomial f is 1.

From Eq. (17) we see that

$$f(x) - y(x) = \frac{\pi_n(x)n!}{n!} = \pi_n(x) \tag{28}$$

Therefore our question asks us to choose the x_i, $i = 0, 1, \ldots, n-1$, to minimize the

$$\max_{-1 \leq x \leq 1} |\pi_n(x)|$$

i.e. find x_i, $i = 0, 1, \ldots, n-1$, to minimize

$$\max_{-1 \leq x \leq 1} |(x - x_0)(x - x_1) \ldots (x - x_{n-1})| \tag{29}$$

This seemingly difficult problem can be solved with the aid of the *Chebyshev polynomials*. These polynomials represent a very important class of polynomials in numerical analysis and we spend some time on their properties here.

Definition The Chebyshev polynomial of degree n is defined as

$$T_n(x) = \cos(n \text{ arc cos } x) \quad \text{for } -1 \leq x \leq 1 \tag{30}$$

which is sometimes written as

$$T_n(x) = \cos n\theta \qquad x = \cos \theta \qquad -\pi \leq \theta \leq \pi$$

The Chebyshev polynomials satisfy a useful recurrence relation which allows for their convenient calculation. By direct calculation we find

$$T_0(x) = \cos 0 = 1$$

and

$$T_1(x) = \cos \theta = x$$

and using the well-known formula for the sum of two cosines, namely

$$\cos(n+1)\theta + \cos(n-1)\theta = 2 \cos n\theta \cos \theta$$

it is easily shown that

$$T_{n+1}(x) = 2xT_n(x) - T_{n-1}(x) \qquad n = 1, 2, \ldots$$

with $x = \cos \theta$ as above and T_0, T_1 defined by Eq. (30).

From the definition and recurrence relation, the following properties of the polynomials can be easily verified:

P(1) The polynomial $T_n(x)$ is a polynomial of degree n with leading coefficient 2^{n-1} $(n > 0)$.

P(2) For n even, the polynomial of degree n is also even (i.e. has only even powers of x). For n odd, the polynomial of degree n is also odd (i.e. has only odd powers of x).

P(3) $T_n(x)$ has exactly n zeros on $[-1, 1]$; the zeros being located at

$$x_i = \cos \frac{(2i+1)}{n} \frac{\pi}{2}, i = 0, 1, \ldots, n-1$$

P(4) The polynomial of degree n attains its maximum and minimum values of ± 1, respectively, alternately at the points

$$x_j^* = \cos \frac{j\pi}{n} \qquad j = 0, 1, \ldots, n$$

The next property is one we require to prove since it is the subject matter of this section.

P(5) $T_n(x)$ is directly related (by a multiplicative constant) to the required polynomial of degree n which satisfies condition (29).

Proof Consider the polynomial $2^{1-n}T_n(x)$ which has a leading coefficient of unity. We know from the properties P(1)–P(4) that this function has $n+1$ extrema situated at $x_j^* = \cos(j\pi/n)$ with alternating values of $\pm 2^{1-n}$, i.e. has values $(-1)^k 2^{1-n}$ for $k = 0, 1, \ldots, n$. Thus let us choose the tabular points in the interpolation problem so that

$$\pi_n(x) = 2^{1-n}T_n(x).$$

To show that this is *the* polynomial which satisfies the min–max property (29), let us assume there does exist another polynomial $Q_n(x)$, say, of degree n, which has a leading coefficient of unity, for which $|Q_n(x)|$ has a smaller maximum value m than 2^{1-n} in $[-1, 1]$. Thus, at the points $\cos(j\pi/n)$, the difference between Q_n and π_n must have the same sign as $T_n(x)$, i.e.

$$D(x) = Q_n(x) - 2^{1-n}T_n(x)$$

satisfies

$$D(x_j) \begin{cases} >0 & \text{if } j \text{ is even} \\ <0 & \text{if } j \text{ is odd} \end{cases}$$

Consequently, $D(x)$ must have n distinct zeros. But D is the difference between polynomials of degree n with leading coefficients of unity and so is a polynomial of degree at most $n-1$. As we know, a polynomial of degree $n-1$ having greater than $n-1$ zeros can only be identically zero. Hence $Q \equiv \pi_n$ and the required property is proved.

Thus the locations of the points $x_0, x_1, \ldots, x_{n-1}$, in order that the polynomial of degree $n-1$ best interpolate a polynomial f of degree n, are at the zeros of the Chebyshev polynomial of degree n, given by property P(3).

To give the next important property we need the following definitions.

Definition A class of functions $\phi_i(x)$, $i = 0, 1, \ldots$, is said to be *orthogonal* over the interval $[a, b]$ with respect to the (positive) weight function $w(x)$ if $\int_a^b w(x)\phi_i(x)\phi_j(x)\,dx = 0$, $i \neq j$.

In the case that the integral is unity for $i = j$, the functions are said to be *orthonormal*.

Orthogonality (and orthonormality) can also be defined on a discrete set. Whereas in the continuous case the space of functions which are orthogonal has dimensions equal to ∞, in the discrete case the dimensionality is equal to the number of points in the interval. Thus if we define the point set $\{x_i\}$, $i = 0, 1, \ldots n$, then the functions $\phi_i(x)$, $i = 0, 1, \ldots, n$, are said to be orthogonal on $\{x_i\} \in [a, b]$ with respect to the weight $w(x)$ if

$$\sum_{i=0}^n w(x_i)\phi_j(x_i)\phi_k(x_i) = 0 \qquad j \neq k, j, k = 0, 1, \ldots, n$$

and for $i = j$ the sum equals one in the case of orthonormality.

Theorem 1 *The Chebyshev polynomials form an orthogonal set over* $[-1, 1]$ *with respect to the weight function* $(1 - x^2)^{-1/2}$.

Proof We consider

$$I_{ij} = \int_{-1}^1 \frac{T_i(x)T_j(x)\,dx}{\sqrt{(1 - x^2)}}$$

Let us transform $x \to \theta$ by $x = \cos\theta$.

$$\therefore \quad I_{ij} = -\int_\pi^0 \frac{\cos i\theta \cos j\theta}{\sin\theta}\,d\theta \sin\theta = \int_0^\pi \cos i\theta \cos j\theta\,d\theta$$

so that

$$I_{ij} = \begin{cases} 0 & i \neq j \\ \dfrac{\pi}{2} & i = j \neq 0 \\ \pi & i = j = 0 \end{cases}$$

and the theorem is proved.

Theorem 2 *The Chebyshev polynomials of degree less than or equal to n form an orthogonal set on the discrete point set*

$$x_i = \cos(1 + 2i)\pi/2n \qquad i = 0, 1, 2, \ldots, n - 1$$

Definition Functions $\phi_i(x)$ belonging to a set $\mathscr{F}$ are said to be *linearly independent* if the linear combination $\sum_{i=0}^n C_i\phi_i(x) = 0$ implies all the C_i are zero. In contrast, a set is *linearly dependent* if $\sum_{i=0}^n C_i\phi_i(x) = 0$ implies $C_i \neq 0$.

Lemma 2   *Orthogonal functions form a linearly independent set.*

Proof   Consider the set of functions $\phi_i(x)$, $i = 0, 1, \ldots, n$, which are orthogonal. Assume for the moment that $\phi_i(x)$, $i = 0, 1, \ldots, n$, are linearly dependent. Then there exist constants C_i such that

$$\sum_{i=0}^{n} C_i \phi_i(x) = 0 \qquad \text{for some nonzero } C_i$$

But $\phi_i(x)$ are orthogonal. Hence

$$\int_a^b w(x)\phi_i(x)\phi_j(x)\,dx = 0 \qquad i \neq j$$

Let us consider the expansion of

$$F(x) = 0 = \sum_{i=0}^{n} C_i \phi_i(x) \tag{31}$$

Multiply both sides of Eq. (31) by $w(x)\phi_j(x)$; integrating, we obtain

$$\int_a^b F(x)w(x)\phi_j(x)\,dx = 0 = \sum_{i=0}^{n} C_i \int_a^b w(x)\phi_i(x)\phi_j(x)\,dx$$

so that

$$\frac{\displaystyle\int_a^b w(x)F(x)\phi_j(x)}{\displaystyle\int_a^b w(x)\phi_j^2(x)\,dx} = C_j \qquad j = 0, 1, 2, \ldots, n$$

But $F(x) = 0$ so all $C_j = 0$ and the assumption was false. Thus the orthogonal functions represent a linearly independent set.

Since the Chebyshev polynomials form an orthogonal set, we can represent any polynomial of degree n as a linear combination of the Chebyshev polynomials up to degree n.

Thus

$$f(x) = \sum_{i=0}^{n} C_i T_i(x) \tag{32}$$

where the C_i are given by

$$C_i = \frac{\displaystyle\int_{-1}^{1} \sqrt{(1-x^2)}f(x)T_i(x)\,dx}{\displaystyle\int_{-1}^{1} \sqrt{(1-x^2)}T_i^2(x)\,dx} \qquad i = 0, 1, \ldots, n \tag{33}$$

Hence we can substitute $f(x)$ given by Eqs. (32) and (33) into Eq. (28), so that

$$y(x) = \pi_n(x) - f(x) = 2^{1-n}T_n(x) - \sum_{i=0}^{n} C_i T_i(x)$$

But by assumption $f(x)$ is a polynomial of degree n with leading coefficient equal to 1, and $T_n(x)$ is the only term to contain x^n. Thus $C_n = 1$

$$\therefore \quad y(x) = \sum_{i=0}^{n-1} C_i T_i(x)$$

is the expression for $y(x)$ i.e. the alternate form of the polynomial $y(x)$ is given in terms of Chebyshev polynomials through degree $n - 1$.

In the derivation of $y(x)$ in terms of the Chebyshev polynomials, we found that by expanding $f(x)$ in terms of the Chebyshev polynomials and omitting the final term we obtain $y(x)$, which is the minimax approximation. By leaving off more than the last term it is often possible to obtain approximations to $f(x)$ which are close to being minimax. Thus since $f(x) = \sum_{i=0}^{n} C_i T_i(x)$ and we now assume $y(x) = \sum_{i=0}^{k} C_i T_i(x)$, then

$$\max_{-1 \le x \le 1} |f(x) - y(x)| = \max_{-1 \le x \le 1} \left| \sum_{i=k+1}^{n} C_i T_i(x) \right| \le \sum_{i=k+1}^{n} |C_i| \qquad (34)$$

because $|T_i(x)| \le 1$ for all i.

Therefore by replacing f by a polynomial of degree k, we have a bound on the error in terms of the calculable coefficients C_i. Such a procedure is called *economization of power series* since we are replacing a power series of high degree by one of lower degree. When the value of the error bound on the right-hand side of Eq. (34) gives a reasonable value, the replacement of f by y can give useful results in terms of the saving in computing time.

In the case where $f(x)$ is *not* a polynomial, then the above procedure can again be applied by expanding $f(x)$ into a Taylor series which is valid on $[-1, 1]$ using, if necessary, the transformation of variables described in Eq. (27) so that $(f(x) - \text{Taylor series expansion}) = \varepsilon$. This power series can be truncated to eliminate the terms of degree greater than n. Then expand this series in terms of the Chebyshev polynomials of degree n. Truncate this to k terms where k is chosen to produce an acceptable error bound from Eq. (34). This produces the required interpolation formula of degree k.

Example 8 For the hyperbolic sine function, construct the polynomial of lowest possible degree such that the error is no greater than 10^{-3} on the interval $[-1, 1]$.

We expand sinh (x) by Taylor series to obtain

$$\sinh(x) = x + \frac{x^3}{3!} + \frac{x^5}{5!} + \cdots$$

For the error term to be less than 10^{-3} we take terms up to and including $x^5/5!$. We therefore require to approximate

$$f(x) = x + \frac{x^3}{3!} + \frac{x^5}{5!}$$

by $y(x) = \sum_{i=0}^{k} C_i T_i(x)$, where C_i are to be determined. Using the definition of $T_i(x)$ we can form the first five terms and hence work out the precise relationship between x^r and the $T_i(x)$. These are given in Table 13.

Table 13

$T_0(x) = 1$	$1 = T_0$
$T_1(x) = x$	$x = T_1$
$T_2(x) = 2x^2 - 1$	$x^2 = (T_0 + T_2)/2$
$T_3(x) = 4x^3 - 3x$	$x^3 = (T_3 + 3T_1)/4$
$T_4(x) = 8x^4 - 8x^2 + 1$	$x^4 = (T_4 + 4T_2 + 3T_0)/8$
$T_5(x) = 16x^5 - 20x^3 + 5x$	$x^5 = (T_5 + 5T_3 + 10T_1)/16$

By replacing powers of x by the T_i from Table 13, we obtain

$$y = T_1 + \frac{(T_3 + 3T_1)}{3! \times 4} + \frac{(T_5 + 5T_3 + 10T_1)}{16 \times 5!}$$

Evaluating the coefficients of T_1, T_3, and T_5, we obtain

$$y = 1.1302T_1 + 0.044\,27T_3 + 0.000\,520\,8T_5$$

Omitting the last term produces the polynomial approximation y_m to our polynomial $f(x)$.
   Thus

$$y_m = 1.1302T_1 + 0.044\,27T_3$$

Because $|T_5| \leq 1$, $|y - y_m| \leq 0.000\,53$, and through our truncation of the Taylor series,

$$\max_{-1 \leq x \leq 1} |\sinh(x) - y| \leq 0.000\,31$$

we have $|\sinh(x) - y_m| \leq 0.000\,31 + 0.000\,53 = 0.000\,84$.

   Thus our polynomial y_m is in error by at most 0.001. Hence by substituting powers of x for the T_i, we can then obtain the required polynomial in normal form. This is found to be

$$y_m = 0.997\,39x + 0.177\,08x^3$$

To see that this polynomial does in fact give the required approximation, we have evaluated y_m and $\sinh(x)$ at points in $[-1, 1]$. These values are given in Table 14.

Table 14

x	$\sinh(x)$	$y_m(x)$	$|\sinh(x) - y_m(x)|$
$(\mp)1.0$	$(\mp)1.175\,20$	$(\mp)1.174\,47$	0.000 73
$(\mp)0.75$	$(\mp)0.822\,31$	$(\mp)0.822\,75$	0.000 44
$(\mp)0.5$	$(\mp)0.521\,10$	$(\mp)0.520\,83$	0.000 27
$(\mp)0.25$	$(\mp)0.252\,61$	$(\mp)0.252\,11$	0.000 50
0.0	0.000 00	0.000 00	0.000 00

Exercises

36. Verify directly, by considering the general form of $Q_2(x) = x^2 + bx + c$, that no choice of b and c can produce a quadratic polynomial with leading coefficient of unity with extrema smaller than $\frac{1}{2}T_2(x)$.
37. Prove Theorem 2.
38. By adding a suitable multiple of the Chebyshev polynomials evaluated at $x = 1$, show that the set of Chebyshev polynomials of degree less than or equal to n forms an orthogonal set.
39. Prove that functions which are orthogonal with respect to the positive weight function $w(x)$ over the finite discrete set $\{x_i\}$, $i = 0, 1, \ldots, n$, form a linearly independent set.
40. For the hyperbolic sine function, construct the polynomial of lowest possible degree such that the error is no greater than 10^{-7} on the interval $[-1, 1]$.
41. Write an ALGOL *recursive* procedure or FORTRAN subroutine to form the Chebyshev polynomial of degree n.
42. If you have a graph plotter in your computing center, use the procedure of Exercise 41 to plot the Chebyshev polynomials for $n = 0, 1, 2, \ldots, 10$.

8 Hermite Interpolation

So far we have considered interpolating a function f from given observed values f_i of the function itself. It is clear that this is not the only case of observed values relating to the function f that can occur. For example, at a point x_0 (say), we may have observed values of the function f and its first n derivatives. In this case, for $x_h = x_0 + h$, h small, we obtain an approximation to $f(x_h) \equiv f_h$ by the use of Taylor series, namely

$$f(x_h) \doteq y(x_0 + h) \equiv y_h$$

$$= f_0 + hf_0^{(1)} + \frac{h^2}{2!}f_0^{(2)} + \frac{h^3}{3!}f_0^{(3)} + \cdots + \frac{h^n}{n!}f_0^{(n)} \qquad (35)$$

The approximation y is in error by an amount equal to the remainder term which in this case is (assuming $f^{(n+1)}(x)$ exists),

$$R_{n+1}(x) = \frac{h^{n+1}}{(n+1)!}f^{(n+1)}(\eta) \qquad x_0 < \eta < x_0 + h$$

Thus approximation (35) can be used for values of the argument for which the finite series converges and for values of h for which $R_{n+1}(x)$ is not so large so as to make the numerical value y_h meaningless.

In practice it is more common to obtain observed values of the function f and just a few of its derivatives (one or two) over a range of values of the abscissa x rather than the function value f and many of its derivatives at a single point. The most simple case is where just the function value is given, as described earlier. Consider now the case where function values f_i and values of the derivative $f_i^{(1)}$ are specified at all the $(n+1)$ tabular points x_i, $i = 0, 1, 2, \ldots, n$. We therefore require to obtain the polynomial which interpolates the values f_i and $f_i^{(1)}$ at x_i, $i = 0, 1, \ldots, n$. With $n + 1$ tabular points and hence $2n + 2$ function and derivative values given, we will require in general a polynomial of degree $2n + 1$ to interpolate the given data. Let us propose a polynomial of degree $2n + 1$ and denote it by $P_{2n+1}(x)$. Let us propose

$$P_{2n+1}(x) = \sum_{j=0}^{n} f_j h_j(x) + \sum_{j=0}^{n} f_j^{(1)} g_j(x) \tag{36}$$

where $h_j(x)$ and $g_j(x)$ are polynomials of degree $2n + 1$. In order that Eq. (36) interpolate the given values, we require

$$h_j(x_i) = \delta_{ij} \qquad i, j = 0, 1, \ldots, n \tag{37}$$

and

$$g_j(x_i) = 0 \qquad i, j = 0, 1, \ldots, n \tag{38}$$

Also, differentiating $P_{2n+1}(x)$ in Eq. (36), we find

$$P_{2n+1}^{(1)}(x) = \sum_{j=0}^{n} f_j h_j^{(1)}(x) + \sum_{j=0}^{n} f_j^{(1)} g_j^{(1)}(x) = f_j^{(1)}$$

If P is to interpolate the derivative values at x_i, $i = 0, 1, \ldots, n$, then

$$h_j^{(1)}(x_i) = 0 \qquad i, j = 0, 1, \ldots, m \tag{39}$$

and

$$g_j^{(1)}(x_i) = \delta_{ij} \qquad i, j = 0, 1, \ldots, n \tag{40}$$

Consequently, the relations (37), (38), (39), and (40) serve to define the required properties of the functions $h_j(x)$ and $g_j(x)$. To obtain their precise form we proceed as follows.

For $h_j(x)$: The polynomials are to be of degree $2n + 1$. The fact that $h_j(x)$ are to vanish at all the tabular points other than $i = j$ and $h_j^{(1)}(x)$ vanish at all the tabular points suggests first that $h_j(x)$ should have n double roots at $x = x_i$, $i = 0, 1, \ldots, n$, $i \neq j$.

Thus $h_j(x)$ is partially made up of a polynomial of degree $2n$ defined by

$$l_j(x) = \prod_{\substack{i=0 \\ i \neq j}}^{n} (x - x_i)^2 \dagger$$

$l_j(x)$ will satisfy all the conditions of Eq. (37) when normalized but, as yet, fails to satisfy just one of the conditions of Eq. (39), namely that $h_i^{(1)}(x_i) = 0$; the condition being satisfied for the other remaining tabular points. In order

---

† Note that $l_j(x)$ is nothing more than $\pi_{nj}^2(x)$ defined in section 3—however, for convenience we have introduced a new symbol.

to make $l_j(x)$ into $h_j(x)$ we must multiply $l_j(x)$ by a linear factor so that $h_j(x)$ is then of degree $2n+1$ and also so that $h_i^{(1)}(x_i)=0$. Let this linear factor be $a(x-x_j)+b$, where a and b are to be determined so that all conditions of Eq. (39) are satisfied. Thus

$$h_j(x) = \{a(x-x_j)+b\} \prod_{\substack{i=0 \\ i \neq j}}^{n} (x-x_i)^2 \tag{41}$$

Using Eqs. (37) and (41), we require

$$h_j(x_j) = [a(x_j-x_j)+b] \prod_{\substack{i=0 \\ i \neq j}}^{n} (x_j-x_i)^2 = 1 \tag{42}$$

and

$$h_j(x_k)_{j \neq k} = [a(x_k-x_j)+b] \prod_{\substack{i=0 \\ i \neq j}}^{n} (x_k-x_i)^2 = 0 \tag{43}$$

Equation (42) defines b explicitly, whilst Eq. (43) is satisfied identically. Hence from Eq. (42),

$$b = \frac{1}{\displaystyle\prod_{\substack{i=0 \\ i \neq j}}^{n} (x_j-x_i)^2} = \frac{1}{l_j(x_j)}$$

Using Eq. (41) in Eq. (39), we require

$$h_j^{(1)}(x_j) = a \prod_{\substack{i=0 \\ i \neq j}}^{n} (x_j-x_i)^2 + [a(x_j-x_j)+b]l_j^{(1)}(x_j) = 0 \tag{44}$$

and

$$h_j^{(1)}(x_k) = a \prod_{\substack{i=0 \\ i \neq j}}^{n} (x_k-x_i)^2 + [a(x_k-x_j)+b]l_j^{(1)}(x_k) = 0 \tag{45}$$

Equation (44) defines a as

$$a = \frac{-bl_j^{(1)}(x_j)}{\displaystyle\prod_{\substack{i=0 \\ i \neq j}}^{n} (x_j-x_i)^2} = \frac{-bl_j^{(1)}(x_j)}{l_j(x_j)} = \frac{-l_j^{(1)}(x_j)}{l_j^2(x_j)}$$

whereas Eq. (45) is satisfied identically, since $l_j^{(1)}(x_k)_{j \neq k} = 0$. Consequently,

$$h_j(x) = \frac{l_j(x)}{l_j(x_j)} \left[1-(x-x_j)\frac{l_j^{(1)}(x_j)}{l_j(x_j)} \right]$$

From section 3, Eq. (14),

$$\frac{l_j(x)}{l_j(x_j)} = L_j^2(x) \tag{46}$$

and

$$\frac{l_j^{(1)}(x_j)}{l_j(x_j)} = 2L_j(x_j)L_j^{(1)}(x_j) = 2L_j^{(1)}(x_j) \tag{47}$$

since $L_j(x_j) = 1$. Consequently,

$$h_j(x) = L_j^2(x)[1 - 2(x - x_j)L_j^{(1)}(x_j)]$$

so that the polynomials of degree $2n + 1$ multiplying the function values f_j in Eq. (36) are defined in terms of Lagrangian interpolating polynomials of degree n and their first derivatives evaluated at the tabular points x_j. $L_j^{(1)}(x_j)$ is simply $\sum_{i=0, i \neq j}^{n} 1/(x_i - x_j)$ and is easy to compute.

For $g_j(x)$: The polynomials multiplying the derivative values in Eq. (36) can be obtained in a similar manner. We leave the actual derivation to Exercise 43. It is found that

$$g_j(x) = (x - x_j)L_j^2(x) \tag{48}$$

so that

$$P_{2n+1}(x) = \sum_{j=0}^{n} f_j L_j^2(x)[1 - 2(x - x_j)L_j^{(1)}(x_j)]$$

$$+ \sum_{j=0}^{n} f_j^{(1)}(x - x_j)L_j^2(x) \tag{49}$$

This result is known as the *Hermite interpolation formula*. It is also called the osculating polynomial.†

Hermite interpolation may be applied in the case where values of the derivative $f^{(1)}(x)$ are not available at all the tabular points. For the case where the function values f_i are given at all the tabular points and the derivative values $f_j^{(1)}$ are given at tabular points x_j, $j = 0, 1, \ldots, r$ $(r < n)$, the Hermite interpolation formula becomes

$$P_{n+r+1}(x) = \sum_{j=0}^{n} h_j(x)f_j + \sum_{j=0}^{r} g_j(x)f_j^{(1)} \tag{50}$$

where

$$h_j(x) = \begin{cases} \{1 - (x - x_j)[L_{jn}^{(1)}(x_j) + L_{jr}^{(1)}(x_j)]\}L_{jn}(x)L_{jr}(x) & j = 0, 1, \ldots, r \\ L_{jn}(x)L_{rn}(x)\dfrac{(x - x_r)}{(x_j - x_r)} & j = r+1, \ldots, n \end{cases} \tag{51}$$

and

$$g_j(x) = (x - x_j)L_{jr}(x)L_{jn}(x) \qquad j = 0, 1, 2, \ldots, r \tag{52}$$

where the second subscript on the Lagrangian polynomal L denotes the degree of the polynomial. This representation clearly reduces to Eq. (49) when $r = n$.

---

† A polynomial is said to be osculating if it agrees with the required function and its first p derivatives for some $p > 0$ at all the tabular points; Eq. (48) is clearly the case $p = 1$.

The interpolation formula given by Eqs. (50), (51), (52) is known as the *modified Hermite interpolation formula*.

The question of uniqueness we asked of the polynomials in the Lagrangian interpolation formula in section 3 can again be asked of the polynomials in the Hermite and modified Hermite interpolation formulae.

In both cases the functions P are unique. The proof of this fact is a simple extension of the proof of uniqueness of the Lagrangian interpolation polynomial and is left as an exercise (Exercise 44).

To find the error associated with Hermite interpolation, we can proceed in a similar manner to that of section 4. As before, consider the function

$$F(z) = f(z) - P(z) - [f(x) - P(x)] \frac{\pi_n(z)\pi_r(z)}{\pi_n(x)\pi_r(x)}$$

where $r \le n$ and $P = P_{n+r+1}$ is the interpolating polynomial of degree $n+r+1$ given by Eq. (50). $\pi_m(\theta)$ is the power polynomial of degree $m+1$ in θ defined in section 4. The function F has $n+r+3$ zeros; double zeros at $x_0, x_1, \ldots, x_r$; single zeros at $x_{r+1}, x_{r+2}, \ldots, x_n$, and x. Consequently a generalization of Rolle's theorem can be applied. Differentiating $F(z)$ $n+r+2$ times, we find that there exists a ξ in the interval spanned by $x_0, x_1, \ldots, x_n$, and x so that

$$0 = F^{n+r+2}(\xi) = f^{(n+r+2)}(\xi) - [f(x) - P(x)] \frac{(n+r+2)!}{\pi_n(x)\pi_r(x)}$$

Thus the error term in the modified Hermite interpolating formula is given by

$$E(x) = \frac{\pi_n(x)\pi_r(x)}{(n+r+2)!} f^{(n+r+2)}(\xi) \tag{53}$$

The interpolating formula with error term thus becomes

$$f(x) = \sum_{j=0}^{n} f_j h_j(x) + \sum_{j=0}^{r} f_j^{(1)} g_j(x) + \frac{\pi_n(x)\pi_r(x)}{(n+r+2)!} f^{(n+2+r)}(\xi) \tag{54}$$

By replacing r by n in Eq. (54) we obtain the Hermite interpolating formula with error term

$$f(x) = \sum_{j=0}^{n} f_j h_j(x) + \sum_{j=0}^{n} f_j^{(1)} g_j(x) + \frac{\pi_n^2(x)}{(2n+2)!} f^{(2n+2)}(\xi)$$

Example 9 Consider interpolating the function $f(x) = x \exp(x)$ using values of f and $f^{(1)}$ at the points $x = ih$, $i = 0, 1, 2, 3$; $h = 0.1$, given in Table 15.

Table 15

x_i	0.0	0.1	0.2	0.3
$f(x_i)$	0.0	0.110 517	0.244 280	0.404 958
$f^{(1)}(x_i)$	1.0	1.215 687	1.465 68	1.754 818

Substituting the values from Table 15 into Eq. (50) we find that $P_7(0.15) =$ 0.174 275 14.

Using the error term, which is bounded in this case by 10^{-11}, we see that $f(x)$ given by Eq. (54) is $0.174\ 275\ 14 \pm 10^{-11}$. The theoretical value of $x \exp (x)$ at $x = 0.15$ is 0.174 275 14 so that to the number of figures quoted the result is exact (as would be expected from the error bound).

Exercises

43. Derive the form of the polynomials $g_i(x)$ given by Eq. (48).
44. Prove that the interpolating polynomials of the Hermite formulas Eq. (49) are unique.
45. Program the Hermite interpolation formula for general n and r. Use your program to interpolate $\sin (x)$ using the data of Table 1 for values of $r = 0, 1, \ldots, 8$; $x = 1.375$.
46. Determine the form of the interpolating polynomial given values of the function f, values of $f^{(1)}$, and values of $f^{(2)}$ at the $(n + 1)$ tabular points $x_0, x_1, \ldots, x_n$. (The resulting formula is known as the *generalized Hermite interpolation formula*.)

9 Inverse Interpolation

In the preceding sections we have assumed values of a function f to be given at the tabular values x_i, $i = 0, 1, \ldots, n$. In many cases, from this data the contrasting question asked is "find the value of the variable x at which the function f takes on a particular value" (zero, say). Later in the book we will devote a chapter to this question, but in the context of interpolation we can answer the question partially now. As we have seen, to interpolate f, we have represented the function f in terms of other functions (polynomials) and substituted the argument x corresponding to that value at which interpolation is desired. Thus

$$f(x) = \sum_{i=0}^{n} a_i \phi_i(x)$$

where $\phi_i(x)$, $i = 0, 1, \ldots, n$, are the basis functions (polynomials) and the a_i are the coefficients dependent upon the known values of the function f_i. If, instead, we interchange the rôles of f and x in our table of values (see Table 1) and then represent x as a sum of polynomials of f, we find

$$x(f) = \sum_{i=0}^{n} b_i \psi_i(f)$$

where ψ_i are polynomials of f. To clarify the situation let us consider an example.

Example 10 The values of the function f for particular x are given symbolically in Table 16.

Table 16

x	x_0	x_1	x_2	x_3	x_4	x_5	x_6	x_7
$f(x)$	f_0	f_1	f_2	f_3	f_4	f_5	f_6	f_7

For this data, the Lagrangian interpolation formula for f is

$$y(x) = \sum_{j=0}^{7} L_j(x) f_j$$

where

$$L_j(x) = \prod_{\substack{i=0 \\ i \neq j}}^{7} \frac{x - x_i}{x_j - x_i} \qquad j = 0, 1, \ldots, 7$$

are the Lagrangian polynomials of section 3. If we now require to interpolate for x at a particular value of f we reverse the rôles of f and x in Table 16 and use the Lagrangian interpolation formula

$$x(f) = \sum_{j=0}^{7} L_j(f) x_j$$

where

$$L_j(f) = \prod_{\substack{i=0 \\ i \neq j}}^{n} \frac{f - f_i}{f_j - f_i} \qquad j = 0, 1, \ldots, 7$$

Such an interpolation problem is called *inverse* interpolation.

Example 11　Find the value of x in $(1.0, 2.0)$ for which the function f, whose values are given in Table 17, equals 0.50.

Table 17

x	1.0	1.2	1.4	1.6	1.8	2.0
$f(x)$	0.000 000	0.127 524	0.302 823	0.517 000	0.764 127	1.039 725

By reversing the rôles of x and $f(x)$ we can then formulate the Lagrangian interpolation formula

$$x = \sum_{j=0}^{5} x_j L_j(f) = \sum_{j=0}^{5} x_j \prod_{\substack{i=0 \\ i \neq j}}^{5} \frac{f - f_i}{f_j - f_i}$$

By substituting the values of $f = 0.5$ and f_j from Table 17 we may evaluate the Langrangian interpolation polynomials as

$$L_0(0.5) = 0.011\ 221\ 05 \qquad L_3(0.5) = 0.939\ 663\ 47$$

$$L_1(0.5) = -0.047\ 255\ 16 \qquad L_4(0.5) = -0.022\ 045\ 92$$

$$L_2(0.5) = 0.116\ 778\ 92 \qquad L_5(0.5) = 0.001\ 637\ 64$$

and then

$$x = \sum_{j=0}^{5} x_j L_j(0.5) = 1.585\,059\,52$$

Thus for $y = f(x)$, what we are doing in inverse interpolation is to interpolate the *inverse function* $x = f^{[-1]}(y)$. Such an interpolation is only possible when x is *a single-valued function* of y in the interval in which the interpolation is to be applied (see Exercise 49).

We may obtain the error term in inverse interpolation in just the same way as it was obtained for the Lagrangian interpolation in section 4. Thus,

$$E(y) = \frac{\pi_{n+1}(y)}{(n+1)!} f^{[-1](n+1)}(\xi)$$

where $f^{[-1](n+1)}(\xi)$ is the $(1+n)$th derivative of the inverse function $f^{[-1]}(y)$ with respect to y evaluated at ξ, a point belonging to the interval of inverse interpolation. The derivatives of the inverse function can be obtained in a straightforward (but tedious) manner as expressions involving derivatives of f. From

$$f^{[-1]}(y) = x$$

we see that differentiation once with respect to x produces

$$f^{[-1](1)}(y)f^{(1)}(x) = 1$$

so that

$$f^{[-1](1)}(y) = \frac{1}{f^{(1)}(x)}$$

Differentiating once again we find that

$$f^{[-1](2)}(y)\{f^{(1)}(x)\}^2 + f^{[-1](1)}(y)f^{(2)}(x) = 0$$

so that

$$f^{[-1](2)}(y) = \frac{-f^{(2)}(x)}{\{f^{(1)}(x)\}^3}$$

By repeated differentiation, it will be found that a power of $f^{(1)}(x)$ always appears in the denominator. Because of the form of the error term, we see that should $f^{(1)}(x)$ become small in the interval of inverse interpolation, the resulting error can become large. In the event that the interpolant x occurs close to a point at which $f^{(1)}(x) = 0$, a variant of iterated linear interpolation can be applied (see [34]).

Exercises

47. Find the zero which lies in the interval $(0, 2)$ of the polynomial

$$f = x^7 - 4x^6 - 14x^5 + 56x^4 + 49x^3 - 196x^2 - 36x + 144$$

using values of f at

$$x = 0.3, 0.5, 0.7, 0.9, 1.1, 1.3, 1.5, 1.6$$

48. Program Aitken's method of linear iterated interpolation for inverse interpolation. Using the values of f in Exercise 47 find the zero in the interval $(0, 2)$ for f of Exercise 47.

49. As stated in page 141 the inverse interpolation can only be applied when x is a single-valued function of y in the interval of interpolation. By evaluating y of Exercise 47 it will quickly become clear that for $x \in [-3, 4]$, x is a many-valued function of y. By choosing the tabular points as the function values corresponding to $x = -3.5, -2.5, -1.5, 0, 1.5, 2.5, 3.5, 4.5$, what point do you find is the zero of the polynomial y of Exercise 47 using Lagrangian interpolation based on these points? Check your result by evaluating y at this value of x.

50. Repeat Exercise 49 with other distributions of points.

10 Piecewise Polynomials

A famous example exists where Lagrangian interpolation is proposed for the Runge function $1/(1 + 25x^2)$ for $x \in [-1, 1]$. If interpolation is used with equi-spaced points x_i, $i = 0, 1, \ldots, n$, then it is observed that as n increases (i.e. we take more equi-spaced points in $[-1, 1]$) the interpolating polynomial does not converge to $1/(1 + 25x^2)$ (try it and see!). However, if the points x_i, $i = 0, 1, \ldots, n$, are placed at the zeros of the Chebyshev polynomial $T_{n+1}(x)$, then as $n \to \infty$ we do get convergence. This is an example of the general situation where placing the interpolating points in a random fashion will cause an erratic behavior of the interpolating polynomial as the degree n increases. The choice of using the zeros of the Chebyshev polynomial is no accident. It may be proved (see [47]) that if the function $f(x)$ being interpolated has a continuous derivative for $x \in [-1, 1]$ then the Lagrangian interpolation polynomial of degree n interpolating $f(x)$ at the zeros of the Chebyshev polynomial of degree $n + 1$ will converge as $n \to \infty$. However, there are still continuous functions $f(x)$ for which this is *not* true (e.g. $f(x)$ which has a discontinuous first derivative). The implication here is that as $n \to \infty$ the error associated with the interpolation polynomial will become small for "nice" $f(x)$ and strategically placed data points. Let us assume we do not want to take large n. If we play safe and use the data points placed at the zeros of the Chebyshev polynomial then for fixed n the error is given by

$$|e_n(x)| \leq \max_{a \leq x \leq b} \frac{|f^{(n+1)}(x)|}{(n+1)!} 2 \left[\frac{b-a}{4} \right]^{n+1}$$

since we can show that

$$\max_{a \leq x \leq b} \left| \prod_{j=0}^{n} (x - y_j) \right| = 2 \left[\frac{b-a}{4} \right]^{n+1}$$

where $\{y_j\}$ are the Chebyshev zeros transformed from $[-1, 1]$ to the interpolating interval $[a, b]$. So, for fixed n the error's value can be reduced only by reducing $b - a$, the length of the interval over which interpolation is taking place. If we proposed interpolating over a small interval then the error would be correspondingly less. So, let us divide up our interval $[a, b]$ into a set of subintervals and propose a different interpolating polynomial on each subinterval.

Example 12 Consider the interval $[a, b]$ subdivided into n subintervals $[x_i, x_{i+1}]$, $i = 0, 1, \ldots, n - 1$, with $a = x_0 < x_1 < \cdots < x_n = b$. On any such subinterval we may define the linear interpolating polynomial

$$y_i(x) = a_i x + b_i \qquad x \in [x_i, x_{i+1}]$$

or equivalently,

$$y_i(x) = \frac{(x_{i+1} - x)f_i + (x - x_i)f_{i+1}}{x_{i+1} - x_i} \qquad x \in [x_i, x_{i+1}], i = 0, 1, \ldots, n - 1 \quad (55)$$

Each $y_i(x)$ is understood to be defined to be nonzero only for $x \in [x_i, x_{i+1}]$, $i = 0, 1, \ldots, n$; for x outside this subinterval, $y_i(x)$ is assumed to be zero. Hence the complete interpolant can be represented as

$$Y(x) = \sum_{i=0}^{n-1} y_i(x) \qquad y_i(x) \text{ given by Eq. (55)} \tag{56}$$

By the nature of derivation, $Y(x)$ is a continuous function of x. However, the derivative of Y is not continuous so that the function has "kinks" at each data point (see Figure 2).

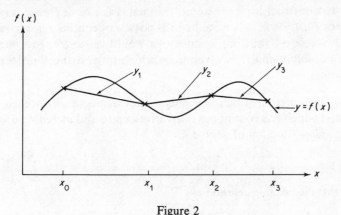

Figure 2

To remove the kinks in Example 12 we would require to make the first derivative of $Y(x)$ continuous at the data points also. We can do this by specifying the first derivative values of $f(x)$ at x_i and proposing cubic Hermite interpolation on each subinterval $[x_i, x_{i+1}]$. These polynomials may easily be

144

seen to be

$$Y_i(x) = f(x_i)\left[1 - \frac{(x-x_i)^2}{(x_i-x_{i+1})^2} - 2\frac{(x-x_i)^2(x-x_{i+1})}{(x_i-x_{i+1})^3}\right]$$

$$+ f(x_{i+1})\left[\frac{(x-x_i)^2}{(x_i-x_{i+1})^2} + 2\frac{(x-x_i)^2(x-x_{i+1})}{(x_i-x_{i+1})^3}\right]$$

$$+ f^{(1)}(x_i)\left[(x-x_i) + \frac{(x-x_i)^2}{(x_i-x_{i+1})} + \frac{(x-x_i)^2(x-x_{i+1})}{(x_i-x_{i+1})^2}\right]$$

$$+ f^{(1)}(x_{i+1})\left[\frac{(x-x_i)^2(x-x_{i+1})}{(x_i-x_{i+1})^2}\right]$$

$$x \in [x_i, x_{i+1}], i = 0, 1, \ldots, n, \text{ (and } = 0 \text{ for } x \in [x_j, x_{j+1}], j \ne i, \tag{57}$$

which can easily be verified to interpolate $f(x_i)$ and $f^{(1)}(x_i)$, $i = 0, 1, \ldots, n$, and hence to produce an interpolating function

$$Y(x) = \sum_{i=0}^{n-1} y_i(x) \tag{58}$$

which is continuous and has continuous first derivative for $x \in [a, b]$. Since the function $Y(x)$ is made up of polynomial pieces, we call such a function a *piecewise polynomial* function.

The function $Y(x)$ defined by Eqs. (57) and (58) still has a discontinuous second derivative at the data points! The question therefore naturally arises "Can we construct a piecewise cubic polynomial $Y(x)$ which interpolates the values $f(x_i)$ at x_i, $i = 0, 1, \ldots, n$, and which has both first and second derivatives continuous at the data points?" In general, the question would be phrased as "Can we construct a piecewise polynomial $Y(x)$ of degree k, say, which interpolates $f(x_i)$, $i = 0, \ldots, n$, and which possesses continuous derivatives of order $1, 2, \ldots, k-1$ (but not k since this would cause $Y(x)$ to degenerate into a single polynomial)?" Polynomials which possess these properties are termed *splines*.

Definition A function which is a polynomial of degree k in each interval $[x_i, x_{i+1}]$ and which has continuous derivatives up to and including order $k-1$ is called a *spline* function of degree k.

Exercises

51. Show that the function defined by

$$S(x) = \begin{cases} 1-2x & x \le -3 \\ 28+25x+9x^2+x^3 & -3 \le x \le -1 \\ 26+19x+3x^2-x^3 & -1 \le x \le 0 \\ 26+19x+3x^2-2x^3 & 0 \le x \le 3 \\ -163+208x-60x^2+5x^3 & 3 \le x \le 4 \\ 157-32x & x \ge 4 \end{cases}$$

is a cubic spline.

52. Plot the graph of the function $S(x)$ and its first *three* derivatives defined in Exercise 51.

In Example 12 the piecewise polynomials are linear and continuous at the tabular points x_i. Thus Y given by Eq. (56) is a spline of degree one or a *linear spline*. If Y were a piecewise quadratic polynomial with continuous first derivative then it would be a spline of degree 2 or a quadratic spline. To determine the particular form of Y in this case we could proceed very simply by assuming

$$y_i(x) = a_i x^2 + b_i x + c_i \qquad x \in [x_i, x_{i+1}], \qquad i = 0, 1, \ldots, n-1, \qquad (59)$$

in each of the intervals where a_i, b_i, c_i are the constants to be determined. By equating $y_i(x_i)$ and $y_i(x_{i+1})$ to the values of f given by $f(x_i)$ and $f(x_{i+1})$ we will have two conditions to determine three coefficients. We also wanted the first derivative to be continuous and hence impose this condition at both x_i and x_{i+1}, $i = 1, 2, \ldots, n-2$ (not $i = 0$ or $i = n-1$). At first sight this might appear to over-determine the coefficients in Eq. (59) but the reader will notice that the continuity of the derivative at the left-hand tabular point for one interval produces the identical relationship as continuity of the derivative at the right-hand tabular point of the previous interval so that in fact, bar the first, we have three conditions for three unknowns in each interval $[x_i, x_{i+1}]$, $i = 1, 2, \ldots, n-1$. For the first interval $[x_0, x_1]$ however, we have just two conditions occurring naturally and hence require an extra condition to be applied in order to obtain the third coefficient. The question naturally arises how to impose this extra condition. The answer, presumably, is so that the resulting interpolant Y is "best." We will only impose an extra condition which is convenient here. For example, we impose the condition that the derivative of $Y_0(x)$ should be zero at $x = x_0$. (We emphasize that this may not even be good let alone best in some cases; merely convenient). Having specified these extra conditions we can now write down the system of linear equations which the coefficients a_i, b_i, and c_i must satisfy. However, there exists a more convenient formulation of the spline function in which only one new coefficient occurs for each new interval. To see how this might arise, let us assume we have progressed from the leftmost interval $[x_0, x_1]$ stepwise to the interval $[x_{i-1}, x_i]$. Let the spline $S_i(x)$ on these intervals possess the required properties that the function values $f_0, f_1, \ldots, f_i$ have been interpolated and the first $k - 1$ derivatives are continuous. Within the interval $[x_i, x_{i+1}]$ let us add to $S_i(x)$, a term $h_i(x)$, a polynomial of degree k, chosen to ensure that we retain continuity of the first $k - 1$ derivatives at x_i. So

$$S_{i+1}(x) = S_i(x) + h_i(x) \qquad (60)$$

Consider

$$S_{i+1}(x_i) = S_i(x_i) + h_i(x_i)$$

By assumption, $S_{i+1}(x_i) = S_i(x_i) = f_i$. This implies

$$h_i(x_i) = 0 \qquad (61)$$

Also, $S_{i+1}^{(j)}(x_i) = S_i^{(j)}(x_i)$, $j = 1, 2, \ldots, k-1$, which further implies

$$h_i^{(j)}(x_i) = 0 \qquad j = 1, 2, \ldots, k-1 \tag{62}$$

Equations (61) and (62) state that the polynomial $h_i(x)$ has a root $(x - x_i)$ of *multiplicity* k. That is,

$$h_i(x) = b_i(x - x_i)^k \tag{63}$$

where b_i is a constant. Since $h_i(x)$ is a polynomial of degree k, Eq. (63) defines $h_i(x)$ uniquely (given b_i). To determine the constant b_i we merely have to interpolate $f(x_{i+1})$, and hence $S_{i+1}(x)$ is defined. We may therefore apply Eq. (60) recursively for $i = 0, 1, 2, \ldots, n-1$, where for $i = 0$,

$$S_0(x) \equiv \sum_{j=0}^{k} a_i x^i$$

is a polynomial of degree k and the coefficients $\{a_i\}$ are to be chosen so that $S_0(x_0) = f(x_0)$; $S_0(x_1) = f(x_1)$ and auxiliary derivative conditions are satisfied. If, for example, we specified

$$S_0^{(j)}(x_0) = 0, j = 1, 2, \ldots, k-1$$

then $k+1$ conditions have been specified which uniquely define $S_0(x)$. Hence for each new interval we introduce just one new parameter b_i through Eqs. (60) and (63). In this case the coefficients b_i are given explicitly—there is no system of linear equations to be solved. Equations (60) and (63) may be combined together in the following convenient notation. The spline $Y(x)$ of degree k on the tabular points $x_0, x_1, \ldots, x_n$ is represented as

$$Y(x) = p_k(x) + \sum_{i=1}^{n-1} b_i(x - x_i)_+^k \tag{64}$$

where $p_k(x)$ is a polynomial of degree k given by $\sum_{i=0}^{k} a_i x^i$ (our function $S_0(x)$ above) and $(x - x_i)_+^k$ is the *truncated power function* defined as

$$(x - x_i)_+^k \equiv \begin{cases} (x - x_i)^k & \text{if } (x - x_i) > 0 \\ 0 & \text{otherwise} \end{cases}$$

Example 13 Let us determine the quadratic spline on the tabular points x_i which interpolates f_i, $i = 0, 1, \ldots, n$, and such that $Y^{(1)}(x_0) = 0$.

For $k = 2$ in Eq. (64), we have

$$Y(x) = p_2(x) + \sum_{i=1}^{n-1} b_i(x - x_i)_+^2$$

$$\equiv a_0 + a_1 x + a_2 x^2 + \sum_{i=1}^{n-1} b_i(x - x_i)_+^2 \tag{65}$$

For $x \in [x_0, x_1]$,

$$Y(x_0) = a_0 + a_1 x_0 + a_2 x_0^2 = f(x_0) \tag{66}$$

$$Y(x_1) = a_0 + a_1 x_1 + a_2 x_1^2 = f(x_1) \tag{67}$$

$$Y^{(1)}(x_0) = a_1 + 2a_2 x_0 = 0 \tag{68}$$

We may solve Eqs. (66), (67), (68) to obtain

$$a_0 = f_0 + a_2 x_0^2; \quad a_1 = -2x_0 a_2$$

where

$$a_2 = \frac{f_1 - f_0}{(x_1 - x_0)^2}$$

For $x \in [x_1, x_2]$, we introduce $b_1(x - x_1)^2$ into Y. Hence Eq. (65) becomes

$$Y(x) = p_2(x) + b_1(x - x_1)^2 \tag{69}$$

Hence using $Y(x_2) = f(x_2)$ in Eq. (69), we find

$$b_1 = [f_2 - p_2(x_2)]/(x_2 - x_1)^2$$

In general, we find for $x \in [x_j, x_{j+1}]$,

$$b_j = \frac{f_{j+1} - p_2(x_{j+1}) - \sum_{i=1}^{j-1} b_i (x_{j+1} - x_i)^2}{(x_{j+1} - x_j)^2} \qquad j = 1, 2, \ldots, n-1$$

So that the coefficients $\{b_j\}$ are defined explicitly.

In a similar manner we may obtain the cubic spline

$$Y(x) = p_3(x) + \sum_{i=1}^{n-1} b_i (x - x_i)_+^3$$

which interpolates f_i at x_i, $i = 0, 1, \ldots, n$, and which satisfies the auxiliary conditions $Y^{(1)}(x_0) = Y^{(2)}(x_0) = 0$. This time a system of four equations in the four unknowns a_0, a_1, a_2, and a_3 must be solved. Then the coefficients $\{b_i\}$ may be obtained explicitly as in Example 13. The details are left to Exercise 53.

In contrast to specifying the auxiliary conditions $Y^{(1)}(x_0) = Y^{(2)}(x_0) = 0$, more "natural" conditions are $Y^{(2)}(x_0) = Y^{(2)}(x_n) = 0$. If we impose such conditions we find that the parameters $\{b_i\}$ are now no longer obtained without solving a system of linear equations.

Example 14 Let us now determine the cubic spline on the tabular points x_i which interpolates f_i, $i = 0, 1, \ldots, n-1$, and such that $Y^{(2)}(x_0) = Y^{(2)}(x_n) = 0$.

Substituting $k = 3$ into Eq. (64), we have

$$Y(x) = p_3(x) + \sum_{i=1}^{n-1} b_i(x - x_i)_+^3$$

$$\equiv a_0 + a_1 x + a_2 x^2 + a_3 x^3 + \sum_{i=1}^{n-1} b_i(x - x_i)_+^3 \tag{70}$$

For $x \in [x_0, x_1]$,

$$Y(x_0) = a_0 + a_1 x_0 + a_2 x_0^2 + a_2 x_0^3 = f(x_0) \tag{71}$$

$$Y(x_1) = a_0 + a_1 x_1 + a_2 x_1^2 + a_3 x_1^3 = f(x_1) \tag{72}$$

$$Y^{(2)}(x_0) = 2a_2 + 6a_3 x_0 = 0 \tag{73}$$

Clearly, Eqs. (71), (72), and (73) do not uniquely specify the four parameters a_0, a_1, a_2, a_3. Solving the equations in terms of a_3 we find $p_3(x) \equiv p_3(x; a_3) = A_3(x)a_3 + A_1(x)$, where $A_i(x)$ are polynomials of degree i, $i = 1, 3$. Hence the interpolation conditions

$$Y(x_j) = A_3(x_j)a_3 + A_1(x_j) + \sum_{i=1}^{j-1} b_i(x_j - x_i)^3 = f_j \qquad j = 2, 3, \ldots, n-1$$

and

$$Y^{(2)}(x_n) = A_3^{(2)}(x_n)a_3 + 6 \sum_{i=1}^{n-1} b_i(x_n - x_i) = 0$$

yield $n - 1$ equations in the $n - 1$ unknowns $\{b_i\}$ and a_3. This system of equations in matrix form is

$$\begin{bmatrix} A_3(x_2) & (x_2 - x_1)^3 & & & \mathbf{0} \\ A_3(x_3) & (x_3 - x_1)^3 & (x_3 - x_2)^3 \\ A_3(x_4) & (x_4 - x_1)^3 & (x_4 - x_2)^3 & (x_4 - x_3)^3 \\ \vdots & \vdots & & & \ddots \\ A_3^{(2)}(x_n) & 6(x_n - x_1) & & & 6(x_n - x_n) \end{bmatrix} \begin{bmatrix} a_3 \\ b_1 \\ b_2 \\ \vdots \\ b_{n-1} \end{bmatrix} = \begin{bmatrix} f_2 - A_1(x_2) \\ f_3 - A_1(x_3) \\ \vdots \\ 0 \end{bmatrix}$$

The coefficient matrix of this system of equations is known as a *lower Hessenberg matrix*. The solution of this system of equations produces the required coefficients a_3, $\{b_i\}$ of the spline $Y(x)$. However, the number of operations (multiplications and divisions) required for the solution is $O(n^2)$. Consequently, particularly for the cubic spline, it is *not* advocated that the spline should be computed in this manner. Rather, a procedure outlined in Conte and de Boor [6, p. 286] allows the computation of the spline in $O(n)$ operations. This procedure is, however, much more cumbersome to describe and the essence of the properties of the spline function yielded by Eq. (64) is lost. Moreover, the generalization of the procedure described in the example for $k = 3$ to arbitrary k, where the auxiliary conditions necessary to describe the kth-degree spline uniquely are specified at the end points x_0 and x_n, is obvious;

the procedure of [6] is extremely tedious. However, many practical applications find cubic splines adequate and consequently the algorithm described in [6] is to be recommended for the *generation* of the interpolating cubic spline; see also [12]

Exercises

53. Determine the coefficients b_i of the cubic spline

$$Y(x) = p_3(x) + \sum_{i=1}^{n-1} b_i(x - x_i)_+^3$$

which interpolates the data $f(x_i)$, $i = 0, 1, \ldots, n$, and which satisfies the auxiliary conditions $Y^{(1)}(x_0) = Y^{(2)}(x_0) = 0$.

54. The Runge function $1/(1 + 25x^2)$ is to be interpolated for $x \in [-1, 1]$ using function values at x_i, $i = 0, 1, \ldots, n$, defined by $x_i = (-1 + ih)$, $i = 0, 1, \ldots, n$, where $(n + 1)h = 2$.

Using the equations for the coefficients of the cubic spline in Exercise 53, program the interpolation problem for increasing n. Plot the Runge function and the cubic spline for $n = 10$, 20, and 100. What do you conclude as n increases compared with the Lagrangian interpolation of the Runge function using the same points?

11 Extrapolation

We conclude this chapter with a brief account of the extrapolation problem. It is often the case that, given a set of readings or measurements corresponding to a given set of abscissas, it is required to determine a value of the function outside this range of abscissas. For example, given the population increase for the years 1970, 71, 72, 73, 74, 75, 76, predict what the population increase will be in 1980 or 1985, etc.

The interpolating formulas which have been discussed could have been employed for extrapolating problems. However, the problem of extrapolation by its very nature must be a more inaccurate technique than the interpolating problems since a bound on the required derivative in the error terms is not usually forthcoming. Consequently it is advisable to try and make sure, in some way, that the point at which approximation is required is an "interval" point so that interpolation exists. In the event that this is impossible, like the example of the population increase, extrapolation should only be performed for very short distances from the end of the interval.

Example 15   In section 2 of Chapter 1 we described the problem of ascertaining oil productions at points intermediate to existing wells. If, as is reasonable to assume, a well is to be drilled outside the existing set of wells, we have the problem of extrapolation if we try to ascertain the production at the specified x, prior to drilling.

CHAPTER 5

Least Squares Approximation

1 The Principle of Least Squares

In the previous chapter we discussed exact approximation in which it was assumed that the *exact* values f_i of the function f were known at $n + 1$ tabular points x_i, $i = 0, 1, \ldots, n$. Furthermore, we assumed that the degree of the approximating polynomial was equal to n. In the present chapter we will consider relaxing these two assumptions. First, we assume the f_i are not known exactly but are themselves approximations to f at $x = x_i$. For example, the values of production read in Example 3 in Chapter 1 are highly likely to be in error owing to the engineer's inability to read values of the production exactly by some physical device. We will denote by f_i, $i = 0, 1, \ldots, n$, the correct function values, and by $\bar{f}_i$, $i = 0, 1, \ldots, n$, the approximate or observed values at the tabular points x_i, $i = 0, 1, \ldots, n$. Second, we will assume that we require to use an approximating function

$$y(x) = \sum_{j=0}^{m} a_j \phi_j(x) \tag{1}$$

where $\phi_j(x)$ are polynomials of degree j in which the number of degrees of freedom (i.e. the number of parameters a_j), $m + 1$, is less than the number of tabular points, $n + 1$. To indicate the reason for this let us consider the following example.

Example 1 Consider the linear function $f(x) \equiv 0.2x + 1$. This function is plotted for $x \in [1, 3]$ in Figure 1. Let us assume we are now given observed values of f at $x = 1$, 2, and 3, namely 1.18, 1.45, and 1.59, instead of 1.2, 1.4 and 1.6 respectively. If we interpolate this data by a quadratic polynomial we obtain the function depicted as $y(x)$ in Figure 1. Clearly this function does not coincide with $f(x)$. However, in this case we might surmise that $|y - f|$ for $x \in [1, 3]$ would not be large. If, in contrast, we require to evaluate the derivative of f at the tabular points then the error is larger. For example,

150

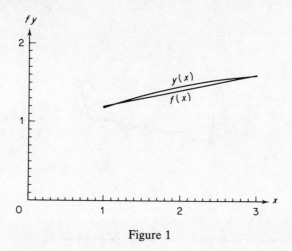

Figure 1

$y^{(1)}(x) = 0.335$, 0.205, and 0.075 at $x = 1$, 2, and 3 respectively, in contrast to the theoretical 0.2. To introduce more data which is in error by the same order of magnitude only causes the situation to worsen. This is shown in Figure 2 using an interpolation polynomial of degree 10 through the eleven points given in Table 1.

Table 1

x	$\bar{f}$	f
1.00	1.18	1.20
1.20	1.26	1.24
1.40	1.23	1.28
1.60	1.37	1.32
1.80	1.37	1.36
2.00	1.45	1.40
2.20	1.42	1.44
2.40	1.46	1.48
2.60	1.53	1.52
2.80	1.59	1.56
3.00	1.59	1.60

The error inherent in the f_i is often called *noise*. Thus in proposing an exact approximation the noise enters into the approximation as well. What we really would like to do is to approximate the function f in the absence of noise. Thus we require to damp or smooth these inherent errors. The damping is achieved by specifying that the degree of the approximating function $y(x)$ be less than the actual function $f(x)$. That this is so can be easily seen from Example 1. If we take a high-degree interpolant for the data $\bar{f}$, the behavior of the interpolant compared with the theoretical f becomes increasingly erratic

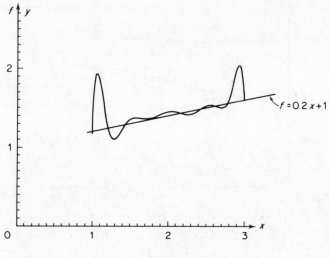

Figure 2

as the number of points increases. If, instead, we allowed the number of data points to increase but specified that the interpolant must be at most degree 2 (say), then this interpolant would look appreciably more like f than $y(x)$ in Figure 2; see Figure 3, where $y_1(x)$ and $y_2(x)$ are two possible quadratic interpolants each passing through some chosen sets of the noisy data. Of course this situation is not satisfactory as the question immediately arises: which of the noisy data should we choose to interpolate and which should we ignore? The reader may be tempted to suggest these points nearest to f; but recall in practice f does not appear on the graph, so "nearest" is not meaningful.

Thus we see that increasing the degree of the approximation $y(x)$ up to (but not exceeding) the number of points can produce an increasingly erratic function. The lower the degree of the approximation, the more *smooth* is the

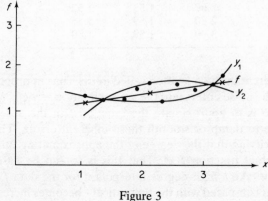

Figure 3

resulting approximation. Thus the example suggests that we keep the degree of $y(x)$ small.

Clearly therefore, we cannot advocate choosing the coefficients a_j through interpolating the given data at the points x_i. Thus we are left with the problem of choosing the coefficients a_j in our approximation

$$y(x) = \sum_{j=0}^{m} a_j \phi_j(x) \qquad m < n \text{ (for smoothing)} \tag{1a}$$

given the data $\bar{f}_i$ at x_i, $i = 0, 1, \ldots, n$.

We have indicated that interpolation is not to be advocated, so that we do not specify $y(x_i) = \bar{f}_i$ for any of the values of i. A clue to the choice of coefficients emerges if we notice that in Figure 3, except at the points we interpolated, the interpolant $y(x)$ and the data $\bar{f}_i$ differed at the tabular points x_i, $i = 0, 1, \ldots, n$. Thus we could choose the coefficients a_j by demanding that the sum of the absolute values of the differences between the approximating function $y(x)$ and the observed data $\bar{f}_i$ be minimized at the tabular points x_i, $i = 0, 1, \ldots, n$. Whereas we can certainly demand this, the mathematics it leads to is not conducive to a simple algorithm. If, however, in contrast we ask that the sum of squares of $y(x_i) - \bar{f}_i$ be minimized then this *does* lead to a simple algorithm.

Thus we will choose the a_j, $j = 0, 1, \ldots, m$ such that

$$\sum_{i=0}^{n} [\bar{f}_i - y(x_i)]^2 \tag{2}$$

is minimized. From calculus we know that a function of several variables has a stationary point when the partial derivatives of this function with respect to the independent variables are each zero. The independent variables in our approximation are the coefficients a_j, $j = 0, 1, \ldots, m$. Hence we require that the partial derivatives of Eq. (1) with respect to each a_j, $j = 0, 1, \ldots, m$ be zero. Because of the form of (2) it can be shown that these conditions also guarantee a minimum rather than just a stationary point. Thus we choose the $\{a_j\}$ such that

$$\frac{\partial}{\partial a_k} \left\{ \sum_{i=0}^{n} [\bar{f}_i - y(x_i)]^2 \right\} = 0 \qquad k = 0, 1, \ldots, m \tag{3}$$

We will have occasion later to discuss particular forms of the polynomials $\phi_j(x)$ used in $y(x)$ in Eq. (1). For the moment, however, it will be sufficient to let $\phi_j(x) = x^j$ so that $y(x) = \sum_{j=0}^{m} a_j x^j$. Hence substituting this $y(x)$ into Eq. (3) and interchanging the differentiation and summation over i, since this is finite, we require to choose $\{a_j\}$ such that

$$\sum_{i=0}^{n} \frac{\partial}{\partial a_k} \left[\bar{f}_i - \sum_{j=0}^{m} a_j x_i^j \right]^2 = 0 \tag{4}$$

Performing the differentiation, this gives

$$\sum_{i=0}^{n} \bar{f}_i x_i^k - \sum_{i=0}^{n} \sum_{j=0}^{m} a_j x_i^j x_i^k = 0 \tag{5}$$

Equation (5) is clearly the system of $m + 1$ equations,

$$\sum_{j=0}^{m} a_j \sum_{i=0}^{n} x_i^j x_i^k = \sum_{j=0}^{m} \bar{f}_i x_i^k \qquad k = 0, 1, \ldots, m \tag{6}$$

Example 2 Equation (6) looks very complicated! Let us see what the system of equations looks like if we have six data points and require the degree of the approximation $y(x)$ to be 2. Then $m = 2$ and $n = 5$, so we require to choose a_j, $j = 0, 1, 2$, such that

$$\frac{\partial}{\partial a_k} \left[\sum_{i=0}^{5} \left[\bar{f}_i - \sum_{j=0}^{2} a_j x_i^j \right]^2 \right] = 0 \qquad k = 0, 1, 2 \tag{7}$$

Differentiate Eq. (7) with respect to a_0, and we obtain

$$\sum_{i=0}^{5} \left[\bar{f}_i x_i^0 - \sum_{j=0}^{2} a_j x_i^j x_i^0 \right] \equiv \sum_{i=0}^{5} \left[\bar{f}_i - \sum_{j=0}^{2} a_j x_i^j \right] = 0 \tag{8}$$

Differentiate Eq. (7) with respect to a_1; then

$$\sum_{i=0}^{5} \left[\bar{f}_i x_i - \sum_{j=0}^{2} a_j x_i^j x_i \right] \equiv \sum_{i=0}^{5} \left[\bar{f}_i x_i - \sum_{j=0}^{2} a_j x_i^{j+1} \right] = 0 \tag{9}$$

Finally, differentiate Eq. (7) with respect to a_2, whence

$$\sum_{i=0}^{5} \left[\bar{f}_i x_i^2 - \sum_{j=0}^{2} a_j x_i^j x_i^2 \right] \equiv \sum_{i=0}^{5} \left[\bar{f}_i x_i^2 - \sum_{j=0}^{2} a_j x_i^{j+2} \right] = 0 \tag{10}$$

Rearranging the summation signs in Eq. (8), (9), and (10), we obtain

$$\sum_{j=0}^{2} a_j \sum_{i=0}^{5} x_i^j = \sum_{i=0}^{5} \bar{f}_i \tag{8.1}$$

$$\sum_{j=0}^{2} a_j \sum_{i=0}^{5} x_i^{j+1} = \sum_{i=0}^{5} \bar{f}_i x_i \tag{8.2}$$

and

$$\sum_{j=0}^{2} a_j \sum_{i=0}^{5} x_i^{j+2} = \sum_{i=0}^{5} \bar{f}_i x_i^2 \tag{8.3}$$

Now we can write Eqs. (8.1), (8.2), (8.3) more familiarly as

$$6a_0 + a_1 \sum_{i=0}^{5} x_i + a_2 \sum_{i=0}^{5} x_i^2 = \sum_{i=0}^{5} \bar{f}_i \tag{11}$$

$$a_0 \sum_{i=0}^{5} x_i + a_1 \sum_{i=0}^{5} x_i^2 + a_2 \sum_{i=0}^{5} x_i^3 = \sum_{i=0}^{5} \bar{f}_i x_i \tag{12}$$

and

$$a_0 \sum_{i=0}^{5} x_i^2 + a_1 \sum_{i=0}^{5} x_i^3 + a_2 \sum_{i=0}^{5} x_i^4 = \sum_{i=0}^{5} \bar{f}_i x_i^2 \tag{13}$$

which is a system of three linear equations in the three unknowns a_0, a_1, a_2.

Choosing the coefficients a_j according to Eq. (4) gives an approximation known as *discrete†* *least squares approximation.*

Exercises

1. Calculate the linear least squares approximation to the data in Table 2.

Table 2

x_i	0.0	0.1	0.2	0.3	0.4	0.5
f_i	0.11	0.251	0.352	0.453	0.54	0.66

   Graph your approximation. What does the Lagrangian interpolant of this data look like?
2. Calculate the quadratic least squares approximation to the data in Table 2. Graph your approximation.
3. The discrete least squares solutions in Exercises 1 and 2 have been obtained by minimizing Eq. (2) for different m. What values of the expression in Eq. (2) do you get for the approximations in Exercises 1 and 2? Which is smaller? Does it follow that the better approximation corresponds to the smaller value of Eq. (2)? Why?
4. Derive the form of the cubic least squares approximation. Would you advocate deriving least squares approximations in this manner for values of m greater than 3? Substitute the data from Table 1 to obtain a precise formula and graph the resulting approximation. What is the value of the expression in Eq. (2) in this case?

The discrepancies $(\bar{f}_i - f_i)$ in the observed values $\bar{f}_i$ have arisen from the inability to measure or observe the function f exactly. These discrepancies will be smaller for some tabular points x_i than others because the nature of the observation allows certain of the data to be measured more accurately than others. For example, if we observed the temperature at a fixed point in a heated metal bar over a given time, we are more likely to make a larger error in measuring the temperature when the temperature varies quickly than when the bar has cooled and the temperature change is much slower. In other words, we attach more reliability to our measurements made when the bar is cooler than when it is very hot. We would like to introduce this certainty into the approximation $y(x)$ so that the more reliable observations have more weight than the less reliable ones. We do this by associating at each tabular

---

† Discrete because we are minimizing the differences at a discrete set of points. If we chose the coefficients so that $\int_a^b [\bar{f}(x) - y(x)]^2 dx$ was minimized, the resulting approximation would be termed continuous least squares. Discrete least squares is most often the situation arising in practice and for this reason we will confine our attention to the discrete data set.

point x_i a nonnegative weight w_i so that the more reliable observations have larger w_i than the less reliable ones.

Thus let us denote the *residual* R_i by

$$R_i = w_i[\bar{f}_i - y(x_i)] \tag{14}$$

Then we propose a *weighted least squares* approximation $y(x)$ given by Eq. (1) where the coefficients a_j, $j = 0, 1, \ldots, m$, are chosen to

$$\text{minimize} \sum_{i=0}^{n} R_i^2 = \min \sum_{i=0}^{n} w_i^2 \left[\bar{f}_i - \sum_{j=0}^{m} a_j \phi_j(x_i) \right]^2 \tag{15}$$

The coefficients $\{a_j\}$ are now determined in an analogous manner to Eq. (3). Because of the presence of the weight function, our system of equations will be slightly modified. For $\phi_j(x) = x^j$ our system becomes

$$\sum_{j=0}^{m} a_j \left(\sum_{i=0}^{n} w_i^2 x_i^j x_i^k \right) = \sum_{j=0}^{m} w_i^2 \bar{f}_i x_i^k \qquad k = 0, 1, \ldots, m \tag{16}$$

The linear equations (16) are called the *normal* equations. In order that the normal equations have a unique solution we require that the functions $\phi_j(x)$ be *linearly independent*.† For $\phi_j = x^j$ this will be the case if the number of points x_i is greater than m.

Given that we can choose convenient linearly independent functions $\phi_j(x)$, the problem of calculating the least squares solution would seem to be solved; simply solve the normal equations. Unfortunately, this turns out to be harder than might at first appear. The assumption of linear independence of $\phi_j(x)$ which guarantees the nonsingularity of the matrix coefficients does not prevent the determinant of this matrix being small. This, as we have indicated in Chapter 3, is characteristic of the system of equations being *ill-conditioned* (see also Chapter 2). As we have indicated, such systems of equations are inherently extremely difficult to solve accurately; the computations carried out in determining the solution are subject to a large growth of rounding error. This in turn means that the numerical values of a_j are badly affected by these errors and the resulting least squares solution is poor. The seriousness of this ill-conditioning is exemplified by the following example.

Example 3 Consider the weight functions w_i to be identically equal to 1. Consider a least squares approximation based on tabular points x_i which are fairly uniformly distributed over the interval $[0, 1]$. Let $\phi_j(x) = x^j$. The normal equations are easily evaluated in this case and we find that the coefficient of a_j in the kth equation is given by $\sum_{i=0}^{n} x_i^{j+k}$, $k, j = 0, 1, \ldots, m$.

† See Chapter 4, section, 7, for a definition.

If the number of tabular points is sufficiently large we may approximate this sum by

$$\sum_{i=0}^{n} x_i^{j+k} \doteq (n+1) \int_0^1 x^{j+k} \, dx$$

Carrying out this integration for each a_j in each equation, we find that the matrix of coefficients is approximately

$$(n+1) \begin{bmatrix} 1 & 1/2 & 1/3 & 1/4 & \cdots & \dfrac{1}{m+1} \\[2mm] 1/2 & 1/3 & 1/4 & 1/5 & \cdots & \dfrac{1}{m+2} \\[2mm] \vdots & \vdots & \vdots & \vdots & & \vdots \\[2mm] \dfrac{1}{m+1} & \dfrac{1}{m+2} & \dfrac{1}{m+3} & \dfrac{1}{m+4} & \cdots & \dfrac{1}{2m+1} \end{bmatrix} \begin{matrix} (k=0) \\[2mm] (k=1) \\[2mm] \\[2mm] (k=m) \end{matrix}$$

This matrix is a principal minor of the Hilbert matrix, which is notoriously ill conditioned. That is, as m gets larger the conditioning of the normal equation deteriorates to such an extent that the numerical method soon determines a solution which has no significant digits of accuracy. Since for small m (2 or 3) the conditioning is not too bad, we obtain reasonable least squares approximations, but for m any larger than 9 the solution is worthless. We therefore need an alternative approach to deriving our least squares solution if it is to be of any practical use.

The last comment in Example 3 suggests we look for another approach to deriving the least squares solution. What is really needed is a method whereby a coefficient matrix arises which is well conditioned. It is clear that there is little we can do when $\phi_i(x) = x^i$, as we have shown in Example 3. However, the monomials are not the only linearly independent polynomials. Thus, perhaps if we chose another set $\{\phi_j\}$ we would end up with a well-conditioned matrix. Guessing at sets of linearly independent polynomials $\phi_j(x)$ is not very scientific and might not produce any better results than the monomials. Whereas there may be several ways of ensuring a matrix is well conditioned, the easiest is to ensure that it is *diagonally dominant* or, what is even better, by ensuring that the matrix is diagonal. This may be simply achieved in the following manner.

If we carry out the differentiation in Eq. (15) where now the $\{\phi_j(x)\}$ is a general sequence of linearly independent functions, we obtain, on setting the resulting expressions to zero, the system of linear equations

$$\sum_{i=0}^{n} w_i^2 \left[\bar{f}_i - \sum_{j=0}^{m} a_j \phi_j(x_i) \right] \phi_k(x_i) = 0 \qquad k = 0, 1, \ldots, m \qquad (17)$$

(If you find this perplexing, compare Eq. (17) with Eq. (16) when $\phi_k(x) = x^k$.) We may rewrite Eq. (17) as

$$\sum_{i=0}^{n} w_i^2 \left\{ \sum_{j=0}^{m} a_j \phi_j(x_i) \phi_k(x_i) \right\} = \sum_{i=0}^{n} w_i^2 \bar{f}_i \phi_k(x_i) \qquad k = 0, 1, \ldots, m$$

which we may once more rearrange to give

$$\sum_{j=0}^{m} a_j \left\{ \sum_{i=0}^{n} w_i^2 \phi_j(x_i) \phi_k(x_i) \right\} = \sum_{i=0}^{n} w_i^2 \bar{f}_i \phi_k(x_i) \qquad k = 0, 1, \ldots, m \qquad (18)$$

Recall that we would like to make the coefficient matrix diagonal. In terms of the coefficients in the system of linear equations (18), this would require us to *choose* the functions $\{\phi_j\}$ so that

$$\sum_{i=0}^{n} w_i^2 \phi_j(x_i) \phi_k(x_i) = 0 \qquad j \neq k \qquad (19)$$

since the coefficient a_k in the kth equation is

$$\sum_{i=0}^{n} w_i^2 \phi_k(x_i) \phi_k(x_i) = \sum_{i=0}^{n} w_i^2 \phi_k^2(x_i) (\neq 0) \qquad (20)$$

and we require the other coefficients in that row of the matrix, namely those given by Eq. (19), to be zero. Equation (19) is familiar if we recall our discussion of *orthogonal functions* given in Chapter 4. That is, the requirement of a diagonal matrix requires us to choose the linearly independent functions $\{\phi_j\}$ to be orthogonal with respect to the weight function w_i^2 on the discrete data set $\{x_i\}$. Assuming we are able to construct such a set of orthogonal functions (we will actually describe the process in the next section) we can then use the property (19) to determine the coefficients $\{a_j\}$. Substituting Eqs. (19) and (20) into Eq. (18), we then obtain

$$a_k \sum_{i=0}^{n} w_i^2 \phi_k^2(x_i) = \sum_{i=0}^{n} w_i^2 \bar{f}_i \phi_k(x_i) \qquad k = 0, 1, \ldots, m$$

or

$$a_k = \sum_{i=0}^{n} w_i^2 \bar{f}_i \phi_k(x_i) \Big/ \sum_{i=0}^{n} w_i^2 \phi_k^2(x_i) \qquad k = 0, 1, \ldots, m \qquad (21)$$

as the *explicit* relations for the least squares coefficients $\{a_k\}$. We merely require a means of constructing orthogonal functions with respect to given weights w_i^2 on a discrete set $\{x_i\}$.

Example 4 Consider constructing the least squares approximation for the data in Table 3 using the orthogonal (on this discrete data with respect to the given weights) functions

$$\phi_0(x) = 1 \qquad \phi_1(x) = x - 0.358\,21 \qquad \phi_2(x) = x^2 - 1.101\,34x + 0.167\,72$$

Table 3

x_i	0	0.2	0.4	0.6	0.8	1.0
$\bar{f}_i$	0	0.3	0.6	0.7	0.6	0.4
w_i^2	1.0	1.1	1.2	1.1	1.3	1.0

Using the data of Table 3 in Eq. (21), we find

$$a_0 = \sum_{i=0}^{5} w_i^2 \bar{f}_i \phi_0(x_i) \Big/ \sum_{i=0}^{5} w_i^2 \phi_0^2(x_i) = \sum_{i=0}^{5} w_i^2 \bar{f}_i \Big/ \sum_{i=0}^{5} w_i^2$$

Hence

$$a_0 = [(1.0) \cdot (0.0) + (1.1) \cdot (0.3) + (1.2) \cdot (0.6) + (1.1) \cdot (0.7) + (1.3) \cdot (0.6)$$
$$+ (1.0) \cdot (0.4)]/[1.0 + 1.1 + 1.2 + 1.1 + 1.3 + 1.0] = 3.00/6.7 = 0.447\,76$$

Similarly,

$$a_1 = \sum_{i=0}^{5} w_i^2 \bar{f}_i (x_i - 0.358\,21) \Big/ \sum_{i=0}^{5} w_i^2 (x_i - 0.358\,21)^2$$

which gives $a_1 = 0.862\,02$.

Finally,

$$a_2 = \sum_{i=0}^{5} w_i^2 \bar{f}_i (x_i^2 - 1.101\,34 x_i + 0.167\,718) \Big/$$

$$\sum_{i=0}^{5} w_i^2 (x_i^2 - 1.101\,34 x + 0.167\,718)^2$$

which yields

$$a_2 = -2.934\,61$$

Thus we have obtained, without having to solve a system of equations, the required coefficients in the quadratic least squares approximation

$$y(x) = a_0 + a_1 \phi_1(x) + a_2 \phi_2(x)$$

so that $y(x) = 0.447\,76 + 0.862\,02(x - 0.358\,21) - 2.934\,61(x^2 - 1.101\,34x + 0.167\,718)$, which can be written

$$y(x) = -2.934\,61 x^2 + 4.094\,02 x - 0.353\,21$$

Exercise

5. Show that the functions $\phi_0(x)$, $\phi_1(x)$, and $\phi_2(x)$ are, indeed, orthogonal on the given discrete data with respect to the weights w_i^2 in Table 3.

If we change any of the data x_i or w_i^2 in Table 3 it can easily be shown that the functions $\phi_0(x)$, $\phi_1(x)$, and $\phi_2(x)$ are no longer orthogonal with respect to the new data. In this case we could not proceed along the lines described above using these ϕ's. What we need, of course, is a method of constructing orthogonal functions in terms of any given data. In the next sections we will describe two approaches to constructing these functions.

2 The Generation of Orthogonal Functions

We will first consider the generation of orthogonal functions using the *Gram–Schmidt* process. Consider linearly independent functions $\phi_j(x)$, $j = 0, 1, \ldots, m$. We want to construct (orthogonal) functions $q_j(x)$ so that

$$\sum_{i=0}^{n} q_j(x_i)q_k(x_i) \left. \right\} \begin{matrix} = 0 & j \neq k \\ \neq 0 & j = k \end{matrix} \qquad j, k = 0, 1, \ldots, m$$

(We will restrict our discussion here to the case where $w_i^2 = 1$; the extension of the algorithm to the case of general $w_i^2 \geq 0$ follows easily.)

Define

$$q_0(x) = \phi_0(x)$$

(the functions $\phi_j(x)$ have been ordered arbitrarily so this is in fact any of the ϕ's).

Take a linear combination of $q_0(x)$ and $\phi_1(x)$ (again the arbitrary ordering means any of the remaining ϕ's) so that the resulting combination is orthogonal to $q_0(x)$.

Thus we want to choose α_{01} in

$$q_1(x) = \phi_1(x) - \alpha_{01}q_0(x) \tag{22}$$

so that

$$\sum_{i=0}^{n} q_1(x_i)q_0(x_i) = 0 \tag{23}$$

Substitute for $q_1(x)$ from (22) into (23) so that we choose α_{01} in order that

$$\sum_{i=0}^{n} \phi_1(x_i)q_0(x_i) - \sum_{i=0}^{n} \alpha_{01}q_0^2(x_i) = 0$$

or

$$\alpha_{01} = \sum_{i=0}^{n} \phi_1(x_i)q_0(x_i) \bigg/ \sum_{i=0}^{n} q_0^2(x_i) \tag{24}$$

With this α_{01} we now have two functions $q_0(x)$ and $q_1(x)$ which are orthogonal.

We can now continue. We take a linear combination of the orthogonal $q_0(x)$ and $q_1(x)$ together with $\phi_3(x)$ (one of the remaining linearly independent functions). Thus choose α_{02} and α_{12} so that

$$q_2(x) = \phi_2(x) - \alpha_{02}q_0(x) - \alpha_{12}q_1(x) \tag{25}$$

is orthogonal to q_0 and q_1. This requires that

$$\sum_{i=0}^{n} \phi_2(x_i)q_0(x_i) - \sum_{i=0}^{n} \alpha_{02}q_0^2(x_i) - \sum_{i=0}^{n} \alpha_{12}q_1(x_i)q_0(x_i) = 0$$

and

$$\sum_{i=0}^{n} \phi_2(x_i)q_1(x_i) - \sum_{i=0}^{n} \alpha_{02}q_0(x_i)q_1(x_i) - \sum_{i=0}^{n} \alpha_{12}q_1^2(x_i) = 0$$

Since $q_0(x)$ and $q_1(x)$ are already orthogonal, we obtain

$$\alpha_{02} = \sum_{i=0}^{n} \phi_2(x_i)q_0(x_i) \Big/ \sum_{i=0}^{n} q_0^2(x_i) \tag{26}$$

and

$$\alpha_{12} = \sum_{i=0}^{n} \phi_2(x_i)q_1(x_i) \Big/ \sum_{i=0}^{n} q_1^2(x_i) \tag{27}$$

Thus (25) together with (26) and (27) serves to define the next orthogonal function $q_2(x)$. We can continue in this mode adding one extra linearly independent ϕ at a time and building up the sequence of orthogonal q's. To obtain the general formula, let us assume we have constructed p orthogonal functions

$$q_j(x) \qquad j = 0, 1, \ldots, p-1$$

We therefore add $\phi_p(x)$ into a linear combination with these orthogonal $q_j(x)$ to produce the next orthogonal $q_p(x)$.

Thus choose coefficients α_{jp}, $j = 0, 1, \ldots, p-1$, so that

$$q_p(x) = \phi_p(x) - \sum_{j=0}^{p-1} \alpha_{jp}q_j(x) \tag{28}$$

is orthogonal to the previous $q_j(x)$, $j = 0, 1, \ldots, p-1$. We now find in a straightforward manner that

$$\alpha_{jp} = \sum_{i=0}^{n} \phi_p(x_i)q_j(x_i) \Big/ \sum_{i=0}^{n} q_j^2(x_i) \qquad j = 0, 1, \ldots, p-1 \tag{29}$$

Equations (28) and (29) serve to define the orthogonal functions for general p. To save unwieldy notation, we introduce the following new notation.

Define the *inner product*

$$(f, g) = \sum_{i=0}^{n} f(x_i)g(x_i) \tag{30}$$

We may then employ this notation in the procedure for generating the orthogonal functions in Eq. (29) so that

$$\alpha_{jp} = (\phi_p, q_j)/(q_j, q_j) \tag{31}$$

The least squares approximation $y(x)$ is then obtained by substituting $q_i(x)$ for $\phi_i(x)$ in Eqs. (1) and (2).

Although the process for generating orthogonal functions just described produces the required orthogonal properties, there is a degree of arbitrariness about the functions q_j. That is, if $\{q_j\}$ are constructed to be orthogonal, then $\{\alpha_j q_j\}$ where α_j are constants are also orthogonal for nonzero α_j, $j = 0, 1, \ldots, m$. That is, we can multiply each of the orthogonal functions by an arbitrary constant without destroying the orthogonalty. This arbitrariness can in fact be used to advantage. For example, in the derivation of the coefficients $\{a_j\}$ in the least squares approximation given by Eq. (21) we see that the computation of these coefficients would be made more convenient if we constructed the orthogonal functions $\{q_j(x)\}$ so that

$$\sum_{i=0}^{n} w_i^2 q_j^2(x_i) = 1 \qquad j = 0, 1, \ldots, m \tag{32}$$

for in this case

$$a_j = \sum_{i=0}^{n} w_i^2 \bar{f}_i q_j(x_i)$$

Functions $\{q_j(x)\}$ which satisfy both

$$\sum_{i=0}^{n} w_i^2 q_j(x_i) q_k(x_i) = 0 \qquad j \neq k, j, k = 0, 1, \ldots, m \tag{33}$$

and

$$\sum_{i=0}^{n} w_i^2 q_j^2(x_i) = 1 \qquad j = 0, 1, \ldots, m$$

are called *orthonormal functions* on the discrete set $\{x_i\}$ and with respect to the weight functions $\{w_i^2\}$. Thus we can construct these orthonormal functions by suitably modifying the Gram–Schmidt method described above: the details are left to Exercise 7.

Exercises

6. Generalize the Gram–Schmidt process method for arbitrary weights w_i^2.
7. Modify your algorithm of Exercise 6 to construct orthonormal functions.
8. Write a computer program which will construct orthonormal functions for a given discrete set $\{x_i\}$ and given weights w_i^2 for a given set of linearly independent functions $\{\phi_i(x)\}$.
9. Use your program of Exercise 8 to compute the least squares approximation to the data in Table 1 assuming a weight function $w_i^2 = 1$ for each data point. The orthogonal functions are to be constructed from the set of linearly independent functions

$$\phi_0 = 1 + x + x^2 + x^3 + x^4 + x^5$$
$$\phi_1 = 1.05 + 0.95x + 0.05x^2 + x^3 - 0.05x^4$$

$$\phi_2 = 2 + x + x^2 + x^3 + x^4 + x^5$$
$$\phi_3 = 1.05 + 0.95x + 0.05x^2 + x^3 - 0.05x^5$$
$$\phi_4 = 1 + 2x + x^2 + x^3 + x^4 + x^5$$
$$\phi_5 = 1 + 3x + x^2 + 2x^3 + x^4 + x^5$$

Has the order in which the ϕ's are numbered, and hence introduced to construct the orthogonal functions, any effect upon you results?

To compute such orthogonal functions q we could (?) use ordered subroutines (in FORTRAN) which assume $q_0, q_1, \ldots, q_{K-1}$ have been calculated prior to the calculation of q_K and for predetermined m call the subroutine for increasing subscript $j = 0, 1, 2, \ldots$; *or* define a *recursive* procedure (ALGOL) to construct recursively the orthogonal q. In either case this approach leads to unnecessary complications and inefficiencies in programming, and the following approach is advocated.

Consider the $(n+1) \times (m+1)$ matrices

$$\Phi = \begin{bmatrix} \phi_0(x_0) & \phi_1(x_0) & \cdots & \phi_m(x_0) \\ \phi_0(x_1) & \phi_1(x_1) & \cdots & \phi_m(x_1) \\ \vdots & \vdots & & \vdots \\ \phi_0(x_n) & \phi_1(x_n) & \cdots & \phi_m(x_n) \end{bmatrix}$$

and

$$Q = \begin{bmatrix} q_0(x_0) & q_1(x_0) & \cdots & q_m(x_0) \\ q_0(x_1) & q_1(x_1) & \cdots & q_m(x_1) \\ \vdots & \vdots & & \vdots \\ q_0(x_n) & q_1(x_n) & \cdots & q_m(x_n) \end{bmatrix}$$

then for the $(m+1) \times (m+1)$ upper triangular matrix

$$R = \begin{bmatrix} 1 & \alpha_{01} & \alpha_{02} & \cdots & \alpha_{0m} \\ & 1 & \alpha_{12} & & \alpha_{1m} \\ & & 1 & & \vdots \\ & \mathbf{0} & & \ddots & 1 \end{bmatrix}$$

with $\{\alpha_{jk}\}$ defined by Eq. (31), by direct computation it is straightforward to verify that

$$\Phi = QR$$

(so that $Q = \Phi R^{-1}$). $\qquad\qquad$ (34)

We can therefore construct the matrices Q and R column by column. For example,

$$\mathbf{q}_0 = \boldsymbol{\phi}_0 \equiv (\phi_0(x_0), \phi_0(x_1), \ldots, \phi_0(x_n))^{\mathrm{T}}$$

is the first column of Q; and

$$\mathbf{q}_1 = \boldsymbol{\phi}_1 - \alpha_{01}\mathbf{q}_0$$

is the second column of Q, where

$$\alpha_{01} = \frac{\boldsymbol{\phi}_0^T \boldsymbol{\phi}_1}{\mathbf{q}_0^T \mathbf{q}_0} \quad \text{and} \quad \boldsymbol{\phi}_1 = (\phi_1(x_0), \phi_1(x_1), \ldots, \phi_1(x_n))^T$$

In general,

$$\mathbf{q}_K = \boldsymbol{\phi}_K - \sum_{j=0}^{K-1} \alpha_{jk}\mathbf{q}_j \qquad (35)$$

with

$$\left. \begin{array}{c} \\ \\ \end{array} \right\} \quad K = 0, 1, 2, \ldots, m$$

$$\alpha_{jk} = \frac{\mathbf{q}_j^T \boldsymbol{\phi}_K}{\mathbf{q}_j^T \mathbf{q}_j} \qquad j = 0, 1, \ldots, K-1 \qquad (36)$$

where Eq. (35) defines the $(1+K)$th column of Q, and Eq. (36) defines the $(1+K)$th column of R (with diagonal elements unity and lower triangular elements zero). Now the least squares approximation

$$p(x) = a_0\phi_0(x) + \cdots + a_m\phi_m(x)$$

produces a vector

$$\mathbf{p}^T \equiv (p(x_0), p(x_1), \ldots, p(x_n)) = \mathbf{a}^T \Phi^T \qquad (37)$$

If, instead of representing $p(x)$ in terms of the bases $\{\phi_i(x)\}$ we write

$$p(x) = b_0 q_0(x) + \cdots + b_m q_m(x)$$

then

$$\mathbf{p}^T = \mathbf{b}^T Q^T \qquad (38)$$

where $\mathbf{b}^T = (b_0, b_1, \ldots, b_m)$ are given by

$$b_j = \frac{(\bar{f}, q_j)}{\mathbf{q}_j^T \mathbf{q}_j} \qquad j = 0, 1, 2, \ldots, m$$

We therefore obtain from Eqs. (37) and (38)

$$\mathbf{b}^T Q^T = \mathbf{a}^T \Phi^T$$

Substitute for Q from (34) so that

$$\Phi R^{-1}\mathbf{b} = \Phi\mathbf{a}$$

or

$$R\mathbf{a} = \mathbf{b} \qquad (39)$$

This triangular system can be solved (simply) for the required vector $\mathbf{a}$ so that the least squares approximation $p(x)$ is calculated (in a *stable manner*) in terms of the given functions $\{\phi_i(x)\}$.

In the above discussion it has been assumed that the functions $\phi_i(x)$ from which the orthogonal $q_i(x)$ are constructed, are linearly independent. In some cases, a set of functions $\{\phi_i(x)\}$ can be linearly independent but can possess some members which are "nearly" linearly dependent on others. For instance, the functions $1+x$, $1+x^2$, $1+\alpha x+\beta x^2$ will be linearly dependent if $\alpha = \beta = 0.5$. If for some reason the functions $\phi_i(x)$ contain these three functions with $\alpha = 0.51$ and $\beta = 0.49$ then the set of orthogonal functions $\{q_i(x)\}$ constructed from the $\{\phi_i(x)\}$ will be poorly defined by the Gram–Schmidt process because $1+0.51x+0.49x^2$ is approximately $\frac{1}{2}(1+x+1+x^2)$, indicating near linear dependence. Orthogonal functions produced from such nearly linearly dependent functions give rise to poor round-off properties in the associated numerical procedure. Alternatively, we may say that this form of the Gram–Schmidt process may exhibit instability with respect to round-off errors when the functions $\phi_i(x)$ are nearly linearly dependent.

A *modified* form of the Gram–Schmidt process has been suggested by Rice [36] which minimizes the effect of such near linear dependence. The reasoning behind this method proceeds as follows.

We choose, at the first stage, any of the linearly independent columns $\boldsymbol{\phi}_j$ as the first of the orthogonal columns $\mathbf{q}_j$, and call it $\mathbf{q}_0$. We then take linear combinations of *each* of the remaining $\boldsymbol{\phi}_j$ with $\mathbf{q}_0$ and choose the associated parameter α_{0j} in each case to ensure the orthogonality of each linear combination to $\mathbf{q}_0$. We call these new columns $\boldsymbol{\phi}_j^{I_1}$ (I_1, for intermediate level 1). Mathematically, therefore, we equate $\mathbf{q}_0$ to $\boldsymbol{\phi}_0$ and then construct

$$\boldsymbol{\phi}_j^{I_1} = \boldsymbol{\phi}_j - \alpha_{0j}\mathbf{q}_0 \qquad j = 1, 2, \ldots, m \tag{40}$$

where α_{0j} is chosen to ensure the orthogonality of each $\boldsymbol{\phi}_j^{I_1}$ to $\mathbf{q}_0$, namely

$$\alpha_{0j} = \boldsymbol{\phi}_j^{\mathrm{T}}\mathbf{q}_0/\mathbf{q}_0^{\mathrm{T}}\mathbf{q}_0 \qquad j = 1, 2, \ldots, m$$

where we have used the notation of Eq. (30). One of these $\boldsymbol{\phi}_j^{I_1}$ is to be chosen as the next orthogonal function; we have a choice since, by construction, all the $\boldsymbol{\phi}_j^{I_1}$ are orthogonal to $\mathbf{q}_0$. The choice is made so that the particular $\boldsymbol{\phi}_j$ introduced by $\boldsymbol{\phi}_j^{I_1}$ is the one which is most linearly independent of $\mathbf{q}_0$.

Viewing the vectors $\boldsymbol{\phi}_j$ as vectors in the m-dimensional Euclidean vector space, this means choosing that $\boldsymbol{\phi}_j$ which makes the largest angle with $\mathbf{q}_0$. This angle can be calculated from the inner product of $\boldsymbol{\phi}_j$ and $\mathbf{q}_0$. Denoting the angle between $\mathbf{q}_0$ and $\boldsymbol{\phi}_j$ by θ_{0j}, we have

$$\cos(\theta_{0j}) = \frac{\boldsymbol{\phi}_j^{\mathrm{T}}\mathbf{q}_0}{(\mathbf{q}_0^{\mathrm{T}}\mathbf{q}_0)^{1/2}(\boldsymbol{\phi}_j^{\mathrm{T}}\boldsymbol{\phi}_j)^{1/2}} \qquad j = 1, 2, \ldots, m \tag{41}$$

Since this may be negative, we consider its square. Consequently, to ensure $\boldsymbol{\phi}_j$ makes as large an angle as possible with $\mathbf{q}_0$ we require

$$\min_j (\cos(\theta_{0j}))^2 = \min_j \frac{(\boldsymbol{\phi}_j^{\mathrm{T}}\mathbf{q}_0)^2}{(\mathbf{q}_0^{\mathrm{T}}\mathbf{q}_0)(\boldsymbol{\phi}_j^{\mathrm{T}}\boldsymbol{\phi}_j)} \qquad j = 1, 2, \ldots, m \tag{42}$$

The value of j for which this quantity takes on its smallest value determines which $\boldsymbol{\phi}_j^{I_1}$ becomes the next orthogonal column $\mathbf{q}_1$. Assume the value of j for which the expressions in (42) is smallest is $j = J$. Interchange the positions, in the sequence of $\boldsymbol{\phi}_1^{I_1}$ and $\boldsymbol{\phi}_J^{I_1}$, so that we now call $\boldsymbol{\phi}_J^{I_1}$, $\boldsymbol{\phi}_1^{I_1}$ and vice versa. The $\boldsymbol{\phi}$ which is most linearly independent of $\mathbf{q}_0$ is now $\boldsymbol{\phi}_1^{I_1}$. (This interchange is implemented only for computational convenience.) Equate $\mathbf{q}_1$ to $\boldsymbol{\phi}_1^{I_1}$.

Now, take a linear combination of $\mathbf{q}_1$ with each $\boldsymbol{\phi}_j^{I_1}$, $j = 2, 3, \ldots, m$, in turn, choosing the parameters to make the resulting vectors, called $\boldsymbol{\phi}_j^{I_2}$, orthogonal to $\mathbf{q}_1$ (and by construction to $\mathbf{q}_0$).

Thus,

$$\boldsymbol{\phi}_j^{I_2} = \boldsymbol{\phi}_j^{I_1} - \alpha_{1j}\mathbf{q}_1 \qquad j = 2, 3, \ldots, m$$

where the α_{1j} are chosen, as before, to ensure orthogonality of $\boldsymbol{\phi}_j^{I_2}$ to $\mathbf{q}_1$. Thus,

$$\alpha_{1j} = (\boldsymbol{\phi}_j^{T I_1}\mathbf{q}_1)/(\mathbf{q}_1^T\mathbf{q}_1) \qquad j = 2, 3, \ldots, m$$

We now have $(m - 1)$ vectors $\boldsymbol{\phi}_j^{I_2}$ which, by construction, are orthogonal to $\mathbf{q}_0$ and $\mathbf{q}_1$. We have once again to make a choice of one of these functions to produce the next next orthogonal function $\mathbf{q}_2$. We do this by choosing the $\boldsymbol{\phi}_j^{I_2}$ which will introduce the $\boldsymbol{\phi}_j$ which is most linearly independent of the orthogonal vectors constructed to date, namely $\mathbf{q}_0$ and $\mathbf{q}_1$. This is determined by making as small as possible the angle between $\boldsymbol{\phi}_j$ and the *normal* direction $\mathbf{n}$ to the plane containing $\mathbf{q}_0$ and $\mathbf{q}_1$. Let this angle be denoted by θ_{2j}. Furthermore, let the angle between $\boldsymbol{\phi}_j$ and $\mathbf{q}_0$ be θ_{0j} and that between $\boldsymbol{\phi}_j$ and $\mathbf{q}_1$ be θ_{1j}. Since $\mathbf{q}_0$, $\mathbf{q}_1$, and the normal $\mathbf{n}$ are orthogonal, from coordinate geometry

$$\cos^2(\theta_{0j}) + \cos^2(\theta_{1j}) + \cos^2(\theta_{2j}) = 1$$

or

$$\cos^2(\theta_{2j}) = 1 - \cos^2(\theta_{0j}) - \cos^2(\theta_{1j})$$

Hence we require to minimize $(\theta_{2j})^2$, or

$$\max_j \cos^2(\theta_{2j}) = \max_j [1 - \cos^2(\theta_{0j}) - \cos^2(\theta_{1j})]$$

$$= -\min_j [\cos^2(\theta_{0j}) + \cos^2(\theta_{1j})] \tag{43}$$

Following Eq. (41), we have

$$\cos^2(\theta_{0j}) = \frac{(\boldsymbol{\phi}_j^T\boldsymbol{\phi}_0)^2}{(\mathbf{q}_0^T\mathbf{q}_0)(\boldsymbol{\phi}_j^T\boldsymbol{\phi}_j)}$$

$$\cos^2(\theta_{1j}) = \frac{(\boldsymbol{\phi}_j^T\boldsymbol{\phi}_1)^2}{(\mathbf{q}_1^T\mathbf{q}_1)(\boldsymbol{\phi}_j^T\boldsymbol{\phi}_j)} \qquad j = 2, 3, \ldots, m \tag{44}$$

If the minimum in (44) is achieved for $j = J$, we interchange $\boldsymbol{\phi}_2^{I_2}$ and $\boldsymbol{\phi}_J^{I_2}$.

Let us assume we have constructed p orthogonal vectors $\mathbf{q}_j$, $j = 0, 1, \ldots, p-1$, by the above process. Then the $(p+1)$th stage may be described as follows:

Calculate

$$\boldsymbol{\phi}_j^{I_p} = \boldsymbol{\phi}_j^{I_{p-1}} - \alpha_{p-1,j}\mathbf{q}_{p-1}$$

$$\boldsymbol{\phi}_j^{I_p} = \boldsymbol{\phi}_j^{I_{p-1}} - \frac{(\boldsymbol{\phi}_j^{T I_{p-1}}\mathbf{q}_{p-1})}{(\mathbf{q}_{p-1}^T\mathbf{q}_{p-1})}\mathbf{q}_{p-1} \tag{45}$$

Choose $\mathbf{q}_p = \boldsymbol{\phi}_j^{I_p}$, where J is the index for which

$$\sum_{l=0}^{p-1} \frac{(\boldsymbol{\phi}_j^T\mathbf{q}_l)^2}{(\mathbf{q}_l^T\mathbf{q}_l)(\boldsymbol{\phi}_j^T\boldsymbol{\phi}_j)} \qquad j = p+1, p+2, \ldots, m \tag{46}$$

takes its smallest value. Repeat the process until the complete set of orthogonal vectors has been constructed.

The construction of the orthogonal vectors $\{\mathbf{q}_j\}$ using this algorithm minimizes the build up of round-off error and thereby gives as accurate as possible orthogonal vectors. The algorithm does not eliminate the near linear dependence of the $\boldsymbol{\phi}_j$ but minimizes its effect upon the orthogonal vectors. Clearly, if all the $\boldsymbol{\phi}_j$ were nearly linearly dependent we could not expect an accurate calculation of orthogonal vectors, no matter which method we used.

To make the algorithms as efficient as possible, it is wise to normalize the vectors $\boldsymbol{\phi}_j$ to 1 before starting the algorithm. Further details may be found in Rice [36].

Exercises

10. Determine the form of the modified Gram–Schmidt method for general weights $\{w_i^2\}$.
11. Modify your algorithm of Exercise 10 so that the sequence of orthogonal vectors is orthonormal.
12. Program the algorithm of Exercise 11 for a given discrete data set $\{x_i\}$ and associated weights $\{w_i^2\}$ and given linearly independent functions $\{\phi_j(x)\}$.
13. Use your program of Exercise 12 to compute the least squares approximation to the data in Table 1 assuming a weight function $w_i^2 = 1$ for each data point. Use as the linearly independent functions the functions given in Exercise 9. What is the order in which the linearly independent vectors are introduced? Compare your results with those obtained in Exercise 9.

3 The Recursive Generation of Orthogonal Polynomials

If we look carefully at the equations required to define the orthogonal vectors by either of the Gram–Schmidt processes in the previous section, we see that the process will involve a considerable amount of work because of the large number of inner products required. This inefficiency motivates us to look further for a method of generating the orthogonal functions. In fact our consideration will now be restricted to the generation of orthogonal *polynomials* since the process we now require to develop does not work in general for other functions. This technique seeks to generate the orthogonal polynomials in a recursive manner. In fact we will show that orthogonal polynomials satisfy a *three-term* recurrence relation. To see this, let us assume

$$q_j(x) \qquad j = 0, 1, \ldots, p+1$$

are orthogonal polynomials of degree j. Consider the polynomials $xq_p(x)$ of degree $p+1$. We can express this polynomial as a linear combination of orthogonal polynomials of degree up to and including ones of degree $p+1$, i.e., of $q_j(x)$, $j = 0, 1, \ldots, p+1$. Let

$$xq_p(x) = \sum_{j=0}^{p+1} \beta_{jp} q_j(x) \tag{47}$$

Multiply both sides of Eq. (47) by $q_k(x)$, $k = 0, 1, \ldots, p+1$. Then on rearrangement we obtain

$$q_p(x)xq_k(x) = \sum_{j=0}^{p+1} \beta_{jp} q_j(x) q_k(x) \qquad k = 0, 1, \ldots, p+1 \tag{48}$$

Now $xq_k(x)$ is a polynomial of degree $k+1$ and can be expressed as a linear combination of orthogonal polynomials of degree up to and including degree $k+1$. Let

$$xq_k(x) = \sum_{l=0}^{k+1} c_l q_l(x)$$

Substituting this expression into Eq. (48) and summing over x_i, $i = 0, 1, \ldots, n$, we obtain

$$\sum_{l=0}^{k+1} c_l \sum_{i=0}^{n} q_p(x_i) q_l(x_i) = \sum_{j=0}^{p+1} \beta_{jp} \sum_{i=0}^{n} q_j(x_i) q_k(x_i) \qquad k = 0, 1, \ldots, p+1 \tag{49}$$

But the $q_j(x)$ are orthogonal and hence the left-hand side is zero except for $k+1 = p$, $p+1$, and $p+2$, i.e., for $k = p-1$, p, $p+1$. Similarly, the coefficients on the right-hand side of Eq. (49) are given by

$$\beta_{kp} \sum_{i=0}^{n} q_k^2(x_i)$$

Hence for $k < p-1$ we have $\beta_{kp} \equiv 0$. For $k = p-1$, p, $p+1$ we have the three-term recurrence relation

$$xq_p(x) = \beta_{p-1\,p} q_{p-1}(x) + \beta_{pp} q_p(x) + \beta_{p+1\,p} q_{p+1}(x) \tag{50}$$

which on rearrangement is

$$\beta_{p+1\,p} q_{p+1}(x) = (x - \beta_{pp}) q_p(x) - \beta_{p-1\,p} q_{p-1}(x) \tag{51}$$

We complete the recurrence relation by obtaining the explicit form of the coefficients. First, we note that $q_{p+1}(x)$ will be orthogonal if the sum over x_i of the product of the right-hand side of Eq. (51) and $q_j(x)$ is zero for $j \neq p+1$ and is nonzero for $j = p+1$, which will occur irrespective of what value (nonzero, of course) is assigned to $\beta_{p+1\,p}$. That is, the coefficient $\beta_{p+1\,p}$ plays the rôle of a normalizing parameter (in some sense). Consequently, to choose the $\beta_{p+1\,p}$ explicitly we have to enforce an extra condition to orthogonality, namely a decision of what form the normalization should take. We will discuss this a little more fully when we have derived the remaining parameters β_{pp} and $\beta_{p-1\,p}$. To do this, multiply both sides of Eq. (51) by $q_k(x)$, $k = p-1, p, p+1$. Sum the resulting expression over x_i, $i = 0, 1, \ldots, n$. We then obtain

$$\sum_{i=0}^{n} x_i q_p(x_i) q_k(x_i) = \beta_{p-1\,p} \sum_{i=0}^{n} q_{p-1}(x_i) q_k(x_i) + \beta_{pp} \sum_{i=0}^{n} q_p(x_i) q_k(x_i)$$

$$+ \beta_{p+1\,p} \sum_{i=0}^{n} q_{p+1}(x_i) q_k(x_i) \qquad k = p-1, p, p+1$$

For $k = p-1$,

$$\sum_{i=0}^{n} x_i q_p(x_i) q_{p-1}(x_i) = \beta_{p-1\,p} \sum_{i=0}^{n} q_{p-1}^2(x_i)$$

where we have used the orthogonality of $q(x)$. Hence

$$\beta_{p-1\,p} = \sum_{i=0}^{n} x_i q_{p-1}(x_i) q_p(x_i) \Big/ \sum_{i=0}^{n} q_{p-1}^2(x_i)$$

or

$$\beta_{p-1\,p} = \sum_{i=0}^{n} \beta_{pp-1} q_p^2(x_i) \Big/ \sum_{i=0}^{n} q_{p-1}^2(x_i)$$

where we have used Eq. (50) with p replaced by $p-1$. We can use our inner product notation to write this as

$$\beta_{p-1\,p} = \beta_{pp-1}(q_p, q_p)/(q_{p-1}, q_{p-1}) \qquad q = 1, 2, \ldots, m \qquad (52)$$

For $k = p$, we easily see that

$$\beta_{pp} = \sum_{i=0}^{n} x_i q_p^2(x_i)/(q_p, q_p) \qquad p = 0, 1, \ldots, m-1 \qquad (53)$$

Thus any three orthogonal polynomials satisfy the recurrence relation (51) with the parameters defined by Eqs. (52) and (53), $\beta_{p+1\,p}$ being arbitrary (at this stage) as described above. We note that β_{pp} and $\beta_{p-1\,p}$ are not dependent on $q_{p+1}(x)$. Consequently we may view this recurrence relation as a means of *defining* the orthogonal function $q_{p+1}(x)$ as a linear combination of the two orthogonal polynomials $q_p(x)$, $q_{p-1}(x)$ with the coefficients in the linear combination chosen to give orthogonality. These are of course just the β_{pp} and $\beta_{p-1\,p}$ defined by Eqs. (50) and (51) respectively. (We leave the proof that $q_{p+1}(x)$ defined in this way is orthogonal to $q_i(x)\,j < p$ to Exercise 14.)

To complete the definition of the recurrence relation, we need to define $q_p(x)$ for $p = 0$ and -1. There is clearly an arbitrariness about such a definition: all we require is that q_1 be orthogonal to q_0 and q_{-1}. Since 0 is orthogonal to all functions including itself, we define $q_{-1}(x) \equiv 0$. We then define $q_0(x)$ to be any polynomial of degree zero, i.e., a constant and denote it by β_{0-1}. Hence our recurrence relation for generating the orthogonal functions $q_{p+1}(x)$ is complete; namely

$$q_{-1}(x) = 0$$

$$q_0(x) = \beta_{0-1} \tag{54}$$

$$\beta_{p+1\,p}q_{p+1}(x) = (x - \beta_{pp})q_p(x) - \beta_{p-1\,p}q_{p-1}(x) \qquad p = 0, 1, \ldots, m-1$$

where $\beta_{p+1\,p}$, $p = -1, 0, 1, \ldots$, are arbitrary and β_{pp}, $\beta_{p-1\,p}$ are given by Eqs. (52) and (53).

Exercises

14. For parameters $\beta_{p-1\,p}$ and β_{pp} defined by Eqs. (52) and (53) respectively, show that $q_{p+1}(x)$ defined by (54) is orthogonal to $q_i(x), j = -1, 0, 1, \ldots, p$, for orthogonal $q_p(x)$, $q_{p-1}(x)$.
15. Derive the orthogonal polynomial for Eq. (54) with $p = 1$.
16. For the orthogonal polynomial $q_1(x)$ defined in Exercise 15 choose β_{10} so that $q_1(x)$ is orthonormal.
17. Derive the recurrence relationship and equations for the coefficients β_{pp}, $\beta_{p-1\,p}$ for functions to be orthogonal over the tabular points $\{x_i\}$ with an associated nonnegative weight function $w^2(x_i)$, $i = 0, 1, \ldots, n$.
18. In the derivation of the recurrence relation (54) we considered the expansion of the polynomial $xq_p(x)$ in terms of orthogonal polynomials $q_j(x), j = 0, 1, \ldots, p+1$. This polynomial has a leading coefficient equal to that of $q_p(x)$. Is there any advantage in introducing an extra parameter, α say, premultiplying this polynomial, and considering the expansion of the resulting polynomial in terms of orthogonal polynomials $q_j(x), j = 0, 1, \ldots, p+1$?
19. Construct a computer program to produce the orthogonal polynomials from Eq. (54) for a general data set $\{x_i\}$ and associated weights $\{w_i^2\}$. The normalization parameters $\beta_{p+1\,p}$ should be produced by a subroutine (FORTRAN) or procedure (ALGOL) which at this stage is undetermined (arbitrary).
20. Check that your program in Exercise 19 works correctly by deriving the orthogonal polynomials ϕ_0, ϕ_1, and ϕ_2 used in Example 4 and specifying the appropriate values of the normalizing coefficients $\beta_{p+1\,p}$, $p = -1, 0, 1$.

We have left to discuss means of choosing the parameters $\beta_{p+1\,p}$. A simple choice would be to choose all $\beta_{p+1\,p}$, $p = -1, 0, 1, \ldots, m-1$, equal to 1. In this way we would obtain from Eq. (54) orthogonal polynomials each of which had a leading coefficient equal to 1, i.e., x^p in $q_p(x)$ has coefficient 1. In this case the recurrence relationship takes on the simplified form

$$q_{-1}(x) = 0$$

$$q_0(x) = 1$$

$$q_{p+1}(x) = (x - \beta_{pp})q_p(x) - \beta_{p-1\,p}q(x)$$

and

$$\beta_{pp} = \sum_{i=0}^{n} x_i q_p^2(x_i)/(q_p, q_p) \qquad p = 0, 1, \ldots, m-1$$

$$\beta_{p-1\,p} = (q_p, q_p)/(q_{p-1}, q_{p-1}) \qquad p = 1, 2, \ldots, m$$

An alternative requirement is that the orthogonal functions $q_i(x)$ should be orthonormal polynomials, i.e.,

$$\sum_{i=0}^{n} q_p(x_i)q_k(x_i) = \delta_{pk}^\dagger \qquad p, k = 0, 1, \ldots, m \tag{55}$$

To obtain the expression for $\beta_{p+1\,p}$ in such a case, let us assume $q_p(x)$ and $q_{p-1}(x)$ in Eq. (54) are orthonormal. Then

$$q_{p+1}(x) = \frac{1}{\beta_{p+1\,p}}[(x - \beta_{pp})q_p(x) - \beta_{p-1\,p}q_{p-1}(x)] \tag{56}$$

is to be orthonormal. Hence

$$\sum_{i=0}^{n} q_{p+1}^2(x_i) = \frac{1}{\beta_{p+1\,p}^2} \sum_{i=0}^{n} \{[(x_i - \beta_{pp})q_p(x_i) - \beta_{p-1\,p}q_{p-1}(x_i)]^2\}^{1/2}$$

Then choose $\beta_{p+1\,p}$ as

$$\beta_{p+1\,p} = \sum_{i=0}^{n} \{[(x_i - \beta_{pp})q_p(x_i) - \beta_{p-1\,p}q_{p-1}(x_i)]^2\}^{1/2} \tag{57}$$

for $p = 0, 1, \ldots, m-1$, with $\beta_{0\,-1} = 1/\sqrt{n+1}$ (why?)

Exercises

21. Verify by direct computation that $q_2(x)$ defined by Eq. (54) with $\beta_{p+1\,p}$ defined by Eq. (57) is orthonormal.
22. Use your program of Exercise 19 to construct orthogonal polynomials up to degree 5 for the data in Table 1. Use these polynomials for the least squares approximation and compare your result with that of Exercise 9 by graphing the associated least squares curves. The normalization constants $\beta_{p+1\,p}$ should be defined to be 1 at each stage.
23. Repeat Exercise 22 using orthonormal polynomials.

---

† δ_{pk} is the Kronecker delta defined by

$$\delta_{pk} = \begin{cases} 1 & p = k \\ 0 & p \neq k \end{cases}$$

CHAPTER 6

Minimax Approximation

1 Introduction

In this chapter we will describe the third criterion for approximating a function $f(x)$ of the real variable x, outlined in Chapter 4. Namely, we require to construct an approximating function $y(x)$ such that

$$\max |y(x) - f(x)| \tag{1}$$

is minimized for x in a prescribed interval. Such an approximation is called minimax approximation of $f(x)$. We have already met a special case of this minimax approximation in Chapter 4, when we approximated a polynomial of degree n by one of degree $n - 1$ so that the maximum error between the two polynomials was minimized over a given interval.

In our previous discussions of approximations of functions in Chapters 4 and 5 we restricted our attention, in the main, to polynomial approximation. In this chapter we could continue in that vein and propose $y(x)$, the minimax approximation, to be a polynomial of specified degree. However, we will generalize our discussion somewhat by considering $y(x)$ to be a rational function, i.e.

$$y(x) = \frac{P_m(x)}{Q_k(x)} \tag{2}$$

where $P_m(x)$ and $Q_k(x)$ are polynomials of degree m and k respectively. This discussion will then contain the polynomial approximation as a special case ($k = 0$). We do this for two reasons. First, it is recognized that for a given amount of computation, rational approximation leads to smaller maximum errors (1) than polynomial approximation. Second, the polynomial approximation will be a special case of Eq. (2) with $k = 0$.

The construction of the minimax approximation will be heavily dependent on the properties of Chebyshev polynomials which we described in Chapter 4. The reader is directed to that chapter to refresh his memory of these properties.

172

2 Padé Approximations

We will attempt to construct the minimax approximations by an iterative process where we start from a near-minimax approximation and attempt to construct a nearer minimax approximation, and then repeat the process until the required approximation is obtained. In our search for such near-minimax approximations we first describe (rational) Padé approximations. These Padé approximations will then be used later as a starting point from which to develop better approximations. We propose a Padé approximation

$$R_{mk}(x) = \frac{P_m(x)}{Q_k(x)} \tag{3}$$

where

$$P_m(x) = \sum_{j=0}^{n} a_j x^j \tag{4}$$

$$Q_k(x) = \sum_{j=0}^{k} b_j x^j \quad \text{and} \quad b_0 = 1† \tag{5}$$

where the $m + k + 1$ coefficients a_j, b_j are chosen so that $R_{mk}(x)$ agrees with $f(x)$ and as many derivatives as possible at $x = \alpha$. To ensure ease in manipulation we will assume $\alpha = 0$ and note that we may always achieve this by a suitable change of variable (see Eq. (16) below). To determine the coefficients a_j, b_j, expand $f(x)$ in terms of its Maclaurin series (see Jones and Jordan [27]), which we assume to exist in a neighborhood of $x = 0$.

Hence

$$f(x) = \sum_{j=0}^{\infty} c_j x^j \qquad c_j = f^{(j)}(0)/j!$$

Then

$$f(x) - R_{mk}(x) = \sum_{j=0}^{\infty} c_j x^j - \frac{\sum_{j=0}^{m} a_j x^j}{\sum_{j=0}^{k} b_j x^j}$$

which we rewrite as

$$f(x) - R_{mk}(x) = \frac{\sum_{j=0}^{\infty} c_j x^j \sum_{j=0}^{k} b_j x^j - \sum_{j=0}^{m} a_j x^j}{\sum_{j=0}^{k} b_j x^j} \tag{6}$$

† In a rational function $R_{mk}(x)$ given by Eq. (3) we usually have $m + k + 2$ coefficients, but since we can divide the numerator and denominator by an arbitrary (nonzero) constant without affecting its value, we in fact have only $m + k + 1$ free coefficients. We therefore specify b_0 to be 1 for convenience.

Example 1   Determine a_0, a_1, and b_1 ($b_0 = 1$) in $R_{11}(x) = P_1(x)/Q_1(x)$ which agrees with $\exp(x)$ and as many of its derivatives as possible at $x = 0$.

We expand $\exp(x)$ in terms of its Maclaurin expansion, namely

$$\exp(x) = \exp(0) + x\exp(0) + \frac{x^2}{2}\exp(0) + \frac{x^3}{6}\exp(0) + \cdots$$

So

$$f(x) - R_{11}(x) = (1 + x + x^2/2 + x^3/6 + \cdots) - (a_1x + a_0)/(b_1x + 1)$$

$$= \frac{(1 + x + x^2/2 + x^3/6 + \cdots)(b_1x + 1) - (a_1x + a_0)}{(b_1x + 1)} \tag{7}$$

Let us write Eq. (7) as

$$f(x) - R_{11}(x) = N(x)/D(x) \tag{8}$$

where, note, $N(x)$ and $D(x)$ are polynomials. Then for agreement at $x = 0$,

$$f(0) - R_{11}(0) = N(0)/D(0) = 0 \tag{9}$$

from which it is clearly sufficient that $N(0) = 0$, namely

$$1 - a_0 = 0$$

Differentiate Eq. (8), so that

$$f^{(1)}(x) - R_{11}^{(1)}(x) = \frac{D(x)\dfrac{\mathrm{d}}{\mathrm{d}x}N(x) - N(x)\dfrac{\mathrm{d}}{\mathrm{d}x}D(x)}{D^2(x)} \tag{10}$$

For agreement at $x = 0$,

$$f^{(1)}(0) - R_{11}^{(1)}(0) = \frac{D(0)\left(\dfrac{\mathrm{d}}{\mathrm{d}x}N(x)\right)_{x=0} - N(0)\left(\dfrac{\mathrm{d}}{\mathrm{d}x}D(x)\right)_{x=0}}{D^2(0)} = 0$$

where the notation $(\ )_{x=0}$ means evaluate the expression in the brackets at $x = 0$. From Eq. (9) we have $N(0) = 0$, so for

$$f^{(1)}(0) - R_{11}^{(1)}(0) = \frac{\left(\dfrac{\mathrm{d}}{\mathrm{d}x}N(x)\right)_{x=0}}{D(0)} = 0 \tag{11}$$

it is sufficient that $(\mathrm{d}/\mathrm{d}x\,N(x))_{x=0} = 0$, where by (9) we have already specified $N(0) = 0$.

Continuing, we differentiate (10) so that

$$f^{(2)}(x) - R_{11}^{(2)}(x) = \left\{ D^2(x)\left[\left(\frac{\mathrm{d}}{\mathrm{d}x}D(x)\right)\left(\frac{\mathrm{d}}{\mathrm{d}x}N(x)\right) + D(x)\frac{\mathrm{d}^2}{\mathrm{d}x^2}N(x) \right.\right.$$

$$\left. - \left(\frac{\mathrm{d}}{\mathrm{d}x}N(x)\right)\left(\frac{\mathrm{d}}{\mathrm{d}x}D(x)\right) - N(x)\frac{\mathrm{d}^2}{\mathrm{d}x^2}D(x)\right]$$

$$\left. - 2D(x)\left[D(x)\frac{\mathrm{d}}{\mathrm{d}x}N(x) - N(x)\frac{\mathrm{d}}{\mathrm{d}x}D(x)\right]\right\} \Big/ D^4(x)$$

For agreement at $x = 0$, on using $N(0) = 0$ and $(d/dx\, N(x))_{x=0}$, we obtain

$$f^{(2)}(0) - R_{11}^{(2)}(0) = \left\{ D^2(0) \left(\frac{d^2}{dx^2} N(x) \right)_{x=0} \right\} / D^4(0) = 0$$

which is satisfied if we choose the coefficients a_0, a_1, and b_1 such that

$$\left(\frac{d^2}{dx^2} N(x) \right)_{x=0} = 0 \tag{12}$$

By direct comparison in Eqs. (9), (11), (12), we find

$$N(0) = 1 - a_0 = 0$$

$$\left(\frac{d}{dx} N(x) \right)_{x=0} = b_1 + 1 - a_1 = 0$$

$$\left(\frac{d^2}{dx^2} N(x) \right)_{x=0} = 2(b_1 + 1/2) = 0$$

which define the coefficients a_0, a_1, and b_1. Solving these equations we obtain $a_0 = 1$, $b_1 = -1/2$, $a_1 = 1/2$, so that

$$R_{11}(x) = \frac{1 + \frac{1}{2}x}{1 - \frac{1}{2}x}$$

We leave to Exercise 1 the verification that $f^{(3)}(0) - R_{11}^{(3)}(0) \neq 0$.

Exercises

1. In Example 1 derive $f^{(3)}(x) - R_{11}^{(3)}(x)$ and hence confirm that $f^{(3)}(0) - R_{11}^{(3)}(0) \neq 0$.
2. Derive the $R_{22}(x)$ approximation to $\exp(x)$. What is the value of the first nonzero derivative of $f(x) - R_{22}(x)$ at $x = 0$?
3. What relation has $R_{mo}(x)$ to the Maclaurin expansion of $f(x)$?
4. Derive the $R_{40}(x)$ and $R_{04}(x)$ approximation to $\exp(x)$. What is the relation between these two approximations? Can you generalize this relation to $R_{mk}(x)$ and $R_{km}(x)$?
5. Compute $\exp(x) - R_{11}(x)$ for $x = (-1, (0.1), 1)$.† Comment on the difference as x varies over the interval.
6. Compute and graph $\exp(x) - R_{mk}(x)$ for $x = (-1, (0.1), 1)$ for $m = 0, 1, \ldots, 4$; $k = 0, 1, \ldots, 4$. Comment on these errors as $m + k$ increases. For fixed $m + k$, which member of the set of approximations has the samller maximum error for $x \in (-1, 1)$ and where in the range is this achieved?

Returning to Eq. (6), following the steps described in Example 1, in order that $f(x)$ and $R_{mk}(x)$ and as many of their derivatives agree at $x = 0$ it is sufficient that

$$\frac{d^r}{dx^r} \left[\sum_{j=0}^{\infty} c_j x^j \sum_{j=0}^{k} b_j x^j - \sum_{j=0}^{m} a_j x^j \right] = 0 \qquad r = 0, 1, \ldots, m + k \tag{13}$$

† x takes values of $-1.0, -0.9, -0.8, \ldots, 1.0$, i.e. increment x from -1 to 1 by steps of 0.1

Equation (13) represents $m + k + 1$ equations in the $m + k + 1$ unknowns a_j, $j = 0, 1, \ldots, m$; b_j, $j = 1, 2, \ldots, k$. We may rearrange Eq. (13) to give

$$\frac{d^r}{dx^r}\left[-\sum_{l=0}^{m}\left(a_l - \sum_{j=0}^{l} c_{l-j}b_j \right) x^l + \sum_{l=m+1}^{m+k}\left(\sum_{j=0}^{k} c_{r-j}b_j \right) x^l + \sum_{j=m+k+1}^{\infty} d_j x^j \right] = 0$$

$$r = 0, 1, \ldots, m + k \quad (14)$$

where $c_j = 0$ if $j < 0$, $b_j = 0$ if $j > k$, and $b_0 = 1$, and d_j are the coefficients of the powers of x greater than $m + k$, which we need not specify. Hence from Eq. (14) we require

$$a_r - \sum_{j=0}^{r} c_{r-j}b_j = 0 \qquad r = 0, 1, \ldots, m$$

$$\sum_{j=0}^{k} c_{r-j}b_j = 0 \qquad r = m+1, m+2, \ldots, m+k \quad (15)$$

For a given m and k the solution of Eq. (15) will produce the required Padé approximation. By simple rearrangement, we may write those equations as

$$\begin{bmatrix} c_{m-k+1} & c_{m-k+2} & \cdots & c_m \\ c_{m-k+2} & c_{m-k+3} & \cdots & c_{m+1} \\ \vdots & \vdots & & \vdots \\ c_m & c_{m+1} & & c_{m+k-1} \end{bmatrix} \begin{bmatrix} b_k \\ b_{k-1} \\ \vdots \\ b_1 \end{bmatrix} = \begin{bmatrix} -c_{m+1} \\ -c_{m+2} \\ \vdots \\ -c_{m+k} \end{bmatrix}$$

and then

$$a_r = \sum_{j=0}^{r} c_{r-j}b_j \qquad r = 0, 1, \ldots, m$$

the latter equations producing the coefficients $\{a_r\}$ *explicitly*.

Practical use of Padé approximations indicate that the largest errors will occur at the end points of the interval (see Exercise 6), i.e., at points furthest from the point where the function and derivatives are equated to $R_{mk}(x)$ and its derivatives ($x = 0$ in our case). For this reason, it is best to ensure that this point is the midpoint of the interval. Assume a Padé approximation on $x \in [-2, 2]$ is proposed using the method of derivation described above. Since the errors are larger the further we move from $x = 0$, let us propose dividing the interval $[-2, 2]$ into four subintervals $[-2, -1]$, $[-1, 0]$, $[0, 1]$, $[1, 2]$. Clearly 0 is never the midpoint of any of these subintervals. Hence let us transform each interval into $[-0.5, 0.5]$ where, of course, 0 will be the midpoint. I.e.

$$x \in [a, b] \to y \in [-0.5, 0.5]$$

where a and b are the generic representations of a left end point and a right end point respectively. The required transformation is

$$y = \frac{1}{2}\left\{ \frac{2x - (a + b)}{(b - a)} \right\} \quad (16)$$

whence

$$y_1(x) = x + 3/2 \qquad x \in [-2, -1]$$
$$y_2(x) = x + 1/2 \qquad x \in [-1, 0]$$
$$y_3(x) = x - 1/2 \qquad x \in [0, 1]$$
$$y_4(x) = x - 3/2 \qquad x \in [1, 2]. \tag{17}$$

Then instead of proposing the Padé approximation for $x \in [-2, 2]$ we can propose a Padé approximation for each $y_i(x)$, $i = 1, 2, 3, 4$. Since the range of the variables $y_i(x)$ will be smaller than $b - a$, the maximum error should be consequently reduced.

Example 2   Consider the Padé approximation $R_{22}(x)$ to $\exp(x)$ for $x \in [-2, 2]$. It is straightforward to show that

$$R_{22}(x) = \frac{12 + 6x + x^2}{12 - 6x + x^2} \qquad x \in [-2, 2] \tag{18}$$

Likewise,

$$R_{22}(y) = \frac{12 + 6y + y^2}{12 - 6y + y^2} \, y \in [-0.5, 0.5]$$

where y will be defined in turn by (17). Now

$$\exp(y) \doteq \frac{12 + 6y + y^2}{12 - 6y + y^2}$$

Thus substituting $y_i(x)$, $i = 1, 2, 3, 4$, from (17) we have

$$\exp(y_1(x)) = \exp(x) \exp(3/2) \doteq \frac{23.25 + 9x + x^2}{5.25 - 3x + x^2} \qquad x \in [-2, -1] \quad (19)$$

$$\exp(y_2(x)) = \exp(x) \exp(1/2) \doteq \frac{15.25 + 7x + x^2}{9.25 - 5x + x^2} \qquad x \in [-1, 0] \quad (20)$$

$$\exp(y_3(x)) = \exp(x) \exp(-1/2) \doteq \frac{9.25 + 5x + x^2}{15.25 - 7x + x^2} \qquad x \in [0, 1] \quad (21)$$

$$\exp(y_4(x)) = \exp(x) \exp(-3/2) \doteq \frac{5.25 + 3x + x^2}{23.25 - 9x + x^2} \qquad x \in [1, 2] \quad (22)$$

Thus, before being able to use these approximations we need a means of evaluating the exponential of the constants $3/2$, $1/2$, $-1/2$, $-3/2$. We may obtain such values using the approximations themselves or, as is more usual, assume that such values are given. For example, in (19) we have

$$\exp(-2) = \exp(-1)^2 = \exp(-3/2) \cdot 37/62 \tag{23}$$

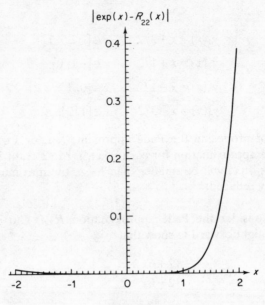

Figure 1

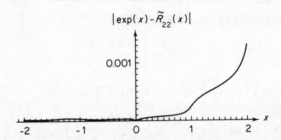

Figure 2   (The scales of Figures 1 and 2 have been adjusted for convenience)

and

$$\exp(-1) = \exp(-3/2) \cdot 61/37. \qquad (24)$$

Combining equations (23) and (24) we find that

$$\exp(-3/2) \doteq (37/61)^3 \; (= 0.22315).$$

In a similar manner we find

$$\exp(-1/2) \doteq 37/61$$
$$\exp(1/2) \doteq 61/37$$
$$\exp(3/2) \doteq 61/37$$

so that our sequence of Padé approximations becomes

$$
\tilde{R}_{22}(x) =
\begin{cases}
(37/61)^3 [23.25 + 9x + x^2]/[5.25 - 3x + x^2] & x \in [-2, -1] \\
(37/61)[15.25 + 7x + x^2]/[9.25 - 5x + x^2] & x \in [-1, 0] \\
61/37[9.25 + 5x + x^2]/[15.25 - 7x + x^2] & x \in [0, 1] \\
61/37[5.25 + 3x + x^2]/[23.25 - 9x + x^2] & x \in [1, 2]
\end{cases}
$$

We have plotted $|\exp(x) - R_{22}(x)|$, $x \in [-2, 2]$ with $R_{22}(x)$ given by Eq. (18), and $|\exp(x) - \tilde{R}_{22}(x)|$ in Figures 1 and 2 respectively. The effect of subdividing the range $[-2, 2]$ is apparent from the smaller errors over the whole range as depicted in Figure 2. Although the value of $|\exp(x) - \tilde{R}_{22}(x)|$ is small for $x \in [1, 2]$ compared with $|\exp(x) - R_{22}(x)|$ in the same interval, to obtain even smaller errors we could repeat the process described by dividing $[1, 2]$ into $[1, 3/2]$, $[3/2, 2]$. This is left as an exercise for the reader.

Exercises

7. The subdivision of the interval over which the Padé approximation is required is clearly not possible (for the whole interval) when this interval is $(-\infty, \infty)$. By writing $\exp(x) = 10^{x \log_{10} \exp(x)}$, where $x \log_{10} \exp(0) = I + F$, I the integer part and F the fractional part, $0 \le F < 1$, show how a Padé approximation may be proposed for $x \in (-\infty, \infty)$ which requires only a Padé approximation to $\exp(y)$, $y \in (0, \ln(0))$ multiplied by 10^I.

8. The approximation in Exercise 7 requires you to find an approximation to $\exp(y)$ for $y \in (0, \ln(0))$. Use a technique analogous to that described in Example 2 and thence program an algorithm for finding an approximation to $\exp(x)$, $x \in (-M, M)$, where M is the largest (in absolute value) machine representable number on your local computer. Graph your results of the difference between $\exp(x)$ and your approximation for a sequence of subdivisions of the interval $(0, \ln(0))$.

3 Rational Approximation and Chebyshev Polynomials

The Padé approximations described in the previous section do not produce a good behavior over the whole interval of approximation, even when we subdivide the interval, and the further away from the midpoint of the interval (or subinterval) the approximation is evaluated, the larger will be the error. In Chapter 4, we discussed the Chebyshev polynomials and showed that, on the interval $[-1, 1]$ of all polynomials of degree n, say, with coefficient of $x^n = 1$, the Chebyshev polynomial of degree n oscillates with minimum/maximum amplitude. Thus it is reasonable to suspect, and we shall subsequently verify numerically, that taking a linear combination of the Chebyshev polynomials instead of monomials should produce an approximation with a more uniform behavior over the interval of approximation. In proposing these approximations we will assume the interval of approximation is $x \in [-1, 1]$; if this is not the case in practice we can again transform the given interval to $[-1, 1]$. In proposing the new approximations we will use the properties P(1)–P(5) of the Chebyshev polynomials described in Chapter 4; we assume the reader has familiarized himself with these. In addition to these properties we require the additional property:

P(6)   The Chebyshev polynomials $T_n(\theta)$ defined by

$$T_n(\theta) = \cos n\theta \qquad -\pi \le \theta \le \pi \qquad n = 0, 1, \ldots$$

satisfy

$$T_{n+m}(\theta) + T_{|n-m|}(\theta) = 2T_n(\theta)T_m(\theta) \tag{25}$$

Proof   From the well-known formula for the sum of two cosines we have

$$\cos(A) + \cos(B) = 2 \cos \frac{(A+B)}{2} \cos \frac{(A-B)}{2} \tag{26}$$

Substitute

$$\frac{A+B}{2} = n\theta \quad \text{and} \quad \frac{A-B}{2} = m\theta \tag{27}$$

Hence solving Eq. (27) for A and B, we find

$$A = (m+n)\theta, \qquad B = (n-m)\theta$$

Thus substitute for A and B into Eq. (26), and then

$$\cos((m+n)\theta) + \cos((n-m)\theta) = 2 \cos n\theta \cos m\theta$$

i.e.

$$T_{m+n}(\theta) + T_{n-m}(\theta) = 2T_n(\theta)T_m(\theta) \tag{28}$$

But Eq. (26) can also be written as

$$\cos(A) + \cos(B) = 2 \cos \frac{(A-B)}{2} \cos \frac{(A+B)}{2} \tag{29}$$

Hence writing $n\theta$ for $(A-B)/2$ and $m\theta$ for $(A+B)/2$ and solving for A and B we find

$$A = (m+n)\theta, \qquad B = (m-n)\theta$$

Thus substitute these A and B into Eq. (29), so that

$$\cos((m+n)\theta) + \cos((m-n)\theta) = 2 \cos n\theta \cos m\theta$$

or

$$T_{n+m}(\theta) + T_{m-n}(\theta) = 2T_n(\theta)T_m(\theta) \tag{30}$$

Thus combining Eqs. (28) and (30) we have

$$T_{n+m}(\theta) + T_{|n-m|}(\theta) = 2T_n(\theta)T_m(\theta)$$

which proves the required result.

Since the Chebyshev polynomials are orthogonal on $[-1, 1]$ we can express any $f(x)$, $x \in [-1, 1]$, as a linear combination of these functions. Hence we propose

$$f(x) = c_0/2 + \sum_{j=1}^{\infty} c_j T_j(x) \tag{31}$$

To determine the coefficients $\{c_j\}$, multiply both sides of Eq. (31) by $(1-x^2)^{-1/2} T_k(x)$ and integrate. We find

$$\int_{-1}^{1} f(x)(1-x^2)^{-1/2} T_k(x) \, dx = \int_{-1}^{1} \frac{c_0}{2} (1-x^2)^{-1/2} T_k(x) \, dx$$

$$+ \int_{-1}^{1} \sum_{j=1}^{\infty} c_j T_j(x) T_k(x)(1-x^2)^{-1/2} \, dx$$

$$k = 0, 1, \ldots, \infty$$

Using the orthogonality of the Chebyshev polynomials, we find

$$\int_{-1}^{1} f(x)(1-x^2)^{-1/2} T_k(x) = c_k^1 \int_{-1}^{1} (1-x^2)^{-1/2} T_k^2(x) \, dx \qquad k = 0, 1, \ldots, \infty \tag{32}$$

where

$$c_k^1 = \begin{cases} c_k & k \neq 0 \\ \dfrac{c_0}{2} & k = 0 \end{cases}$$

Integrating the right-hand side of (32), we obtain

$$c_k = \frac{2}{\pi} \int_{-1}^{1} f(x)(1-x^2)^{-1/2} T_k(x) \, dx \qquad k = 0, 1, \ldots \tag{33}$$

Equation (33) therefore defines the coefficients for the expression of $f(x)$ in terms of Chebyshev polynomials. In many cases, however, this evaluation can be intractible since $f(x)$ may be (sufficiently) complicated. In such cases a change of variable $x = \cos(\theta)$ may facilitate the evaluation when the fast Fourier transform is used—see Hamming [19]. Using Eq. (33) and Eq. (31) we have expressed $f(x)$ as a infinite expansion of Chebyshev polynomials. We clearly cannot propose this as a means of evaluating $f(x)$. However, we may propose tö take a finite number of terms, say the first $(m+1)$ terms. Hence propose

$$f(x) \doteq \tfrac{1}{2}c_0 + \sum_{j=1}^{m} c_j T_j(x) = T_{m,0}(x) \qquad \text{(say)} \tag{34}$$

Example 3 Compare the approximation of $\exp(x)$ given by $m = 4$ in Eq. (34) with the Padé approximations $R_{22}(x)$ and $\tilde{R}_{22}(x)$ defined in Example 2 for $x \in [-1, 1]$.

For $f(x) = \exp(x)$, the coefficients c_k are given by

$$c_k = \frac{2}{\pi} \int_{-1}^{1} \exp(x)(1-x^2)^{-1/2} T_k(x) \, dx \qquad k = 0, 1, \ldots, 4$$

The integrals can be evaluated in terms of the modified Bessel functions of the first kind (see Watson [46]). Evaluating the coefficients we find

$$T_{4,0}(x) = 1.266\ 066 + 1.130\ 318 T_1(x) + 0.271\ 495 T_2(x)$$
$$+ 0.044\ 33 T_3(x) + 0.005\ 474 T_4(x)$$

which may be written as

$$T_{4,0}(x) = 1.000\ 044 + 0.997\ 310x + 0.499\ 200x^2 + 0.177\ 344x^3 + 0.437\ 92x^4$$

when $T_j(x)$ are written in terms of polynomials of x. The approximations based on $R_{22}(x)$, $\tilde{R}_{22}(x)$, and $T_{4,0}(x)$ are depicted in Figures 3, 4, and 5

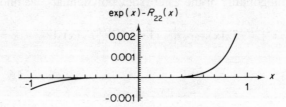

Figure 3

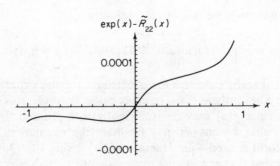

Figure 4

respectively. We have once again adjusted the scales in the three figures in order to show the complete graphs in each case. From Figure 5 it is clear that the single approximation $T_{4,0}(x)$ has a larger maximum error than the sequence of approximations $\tilde{R}_{22}(x)$ shown in Figure 4 but a considerably smaller maximum than the single approximation $R_{22}(x)$ shown in Figure 3. Further, note that the error oscillates between maximum values with alternating signs and that these maximum values are "nearly" equal in magnitude. We have

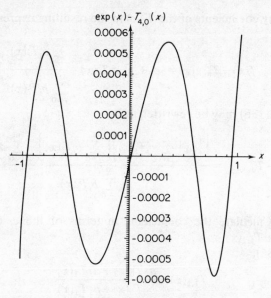

$\exp(x) - \tilde{T}_{4,0}(x)$

Figure 5

therefore obtained two alternative methods for obtaining approximations which, for this example, appear to give errors of similar magnitude, namely $\tilde{R}_{22}(x)$ and $\tilde{T}_{4,0}(x)$. The construction of $\tilde{R}_{22}(x)$ has required us to determine different approximations over different intervals—for many subintervals this may be time consuming. In contrast, the single approximation $\tilde{T}_{4,0}(x)$ requires the evaluation of the coefficients $\{c_j\}$ which in many cases (of $f(x)$) is impracticable. An alternative approach of calculating the coefficients using trigonometric interpolation is given in Ralston [34]. Thus the particular approximation chosen from $\tilde{T}_{4,0}(x)$ and $\tilde{R}_{22}(x)$ will clearly depend upon the particular function $f(x)$ being approximated and the interval over which the approximation is to be applied.

Following our remark in section 1 concerning the motivation for using rational approximation instead of polynomial approximation (the former lead to smaller maximum errors for a given amount of computation than the latter), we will generalize our discussion of approximating $f(x)$ by a linear combination of Chebyshev polynomials to approximating $f(x)$ by a ratio of linear combinations of these polynomials. Namely, we propose

$$f(x) = \frac{\displaystyle\sum_{j=0}^{m} a_j T_j(x)}{\displaystyle\sum_{j=0}^{k} b_j T_j(x)} = T_{m,k}(x) \tag{35}$$

The coefficients $\{a_j\}$ and $\{b_j\}$ are to be chosen in a manner analogous to that process described in section 2. Thus we substitute for $f(x)$ from Eq. (31) and choose as many coefficients of the $T_j(x)$ in the resulting expression

$$f(x) - T_{m,k}(x) = \tfrac{1}{2}c_0 + \sum_{j=1}^{\infty} c_j T_j(x) - \frac{\sum\limits_{j=0}^{m} a_j T_j(x)}{\sum\limits_{j=0}^{k} b_j T_j(x)} \tag{36}$$

to be zero; Eq. (36) may be rewritten as

$$f(x) - T_{m,k}(x) = \frac{\left[\tfrac{1}{2}c_0 + \sum\limits_{j=1}^{\infty} c_j T_j(x)\right]\left[\sum\limits_{j=0}^{k} b_j T_j(x)\right] - \sum\limits_{j=0}^{m} a_j T_j(x)}{\sum\limits_{j=0}^{k} b_j T_j(x)} \tag{37}$$

Example 4 Calculate the coefficients in terms of the c_j of the rational approximation $T_{1,1}(x)$.

From Eq. (35),

$$T_{1,1}(x) = \frac{a_0 T_0(x) + a_1 T_1(x)}{b_0 T_0(x) + b_1 T_1(x)}$$

b_0 is a redundant parameter which is usually chosen to be 1, where this does not lead to an insoluble system of equations in what follows. Hence

$$f(x) - T_{1,1}(x) = \frac{\left[\tfrac{1}{2}c_0 + \sum\limits_{j=1}^{\infty} c_j T_j(x)\right][b_0 T_0(x) + b_1 T_1(x)] - [a_0 T_0(x) + a_1 T_1(x)]}{b_0 T_0(x) + b_1 T_1(x)}$$

We therefore require to choose the parameters a_0, a_1, (b_0), b_1 so that as many coefficients in

$$\tfrac{1}{2}c_0[b_0 T_0(x) + b_1 T_1(x)] + \sum_{j=1}^{\infty} c_j b_0 T_j(x)$$

$$+ \sum_{j=1}^{\infty} c_j T_j(x) T_1(x) - (a_0 T_0(x) + a_1 T_1(x)) \tag{38}$$

are zero. Now,

Coefficient of $T_0(x)$ ($\equiv 1$) is

$$1/2 c_1 b_1 + 1/2 c_0 b_0 - a_0 = 0 \tag{39}$$

where we have used $T_1 T_1 = 1/2(T_2 + T_0)$ from property P(6).

Coefficient of $T_1(x)$ is

$$1/2 c_0 b_1 + c_1 b_0 + \tfrac{1}{2}c_2 b_1 - a_1 = 0 \tag{40}$$

where we have used $T_2 T_1 = \tfrac{1}{2}(T_3 + T_1)$.

Coefficient of $T_2(x)$ is

$$c_2 b_0 + 1/2 c_3 b_1 = 0$$

where we have used

$$T_3 T_1 = \tfrac{1}{2}(T_4 + T_2) \tag{41}$$

Coefficient of $T_3(x)$ is

$$c_3 b_0 + \tfrac{1}{2} c_2 b_1 \stackrel{?}{=} 0 \qquad (42)$$

using

$$T_2 T_1 = \tfrac{1}{2}(T_3 + T_1), \qquad (42)$$

where this last result is in doubt.

Solving Eqs. (39)–(41) we obtain, on substituting $b_0 = 1$,

$$b_1 = -2c_2/c_3; \quad a_1 = c_1 - c_2/c_3(c_0 + c_2); \quad a_0 = c_0 - c_1 c_2/c_3$$

The substitution of these expressions into the left-hand side of Eq. (42) gives $c_3 - c_2^2/c_3 = 0$ only if $c_3^2 = c_2^2$, which for general $f(x)$, will not be true. Hence the first nonzero coefficient in Eq. (38) is the coefficient of $T_3(x)$, namely $(c_3 - c_2^2/c_3)T_3(x)$. Since $|T_3(x)| < 1$, $(c_3 - c_2^2/c_3)$ will often be a reasonable estimate of the error in approximating $f(x)$ by $T_{1,1}(x)$.

Returning to Eq. (37), we may rewrite the right-hand side of this equation, using property P(6), as

$$f(x) - T_{m,k}(x) = \frac{\tfrac{1}{2}c_0 \sum\limits_{j=0}^{k} b_j T_j(x) + \sum\limits_{j=1}^{\infty} \sum\limits_{i=0}^{k} b_i c_j [T_{i+j}(x) + T_{|i-j|}(x)] - \sum\limits_{j=0}^{m} a_j T_j(x)}{\sum\limits_{j=0}^{k} b_j T_j(x)}$$

Collecting coefficients of like terms and setting the resulting coefficients to zero, we find

$$a = \tfrac{1}{2} \sum_{i=0}^{k} b_i c_i$$

and $\qquad (43)$

$$a_r = \tfrac{1}{2} \sum_{i=0}^{k} (c_{|i-r|} + c_{i+r}) \qquad r = 1, 2, \ldots, m+n+1$$

where $a_r = 0$ if $r > m$ and, as remarked in Example 4, b_0 is chosen to be 1 unless doing so produces an insoluble system of equations.

Example 5 Calculate the approximation $T_{2,2}(x)$ to exp (x) and compare it with the approximations for exp (x), $x \in [-1, 1]$ given in Figures 3, 4, and 5.

Substitute $m = k = 2$ into Eq. (43) and put $b_0 = 1$. We obtain the system of equations

$$a_0 = \tfrac{1}{2}(c_0 + b_1 c_1 + b_2 c_2)$$

$$a_1 = c_1 + \tfrac{1}{2}(b_1(c_0 + c_2) + b_2(c_1 + c_3))$$

$$a_2 = c_2 + \tfrac{1}{2}(b_1(c_1 + c_3) + b_2(c_0 + c_4))$$

$$0 = c_3 + \tfrac{1}{2}(b_1(c_2 + c_4) + b_2(c_1 + c_5))$$

$$0 = c_4 + \tfrac{1}{2}(b_1(c_3 + c_5) + b_2(c_2 + c_6)) \qquad (44)$$

186

The coefficients c_i are once again calculated using the modified Bessel function of the first kind. Here

$c_0 = 1.266\ 066$ $c_1 = 1.130\ 318$ $c_2 = 0.271\ 495$ $c_3 = 0.443\ 37$

$c_4 = 0.005\ 474$ $c_5 = 0.000\ 543$ $c_6 = 0.000\ 045$

Substituting these values into Eqs. (44) and solving the resulting system of equations, we obtain

$a_0 = 1.000\ 987\ 5$ $a_1 = 0.482\ 530\ 6$ $a_2 = 0.039\ 709\ 6$

$b_1 = -0.478\ 338\ 7$ $b_2 = 0.038\ 714\ 8$

Substituting these coefficients into Eq. (35) with $m = k = 2$ and writing the Chebyshev polynomials in terms of powers of x (see Table 13, Chapter 4), we find

$$T_{2,2}(x) = \frac{1.000\ 020\ 5 + 0.501\ 978\ 1x + 0.086\ 200x^2}{1.0 - 0.497\ 617\ 3x + 0.080\ 606\ 4x^2}$$

In Figure 6 we have plotted the error $\equiv \exp(x) - T_{2,2}(x)$ for $x \in [-1, 1]$. The error behavior over the interval is clearly now very much superior to the errors of $R_{2,2}(x)$ and $T_{4,0}(x)$ depicted in Figures 3 and 5 respectively. The behavior of $R_{2,2}(x)$ in Figure 4 still compares very well with $T_{2,2}(x)$ in Figure 6; it would be difficult from these figures to pronounce on which of the two approximations is better. However, the maximum error for Figure 6 is 0.000 189 whereas that for Figure 4 is 0.000 239, so that in the context of the basis of approximations described at the start of this chapter $T_{2,2}(x)$ is a better approximation to $\exp(x)$ for $x \in [-1, 1]$ than $\tilde{R}_{2,2}(x)$.

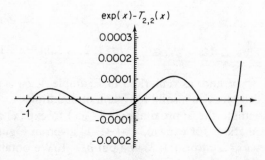

$\exp(x) - T_{2,2}(x)$

Figure 6

Exercises

9. Construct a $(2, 2)$ Padé approximation to $\cos(x)$ for $x \in [-\pi, \pi]$. Program the approximation and compute $\cos(x)$ for values of x in the interval. Compare your results with those produced by the standard function on your local computer.
10. As an alternative to the approximation to $\cos(x)$ in Exercise 9, it is proposed to divide the interval $[-\pi, \pi]$ into two subintervals $[\pi, 0]$ and $[0, \pi]$ and then propose

(2, 2) Padé approximations to cos (x) in each of these intervals. By transforming these intervals, in turn, to the intervals $[-\pi/2, \pi/2]$, derive the (2, 2) Padé approximations for each subinterval and hence obtain an approximation to cos (x) for $x \in [-\pi, \pi]$. Program your algorithm and hence compare the errors in this approximation with those arising from the approximation in Exercise 9.

11. Construct a (2, 2) Padé approximation to $(1/\sqrt{2}) (2-x)^{1/2}$ for $x \in [-2, 2]$. Program the approximation and compute $(1/\sqrt{2}) (2-x)^{1/2}$ for values of x in the interval. Compute your results with those produced by the standard function on your computer.

12. Derive a composite (2, 2) Padé approximation to $(1/\sqrt{2}) (2-x)^{1/2}$ for $x \in [-2, 2]$ by dividing the interval $[-2, 2]$ into four subintervals $[-2, -1]$, $[-1, 0]$, $[0, 1]$, and $[1, 2]$ and, in turn, transforming each interval to $[-1/2, 1/2]$. Calculate approximations to $(1/\sqrt{2}) (2-x)^{1/2}$ for $x \in [-2, 2]$ and compare the resulting errors with those in Exercise 11.

13. Derive the $T_{4,0}$ approximation to $(1/\sqrt{2}) (2-x)^{1/2}$, $x \in [-2, 2]$, and compute approximations for $x \in [-2, 2]$. Compare your results with Exercises 11 and 12.

14. Derive the $T_{2,2}$ approximation to $(1/\sqrt{2}) (2-x)^{1/2}$, $x \in [-2, 2]$ and compute approximations for $x \in [-2, 2]$. Compare your results with Exercises 11, 12, and 13.

4 The Construction of Minimax Approximations: The Algorithm of Remes

In this final section we describe an algorithm due to Remes [35], for constructing minimax approximations. The algorithm is dependent upon an important theorem due to Chebyshev which we will now state; a proof may be found in Ralston [34].

Chebyshev's theorem   Let $f(x)$ be continuous in the finite interval $[a, b]$. Let $y(x)$ be any rational approximation to $f(x)$ of the form (2).
   Let

$$r_{mk} = \max_{a \leq x \leq b} |f(x) - y(x)|$$

Then there exists a unique rational function $y*(x)$ which minimizes r_{mk}.†
   If we denote this unique approximation by

$$y*(x) = \frac{\sum\limits_{j=0}^{m-\nu} a_{j+\nu}^* x^j}{\sum\limits_{j=0}^{k-\mu} b_{j+\mu}^* x^j} \ddagger$$

---

† We can multiply this unique function $y*(x) = P_m *(x)/Q_k *(x)$ by a function $s(x)/s(x)$ where $s(x)$ is assumed not to vanish for x in $[a, b]$ without changing the rational function. We may be tempted to suggest this as an alternative approximation. Hence we assume in the statement of uniqueness that any common factors existing in $y*(x)$ have been divided out so that $P_m *(x)/Q_k *(x)$ cannot be further reduced.

‡ In this formulation the coefficients $a_{j+\nu}^*$ and $b_{j+\mu}^*$ are to be calculated by the algorithms we will describe in this section. In this calculation some of the coefficients may be zero. We have taken cognizance of this fact by allowing the possibility of the coefficients of the highest powers of x in the numerator being zero and the coefficients of the highest powers of x in the denominator being zero.

where $0 < \mu < k$, $0 < \nu < m$, and a_m^*, $b_m^* \neq 0$, and where we assume $y^*(x)$ cannot be reduced, then if $r_{mk}^* = |f(x) - y^*(x)| \neq 0$ the number of consecutive points of $[a, b]$ at which $f(x) - y^*(x)$ takes on its maximum value of magnitude r_{mk}^* with alternating changes of sign is not less than $m + k + 2 - d$ where $d = \min(\mu, \nu)$.

Example 6 For $m = k = 2$, $d = 0$, the best minimax rational approximation

$$y^*(x) = \frac{a_0^* + a_1^* x + a_2^* x^2}{1 + b_1^* x + b_2^* x^2}$$

takes on its maximum difference from the function $f(x)$ with alternating sign at precisely 6 points $x_0, x_1, \ldots, x_5$ (say). In Example 3 we constructed $T_{4,0}(x)$ which approximated $\exp(x)$. For $T_{4,0}(x)$, $m = 4$, $k = 0$, $d = 0$, so $m + k + 2 = 6$. If the reader consults Figure 5 depicting the difference between $\exp(x)$ and $T_{4,0}(x)$, he will note that the error alternates as x increases from -1 to 1. The extreme values of the error occur with alternating sign taking on the values summarized in Table 1.

Table 1

x	-1.0	-0.8	-0.3
max errors $\exp(x) - T_{4,0}(x)$	$-0.000\,50$	$0.000\,51$	$-0.000\,53$
x	0.3	0.8	1.0
max errors $\exp(x) - T_{4,0}(x)$	$0.000\,55$	$0.000\,58$	$0.000\,59$

The extrema are clearly not equal in magnitude. However, they differ in magnitude only in the fifth decimal place. Hence we would surmise that such an approximation is "near" minimax and might prove a useful starting point from which to generate a better minimax approximation.

Likewise, considering Figure 6 where the difference between $\exp(x)$ and $T_{2,2}(x)$ is graphed (again $m + k + 2 = 6$), we note that the error alternates as x increases from $x = -1$ to $x = 1$. The extrema now occur in the manner summarized in Table 2.

Here again the extrema are not equal but the error does take on its maximum with alternating sign at 6 points. The extrema now take on smaller values than those summarized in Table 1. We may surmise that $T_{2,2}(x)$ has produced a nearer minimax approximation than the approximation $T_{4,0}(x)$. However, neither $T_{2,2}(x)$ nor $T_{4,0}(x)$ has produced the minimax approximation since, as Chebyshev's theorem states, the best approximation has alternating extrema which are equal (and the number of extrema is 6 in this example). Thus although $T_{2,2}(x)$ and $T_{4,0}(x)$ are not the minimax approximations, we will use them, in turn, as starting points from which to generate a sequence of better

Table 2

x	-1.0	-0.8	-0.2
max error $\exp(x) - T_{2,2}(x)$	$-0.000\,04$	$0.000\,05$	$-0.000\,07$
x	0.4	0.8	1.0
max error $\exp(x) - T_{2,2}(x)$	$0.000\,11$	$-0.000\,16$	$0.001\,9$

near minimax approximations which ultimately will converge to the required minimax approximation.

For simplicity we assume that $d = 0$ so that there are at least $m + k + 2$ points at which the extreme value of the error is attained. We further assume that we have transformed the interval on which the approximation is required into $x \in [-1, 1]$.

Let us assume we have a starting approximation $y^{[0]}(x)$, say, to $f(x)$ for $x \in [-1, 1]$ which alternates $m + k + 2$ times (for $m + k + 2 = 6$, $T_{4,0}(x)$ or $T_{2,2}(x)$ would be appropriate, following Example 6). Let points at which the error function $f(x) - y^{[0]}(x)$ attains its extrema be denoted by

$$x_0^{[0]} < x_1^{[0]} < \cdots < x_{m+k+1}^{[0]}$$

We call this set of points the critical set for $y^{[0]}(x)$. On this set of points we now determine the rational function which interpolates the points $(x_0^{[0]}, E^{[1]})$, $(x_1^{[0]}, -E^{[1]}), \ldots, (x_{m+k+1}^{[0]}, (-1)^{m+k+1}E^{[1]})$, namely the rational function which takes on the value $E^{[1]}$, with alternating sign at the points $(x_0^{[0]}, x_1^{[0]}, \ldots, x_{m+k+1}^{[0]})$. $E^{[1]}$ is as yet unknown.

Thus we seek to determine the new approximation

$$y^{[1]}(x) = \frac{\sum_{j=0}^{m} a_j x^j}{\sum_{j=0}^{k} b_j x^j} \tag{45}$$

such that

$$f(x_i^{[0]}) - y^{[1]}(x_i^{[0]}) = (-1)^i E^{[1]}, \quad i = 0, 1, \ldots, m+k+1 \tag{46}$$

where as before we assumed $b_0 = 1$. In Eq. (45) we have $(m + k + 1)$ unknowns (the a_j and b_j). Equation (46) represents a system of nonlinear equations (see Chapter 9) which is, in general, a difficult problem to solve. However, with a little ingenuity we will be able to solve this system without recourse to the more sophisticated techniques described in Chapter 9 for such problems. Let us denote the coefficients a_j, b_j which satisfy Eq. (46) by $a_j^{[1]}$, $b_j^{[1]}$. Thus from

Eqs. (45) and (46) we have

$$f(x_i^{[0]}) - \frac{\sum\limits_{j=0}^{m} a_j^{[1]}(x_i^{[0]})^j}{\sum\limits_{j=0}^{k} b_j^{[1]}(x_i^{[0]})^j} = (-1)^i E^{[1]}$$

which can be rewritten as

$$\sum_{j=0}^{m} a_j^{[1]}(x_i^{[0]})^j - (f(x_i^{[0]}) - (-1)^i E^{[1]}) \sum_{j=0}^{k} b_j^{[1]}(x_i^{[0]})^j = 0,$$

$$i = 0, 1, \ldots, m+k+1$$

namely

$$\sum_{j=0}^{m} a_j^{[1]}(x_i^{[0]})^j - (f(x_i^{[0]}) - (-1)^i E^{[1]}) \sum_{j=1}^{k} b_j^{[1]}(x_i^{[0]})^j$$
$$= f(x_i^{[0]}) - (-1)^i E^{[1]} \qquad i = 0, 1, \ldots, m+k+1 \qquad (47)$$

where we have used $b_0^{[1]} = 1$. If we knew $E^{[1]}$ on the left-hand side of Eq. (47) the system of equations would be linear. We could then employ one of the methods for such problems described in Chapter 3. Let us therefore assume we have an *estimate* for $E^{[1]}$ on the left-hand side, say $E_{(0)}^{[1]}$. We can then use this value on the left-hand side to solve for the $a_j^{[1]}$ and $b_j^{[1]}$ on the left-hand side and the value of $E^{[1]}$ on the right-hand side of Eq. (47). Since $E_{(0)}^{[1]}$ is only a guess, this solution will only be an approximation to the correct values of $a_j^{[1]}$, $b_j^{[1]}$, and $E^{[1]}$. Hopefully, we can repeat the process using the *new* value of $E^{[1]}$ in the left-hand side and solve once again for the unknown coefficients. We are therefore proposing solving the system of equations (47) iteratively. We can denote this process as follows:

$$\sum_{j=0}^{m} a_{j(l)}^{[1]}(x_i^{[0]})^j - (f(x_i^{[0]}) - (-1)^i E_{(l-1)}^{[1]}) \sum_{j=1}^{k} b_j^{[1]}(x_i^{[0]})^j$$
$$= f(x_i^{[0]}) - (-1)^i E_{(l)}^{[1]} \qquad i = 0, 1, \ldots, m+k+1; l = 1, 2, \ldots \qquad (48)$$

where $E_{(0)}^{[1]}$ is an initial estimate for $E^{[1]}$.

The number of superscripts and subscripts is sufficiently large so as to be confusing so a short digression on each one's rôle will be in order.

The subscript l is used to denote the iteration count in solving the system of Eq. (47).

The subscript i corresponds to the ith point at which the error of the approximation takes on its extreme value.

The subscript j has been used in the definition of our numerator and denominator of $y(x)$ in Eq. (45).

Finally, the superscript [1] indicates the first stage in our procedure for producing a sequence of approximations $y(x)$ which will ultimately converge to the best approximation. Note that the superscript in [ ] will remain the same until all of the iterations denoted by the subscript l have been completed.

The iterations in Eq. (48) are continued (for increasing l) until $|E^{[1]}_{(l)} - E^{[1]}_{(l-1)}| \leq \varepsilon$ is satisfied, where ε is some preassigned tolerance.

The solution of Eq. (47) by the iterative procedure (48) will have produced the coefficients $\{a^{[1]}_j\}$ and $\{b^{[1]}_j\}$ and hence the new rational approximation $y^{[1]}(x)$, say. In terms of this approximation define

$$h^{[1]}(x) = f(x) - y^{[1]}(x) \qquad (49)$$

Then provided we started with a $y(x)$ which is a near best approximation (in the way $T_{4,0}(x)$ and $T_{2,2}(x)$ were) the locations at which the extrema of $h^{[1]}(x)$ are attained with alternating sign will be close to the points $x^{[0]}_i$. To obtain a new approximation $y^{[2]}(x)$, we first require to determine the points $x^{[1]}_i$ at which $h^{[1]}(x)$ given by Eq. (49) attains these extreme values, i.e., we require

$$\max_{x \in N^{[0]}_i} |f(x) - y^{[1]}(x)| \qquad (50)$$

where $N^{[0]}_i$ is a small neighborhood about $x^{[0]}_i$, $i = 0, 1, \ldots, m+k+1$. In general it is not possible to determine this point exactly, so a numerical algorithm has to be used here, too. Since the intervals $N^{[0]}_i$ are small we will be able to use a simple enumeration of Eq. (49) in $N^{[0]}_i$ and use inverse interpolation as advocated in Chapter 4 to determine where the derivative of $h^{[1]}(x)$ given by Eq. (49) is zero. However, if this is not sufficient we can propose more sophisticated techniques which are described in Chapter 9. The required points are denoted by $x^{[1]}_i$. A situation is possible (see Figure 7) where this set of points does not include the point at which $h^{[1]}(x)$ attains its largest extreme value, $\bar{x}$, say. (This would correspond to a situation where we had initially taken a $N^{[0]}_i$ too small and hence missed the global maximum-modulus value.) In this event we choose to replace that $x^{[1]}_i$ next to $\bar{x}$ by the value of $\bar{x}$ so that the error still achieves its maximum modulus value at successive $x^{[1]}_i$ with alternating sign.

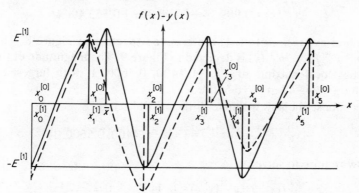

Figure 7 The extreme points of $y^{[0]}(x)$ are denoted by $x^{[0]}_i$, $i = 0, 1, \ldots, 5$. The extreme points, initially chosen of $y^{[1]}(x)$ are denoted by $x^{[1]}_i$ $i = 0, 1, \ldots, 5$. Note that the actual extreme value of $f(x) - y^{[1]}(x)$ occurs at $\bar{x}$ which is not one of the points $x^{[1]}_i$. Hence we replace $x^{[1]}_i$ by $\bar{x}$ thereby preserving the alternating property

The process of generating successive approximations $y^{[p]}(x)$, $p = 0, 1, \ldots$, is repeated until the errors $E^{[p]}$ and $E^{[p-1]}$ satisfy some prescribed condition, for example,

$$|E^{[p]} - E^{[p-1]}| \leq \text{tol}$$

where tol is a prescribed tolerance.

Example 7 Consider finding the minimax approximation to $f(x) = \exp(x)$ on the interval $[-1, 1]$ by a polynomial of degree 2 (i.e. a rational function, $m = 2$, $k = 0$). Let this polynomial be

$$p_2(x) = a_2 x^2 + a_1 x + a_0$$

Let the initial critical set be given as

$$\{x_0^{[0]}, x_1^{[0]}, x_2^{[0]}, x_3^{[0]}\} = \{-1, -\tfrac{1}{2}, \tfrac{1}{2}, 1\}$$

Then we require to solve the 4×4 system of linear equations

$$f(x_i^{[0]}) - p_2^{[1]}(x_i^{[0]}) = (-1)^i E \qquad i = 0, 1, 2, 3$$

That is

$$a_0 - a_1 + a_2 + E = 1/e = 0.3679$$

$$a_0 - 1/2 a_1 + 1/4 a_2 - E = 1/\sqrt{e} = 0.6065$$

$$a_0 + 1/2 a_1 + 1/4 a_2 + E = \sqrt{e} = 1.6487$$

$$a_0 + a_1 + a_2 - E = e = 2.7183$$

The solution of this system is

$a_0 = 0.989\,14 \qquad a_1 = 1.130\,86 \qquad a_2 = 0.553\,494 \quad \text{and} \quad E = -0.044\,34$

Thus

$$p_2^{[1]}(x) = 0.989\,14 + 1.130\,86x + 0.553\,494x^2$$

$p_2^{[1]}(x)$ and $\exp(x)$, $x \in [-1, 1]$ are grouped together in Figure 8. The error curve $e^{[1]}x = f(x) - p_2^{[1]}(x)$ is depicted in Figure 9. The magnitude of the error curve has local maxima at -1, -0.4400, 0.5600, 1, with largest error of magnitude $0.045\,47$ at $x = 0.5600$.

Thus choose the new estimate of the critical set

$$(x_0^{[1]}, x_1^{[1]}, x_2^{[1]}, x_3^{[1]}) = (-1, -0.4400, 0.5600, 1)$$

We now require to solve

$$f(x_i^{[1]}) - p_2^{[2]}(x_2^{[1]}) = (-1)^i E \qquad 0 \leq i \leq 3$$

Solving this system of equations, we find

$$p_2^{[2]}(x) = 0.989\,04 + 1.130\,18x + 0.554\,04x^2, \text{ and } E = -0.04502.$$

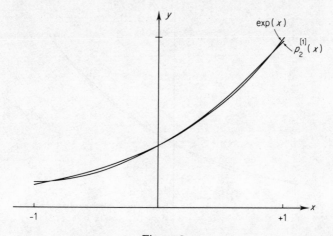

Figure 8

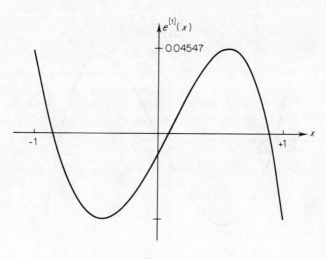

Figure 9

In this case a search for the maxima of the magnitude of the error curve $|f(x) - p_2^{[2]}(x)|$ indicates that the error curve $e^{[2]}(x)$ oscillates with maxima = 0.045 02 so that $p_2^{[2]}(x)$ can be taken as the required minimax polynomial. $p^{[2]}(x)$ and exp (x) are depicted in Figure 10 and the error curve is shown in Figure 11.

Example 8   Use the algorithm of Remes to construct the minimax approximation to exp (x), $-1 \le x \le 1$, using alternatively the extrema produced from the approximations $T_{4,0}(x)$ and $T_{2,2}(x)$ summarized in Tables 1 and 2 respectively.

In constructing the best approximation to exp (x) using the Remes algorithm, we only require, as data, the approximate positions of the extrema

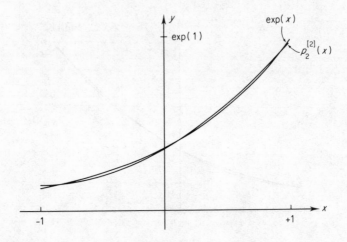

Figure 10

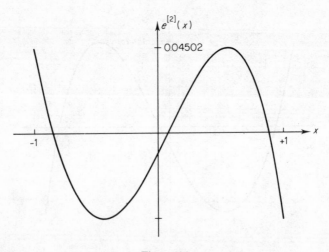

Figure 11

in the final approximation and an initial guess for $E^{[1]}_{(0)}$; the algorithm will then proceed to produce a sequence of approximations converging to the best approximation. Since the error in the approximations $T_{4,0}(x)$ and $T_{2,2}(x)$ alternated in sign with near-equal magnitudes, we conclude that the extrema of these approximations will be "near" the extrema of the best approximation. Furthermore, the initial guess $E^{[1]}_{(0)}$, which could equally well be zero, is chosen as the maximum error in the approximation $T_{4,0}(x)$ and $T_{2,2}(x)$ in turn. The other pieces of data required are the stopping criterion for the iterative solution of the nonlinear equations (47); we specified that iteration should continue until successive estimates for E differed in magnitude by less than 1×10^{-6} or the number of iterations exceed 10. Also we deemed that the sequence of iterations had converged when the minimax errors of succeeding approximations $y^{[n]}$ and $y^{[n+1]}$ differed in magnitude by less than 1×10^{-6}.

For the extrema in Table 1, given by $T_{4,0}(x)$, the sequence of approximations and extreme points are given below:

$$y^{[1]}(x) = \frac{1.000\ 066\ 4 + 0.507\ 308\ 79x + 0.085\ 128\ 42x^2}{1 - 0.492\ 348\ 35x + 0.078\ 212\ 76x^2}$$

The extreme points are located at

$$-1.00, -0.746, -0.153, 0.450, 0.873, 1.00$$

Three iterations were required to solve the system of nonlinear equations.

$$y^{[2]}(x) = \frac{1.000\ 072\ 7 + 0.508\ 596\ 93x + 0.085\ 807\ 58x^2}{1 - 0.491\ 129\ 13x + 0.077\ 723\ 25x^2}$$

with extrema at $-1.000, -0.727, -0.120, 0.473, 0.866, 1.000$. Finally,

$$y^{[3]}(x) = \frac{1.000\ 072\ 3 + 0.508\ 603\ 59x + 0.085\ 813\ 35x^2}{1 - 0.491\ 1236\ 9x + 0.077\ 722\ 35x^2}$$

The error in the final minimax approximation is depicted in Figure 12. The maximum modulus error was found to be $0.000\ 86\ 85$.

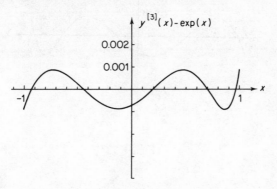

Figure 12

In contrast, the sequence of approximations starting from the extreme points of $T_{2,2}(x)$ given in Table 2 were

$$y^{[1]}(x) = \frac{1.000\ 069\ 6 + 0.508\ 258\ 27x + 0.085\ 639\ 23x^2}{1 - 0.491\ 465\ 77x + 0.077\ 870\ 44x^2}$$

The extreme points are located at

$$-1.000, -0.730, -0.123, 0.465, 0.866, 1.00$$

Three iterations were required to solve the nonlinear system of equations.

$$y^{[2]}(x) = \frac{1.000\ 072\ 3 + 0.508\ 606\ 76x + 0.085\ 814\ 83x^2}{1 - 0.491\ 120\ 51x + 0.077\ 720\ 86x^2}$$

with extrema at

$$-1.000, -0.726, -0.120, 0.473, 0.866, 1.000$$

where just two iterations were required for the nonlinear equations. The starting value for E in this solution was chosen to be the value of E produced in the nonlinear equation solution for $y^{[1]}(x)$.

Finally,

$$y^{[3]}(x) = \frac{1.000\ 072\ 5 + 0.508\ 632\ 62x + 0.085\ 827\ 6x^2}{1 - 0.491\ 0954\ 0x + 0.077\ 709\ 98x^2}$$

where convergence was deemed to have occurred. The error in the final minimax approximation is depicted in Figure 13. The maximum modulus error was found to be 0.000 86 89. To all intents and purposes the two minimax approximations are indistinguishable.

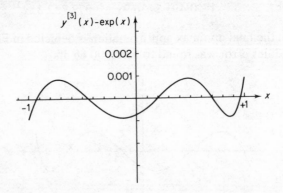

Figure 13

Exercises

15. Program the Remes algorithm described in this section assuming $m = k = 2$. Construct the minimax approximation to exp (x), $-1 \le x \le 1$, assuming the extreme points are situated at

    $$-1.0, -0.8, -0.27, 0.35, 0.82, 1.00$$

    Use -2×10^{-4} as the initial estimate of $E^{[1]}_{(0)}$ and 1×10^{-6} as the criterion for convergence in the solution of the nonlinear equations and successive minimax errors. Compare your result with Figures 12 and 13.

    Note: Your program should use the routine for solving linear equations described in Chapter 3.

16. Combine your program of Exercise 15 with the technique of Exercise 7. Hence produce a minimax approximation for $x \in (-M, M)$ where M is the largest (in absolute value) machine representable number on your local computer. Compare your result with the errors produced in Exercise 8.

CHAPTER 7

Numerical Integration

1 Introduction

Often in practice one meets the problem of requiring to integrate a given function of a single variable over a given interval. In contrast to the school-day exercises where a closed-form solution was usually guaranteed, in practice it is seldom the case that a closed-form solution for our definite integral can be obtained. In this situation, therefore, we are saying that the given function is too difficult to integrate. Alternatively, we may require to evaluate the integral of a function whose analytical form is not known but rather it is represented by a discrete set of values as in our problem of interpolation. In this case we clearly cannot evaluate the function because we do not know its analytical form. These are just two reasons why we may require to determine an algorithm for *numerical integration*.

Example 1 Consider one of the new oil fields in the North Sea. Let us assume the oil company requires to determine the yearly oil production from each of 100 wells. To obtain this production one of several methods could be employed; the following is a simple one. Consider at well i ($i = 1, 2, \ldots, 100$) that at 12 o'clock on the jth day a record is taken of the oil which is produced over the next hour. Using this hourly rate, the rate for the day is assumed to be given by 24 times this rate. Let this be denoted by r_{ij}, $j = 1, 2, \ldots, 365$, $i = 1, 2, \ldots, 100$. If these daily rates are graphed for the 365 days for a particular well i we might obtain a figure like that depicted in Figure 1. If we assumed the production rate was governed by the function $r_i(t)$ denoted by the dotted line in Figure 1, then we would conclude that the year's production at well i would be given by

$$\int_{t_0}^{t_1} r_i(t) \, dt \qquad t_0 = \text{day 1}, \ t_1 = \text{day 365}$$

198

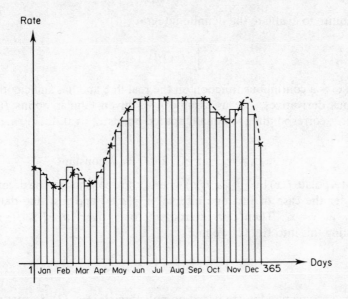

Figure 1

However, we do not know $r_i(t)$ but rather discrete values of it. We could therefore obviously just add each r_{ij} and obtain

$$\text{production} = \sum_{j=1}^{365} r_{ij} \tag{1}$$

If, however, we assume that only one man can take the readings of daily rates and he has to travel over the 100 wells, it may become more convenient for him to take rates of production just once a month. Hence for the well in Figure 1 let us assume the "measuring-man" comes on the first day of each month. Then taking his measurements he will obtain the rates indicated by the Xs in Figure 1. Clearly now a corresponding sum to Eq. (1) is no longer feasible. What we want is an approximation to $r_i(t)$ based on just the given data given by X. Clearly such an approximation is given by interpolating through the data points. The required approximation to the yearly production is then given by integrating the resulting approximation to $r_i(t)$. Thus if $y_i(t)$ is the interpolant of the data in Figure 1,

$$\int_{t_0}^{t_1} r_i(t)\, \mathrm{d}t \doteq \int_{t_0}^{t_1} y_i(t)\, \mathrm{d}t$$

In considering our production rates in Example 1 it is already clear that we intend to propose approximations first for the integrand and then integrate the resulting approximation over the required interval. As in our discussion of interpolation, polynomials once again suggest themselves as the basis of such approximations since clearly these can be easily integrated. To be precise, let us require to evaluate the definite integral

$$\int_a^b f(x)\,\mathrm{d}x \tag{2}$$

where $f(x)$ is a continuous function on the real line and has sufficiently many continuous derivatives. Let us assume $n+1$ given tabular points $f(x_i)$, $i = 0, 1, 2, \ldots$, corresponding to equally spaced points x_i, $i = 0, 1, \ldots, n$, distance h apart, i.e.

$$x_i - x_{i-1} = h \qquad i = 1, 2, \ldots, n, h \text{ constant}$$

Let us interpolate $f(x)$ by $\sum_{i=0}^n \alpha_i f(x_i)$ where α_i are constants to be determined.

Consider the case of just one interval $h = [a, b]$ and just one data value $f(x_0) = f_0$ at $x = x_0$. Then our interpolant for f over $[a, b]$ is $f(x) = \alpha_0 f_0$. Substituting this into Eq. (2) we find

$$\int_a^b f(x)\,\mathrm{d}x \doteq \int_a^b \alpha_0 f_0 \tag{3}$$

Since our approximations are based on polynomials, Eq. (3) if exact must be at least exact for polynomials of degree zero, i.e. constants. Hence if we assume $f(x) = 1$ then Eq. (3) becomes

$$\int_a^b \mathrm{d}x = \alpha_0$$

or $\alpha_0 = b - a = h$. Hence our integrating formula is

$$\int_a^b f(x)\,\mathrm{d}x = hf_0 \tag{4}$$

Clearly this equality holds only for the case $f(x) = $ constant. For other functions, Eq. (4) must be augmented by an error term. Thus in general

$$\int_a^b f(x)\,\mathrm{d}x = hf_0 + E_I \tag{5}$$

where E_I is the error term. (We will consider the general form of the error term in section 3.) The integration formula (5) is known as the *rectangular rule*. Its geometric interpretation is shown in Figure 2, where the required integral represents the area under the curve between $x = a$ and $x = b$, i.e. the area ABCD. The rectangular rule represents $f_0 h = f_0(b-a)$ which is the shaded area ABED. It is clear that the area DEC represents the error involved when approximating the given integral by the integration formula (5), i.e. $E_I = $ area DEC. The rectangular rule is a *zero-order accurate* method. *In general, an integrating formula has order of accuracy p if the integrating formula is exact for polynomials up to, and including, those of degree p.* If, instead, we consider an integrating formula which utilizes values of the function $f(x)$ at $x = x_0$ and $x = x_1$ (we call these points quadrature points), namely $f(x_0)$ and

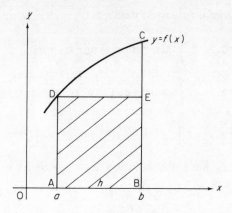

Figure 2

$f(x_1)$, respectively, we propose an integrating formula whose form is

$$\int_a^b f(x)\,dx = \sum_{i=0}^{1} \alpha_i f_i = \alpha_0 f_0 + \alpha_1 f_1 \tag{6}$$

where we have written $f_i \equiv f(x_i)$. The two weights α_0 and α_1 can be determined by setting $f(x)$ equal to a constant and equal to x, integrating the resulting expression, and solving for α_0 and α_1. Thus for $f(x) = 1$,

$$\int_a^b 1\,dx = b - a = \alpha_0 + \alpha_1 \tag{7}$$

and for $f(x) = x$

$$\int_a^b x\,dx = \tfrac{1}{2}(b^2 - a^2) = \alpha_0 a + \alpha_1 b \tag{8}$$

Solving Eqs. (7) and (8), we find $\alpha_0 = \alpha_1 = h/2$. The integrating formula is thus

$$\int_a^b f(x)\,dx = \frac{h}{2}(f_0 + f_1) + E_I \tag{9}$$

and is known as the *trapezoidal rule*. We have augmented our formula with an error term for the cases where $f(x)$ is of degree greater than one. By the derivation, Eq. (9) is exact for polynomials f up to, and including, those of degree one, i.e. it is first-order accurate.

To derive an integrating formula using more than two points, we must redefine our interval. For a formula using three points we define

$$2h = b - a$$

The quadrature points are x_0, x_1, and x_2 and function values $f(x_0)$, $f(x_1)$, and $f(x_2)$ are to be used. In a similar manner to the derivation of the rectangular rule and the trapezoidal rule we obtain:

for $f(x) = 1$,
$$\int_a^b dx = b - a = \alpha_0 + \alpha_1 + \alpha_2 \tag{10}$$

for $f(x) = x$,
$$\int_a^b x \, dx = \tfrac{1}{2}(b^2 - a^2) = \alpha_0 a + \alpha_1 x_1 + \alpha_2 b \tag{11}$$

for $f(x) = x^2$,
$$\int_a^b x^2 \, dx = \tfrac{1}{3}(b^3 - a^3) = \alpha_0 a^2 + \alpha_1 x_1^2 + \alpha_2 b^2 \tag{12}$$

Equations (10)–(12) have the solution $\alpha_0 = \alpha_2 = h/3$; $\alpha_1 = 4h/3$, and our integrating formula is

$$\int_a^b f(x) \, dx = \frac{h}{3}(f_0 + 4f_1 + f_2) + E_I \tag{13}$$

Equation (13) is known as *Simpson's $\frac{1}{3}$-rule*. By its derivation it is exact for polynomials of degree up to and including those of degree two: however, see the next section.

Finally, if we consider $(b - a)$ to be divided into three equal intervals, $3h = (b - a)$, we can in a similar fashion derive the integrating rule

$$\int_a^b f(x) \, dx = \frac{3h}{8}(f_0 + 3f_1 + 3f_2 + f_3) + E_I \tag{14}$$

E_I being the error term. Equation (14) is *Simpson's $\frac{3}{8}$-rule* and is exact for polynomials of degree up to and including those of degree three.

Example 2   Evaluate

$$I = \int_0^{1/2} (1 + x)^{1/2} \, dx$$

using the rectangular rule. Compare your answer with the result obtained by integrating $(1 + x)^{1/2}$ exactly.

Using Eq. (4) with $h = \tfrac{1}{2}$ we evaluate $(1 + x)^{1/2}$ at $x = 0$. Thus

$$I_R = (1)^{1/2} \times 0.5 = 0.5$$

Integrating $(1 + x)^{1/2}$ exactly, we obtain

$$I = \int_0^{1/2} (1 + x)^{1/2} \, dx = \left[\tfrac{2}{3}(1 + x)^{3/2} \right]_0^{1/2} = 0.558$$

Example 3   Evaluate I of Example 2 using the trapezoidal rule. In this case we need $(1 + x)^{1/2}$ evaluated at $x = 0$ and $x = 0.5$. Using Eq. (9) with $h = \tfrac{1}{2}$, we obtain

$$I_T = \tfrac{1}{2} \times \tfrac{1}{2} \times ((1)^{1/2} + (1.5)^{1/2}) = 0.556$$

which is clearly more accurate than I_R.

Exercises

1. Use the rectangular rule (4) to evaluate $\int_1^{1.2} \sin(x)\,dx$. Compare your computed result with the result you obtain when deriving the integral analytically.
2. Use the trapezoidal rule (9) to evaluate the integral of Exercise 1.
3. Evaluate the integral of Exercise 1 using Simpson's $\frac{1}{3}$-rule.
4. Use Simpson's $\frac{3}{8}$-rule to evaluate the integral of Exercise 1.
5. Develop computer programs for evaluating the integrals of the functions $\cos(x)$, $\tan(x)$, $\ln(x)$ between limits a and b, which are read as data, by: (i) the rectangular rule; (ii) the trapezoidal rule; (iii) the Simpson's $\frac{1}{3}$-rule; (iv) the Simpson's $\frac{3}{8}$-rule.
6. Evaluate the integrals of Exercise 5 for: (i) $a = 1$, $b = 2$; (ii) $a = \pi/2$, $b = \pi$. In each case determine which method is the most accurate by comparing your computer results with the values obtained from the analytical derivation of the definite integrals. Does your conclusion coincide with expectation?
7. Derive Simpson's $\frac{3}{8}$-rule.
8. By an analogous technique to that used for the integrating formulas in this section, derive an integrating formula which employs the values of f_0, f_1, f_2, f_3, and f_4.
9. In the derivation of the integrating formulas we considered only functions of the form x^k, $k = 0, 1, \ldots$ Why is it sufficient to consider polynomials of this form rather than the general polynomials $\sum_{i=0}^k a_i x^i$ for successive values of k?

The approach adopted above to determine the integrating formulas is often termed the method of *undetermined coefficients*. As an alternative method for deriving the integrating formulas, we take up the suggestion made in Example 1 where we substitute the Lagrangian interpolating polynomial of Chapter 4 (see Eqs. (11) and (15) of Chapter 4) for the integrand $f(x)$. For just one point we have

$$\int_a^b f(x)\,dx \doteq \int_a^b L_0(x) f_0\,dx = \int_a^b f_0\,dx = hf_0$$

the rectangular rule. For two points we have

$$\int_a^b f(x)\,dx \doteq \int_a^b \{L_0(x) f_0 + L_1(x) f_1\}\,dx = \int_a^b \left\{ \frac{(x - x_1)}{(x_0 - x_1)} f_0 + \frac{(x - x_0)}{(x_1 - x_0)} f_1 \right\} dx$$

$$= \frac{1}{2h}[-(x - x_1)^2 f_0 + (x - x_0)^2 f_1]_a^b = \frac{h}{2}(f_0 + f_1)$$

the trapezoidal rule. For three points,

$$\int_a^b f(x)\,dx \doteq \int_a^b \{L_0(x) f_0 + L_1(x) f_1 + L_2(x) f_2\}\,dx$$

$$= \int_a^b \left\{ \frac{(x - x_1)(x - x_2)}{(x_0 - x_1)(x_0 - x_2)} f_0 + \frac{(x - x_0)(x - x_2)}{(x_1 - x_0)(x_1 - x_2)} \times f_1 \right.$$

$$\left. + \frac{(x - x_0)(x - x_1)}{(x_2 - x_0)(x_2 - x_1)} f_2 \right\} dx$$

$$= \frac{1}{2h^2}\left[\frac{2h^3}{3} f_0 + \frac{8h^3}{3} f_1 + \frac{2h^3}{3} f_2 \right] = \frac{h}{3}[f_0 + 4f_1 + f_2]$$

Simpson's $\frac{1}{3}$-rule.

As the number of tabular points increases, the algebra involved in calculating the coefficients becomes more and more tedious. Therefore if we are to generate more accurate integrating formulas involving more points we require a general approach of generating the coefficients arising from the integration of the Lagrangian polynomials. We will describe this general method in the next section. The integrating formulas based upon equally spaced tabular points are called *Newton–Cotes* integrating formulas. If the end points a, b of the interval $[a, b]$ are included as tabular points and the remaining tabular points are contained strictly within the interval, the methods are termed *closed* Newton–Cotes formulas. If the end points are not tabular points and the tabular points are placed symmetrically† with respect to the end points, then we have *open* Newton–Cotes integrating formulas. The integrating formulas described above are clearly examples of closed Newton–Cotes formulas. The following questions will already have occurred to the reader: (i) What integrating formulas do we get if we allow the given tabular points to be unevenly spaced? (ii) What if we allow the *positions* of the tabular points to be themselves parameters of the integrating method? (iii) Need we necessarily restrict the tabular points to be contained within the interval $[a, b]$?

The use of these resulting formulas will depend, in general, upon the availability of function values $f(x_i)$. The choice of one (of the many) integrating method as opposed to another will depend upon its accuracy and relative ease of application in addition to the availability of the function values.

Exercises

10. Consider the interval $[a, b]$ divided into the intervals $a = x_0$, x_1, $x_2 = b$, where $x_1 - x_0 = h$, $x_2 - x_1 = 2h$, where $b - a = 3h$. Derive the integrating formula for $\int_a^b f(x)\, dx$ using tabular values of f at x_0, x_1, x_2 which is exact for polynomials $f(x)$ of degree up to and including those of degree 2. Are all the coefficients nonzero?

11. Repeat Exercise 10 with $a = x_1$, $x_2 - x_0 = 2h$, $x_2 - x_1 = h$, $x_2 = b$, where once again $(b - a)/3 = h$.

12. The schemes derived in Exercise 10 and 11 may be considered as special cases of the integrating formula

$$\int_a^b f(x)\, dx \doteq \alpha_0 f_0 + \alpha_1 f_1 + \alpha_2 f_2 + \alpha_3 f_3$$

where $f_i = f(x_i)$, $x_i = x_0 + ih$, $i = 0, 1, 2, 3$, $b - a = 3h$, where *one* of the α's is retained as a parameter instead of requiring the maximal accuracy as attained by the Simpson $\frac{3}{8}$-rule. Derive this parametrized method and comment on the form of the coefficients. What values of the remaining parameter produce the schemes of Exercises 10 and 11?

13. To approximate the definite integral $\int_a^b f(x)\, dx$ consider the integrating formula

$$\int_a^b f(x)\, dx \doteq \alpha_0 f_0 + \alpha_1 f(x_1) + \alpha_2 f_2$$

---

† This is tantamount to labeling the end points as $a = x_1$ and $b = x_{n-1}$ in our notation.

where α_0, α_1, and α_2 are parameters, $x_0 = a$, $x_2 = b$, and x_1 is an additional parameter, i.e. the position of the second tabular point is itself a parameter. Determine the form of the integrating formula if α_0, α_1, and α_2 are specified (in terms of x_1) to produce as high accuracy as possible with just three parameter values. What is the position of x_1 in order to achieve maximal accuracy? Is it unique? What is the resulting scheme?

14. Consider approximating the definite integral $\int_a^b f(x)\,dx$ by an integrating formula based on the points $x_0 = a - h$, $x_1 = a$, $x_2 = b$, where $b - a = h$. What is the precise form of the integrating formula?

In the remainder of this section we will indicate the forms of the error terms E_I in the simple cases of the rectangular rule and the trapezoidal rule. We will leave a general discussion of the error term until section 3. To derive the error term we require the *second mean value theorem*, which states that:

If $f(x)$ and $g(x)$ are continuous and integrable on the interval $a \le x \le b$ and if $g(x)$ is of constant sign in $a \le x \le b$, then

$$\int_a^b f(x)g(x)\,dx = f(\eta)\int_a^b g(x)\,dx$$

for some η satisfying $a < \eta < b$.

To derive E_I we will use the error term $E(x)$ obtained for the interpolating formulas in Chapter 4. Consider, with the notation of Chapter 4, the interpolating polynomial $y(x)$ when

$$E(x) = f(x) - y(x) \tag{15}$$

is given by Eq. (17) of Chapter 4. If we integrate Eq. (15) between the points a and b, we obtain

$$\int_a^b E(x)\,dx = -\int_a^b y(x)\,dx + \int_a^b f(x)\,dx$$

Substituting $E(x) = f^{(n+1)}(\xi)\pi_{n+1}(x)/(n+1)!$, where $a < \xi < b$, we obtain

$$\int_a^b f(x)\,dx = \int_a^b y(x)\,dx + \int_a^b f^{(n+1)}(\xi)\pi_{n+1}(x)/(n+1)!\,dx \tag{16}$$

Comparing Eq. (16) with the integrating formulas (5), (9), (13), (14), we see that

$$E_I = \int_a^b f^{(n+1)}(\xi)\pi_{n+1}(x)/(n+1)!\,dx \tag{17}$$

for the appropriate n, where we have assumed f has $(n+1)$ continuous derivatives on $[a, b]$. For $n = 0$, we obtain the error term for the rectangular rule (5), namely

$$E_I = \int_a^b f^{(1)}(\xi)\pi_1(x)\,dx = \int_a^b f^{(1)}(\xi)(x - x_0)\,dx$$

with $x_1 = b$, $x_0 = a$. In $[a, b]$, $(x - x_0)$ is of constant sign and hence using the second mean value theorem we have for the rectangular rule,

$$E_I = f^{(1)}(\eta) \int_a^b (x - a) \, dx = \frac{h^2}{2} f^{(1)}(\eta) \qquad a < \eta < b \qquad (18)$$

Similarly, for the trapezoidal rule,

$$E_I = \frac{1}{2} \int_a^b f^{(2)}(\xi) \pi_2(x) \, dx = \frac{1}{2} \int_a^b f^{(2)}(\xi)(x - x_0)(x - x_1) \, dx$$

where $x_0 = a$, $x_1 = b$. In $[a, b]$, $(x - x_0)(x - x_1)$ is of constant sign and hence we can again employ the second mean value theorem to obtain

$$E_I = f^{(2)}(\eta) \frac{h^3}{12} \qquad a < \eta < b \qquad (19)$$

Finally, for the Simpson's $\frac{1}{3}$-rule, we find

$$E_I = \int_a^b \frac{1}{6} f^{(3)}(\xi) \pi_3(x) \, dx = \int_a^b \frac{1}{6} f^{(3)}(\xi)(x - x_0)(x - x_1)(x - x_2) \, dx$$

where $x_0 = a$, $x_1 = a + h$, $x_2 = b$. However, $\pi_3(x)$ is no longer of constant sign in $[a, b]$ and consequently we cannot use the second mean value theorem directly; a more careful analysis is necessary—see section 3. However, see Exercise 22.

Example 4   As an illustration of the validity of the expressions for the error terms E_I let us evaluate the E_I for the rectangular rule and the trapezoidal rule for the integrals of Examples 2 and 3. From Eq. (18) we have

$$E_I = \frac{h^2}{2} f^{(1)}(\eta)$$

where $h = \frac{1}{2}$, and $0 < \eta < \frac{1}{2}$. Also $f^{(1)}(\eta) = \frac{1}{2}(1 + \eta)^{-1/2}$ and

$$E_I = \frac{1}{16} \times 1/(1 + \eta)^{1/2}$$

Since η can lie anywhere between 0 and $\frac{1}{2}$, our actual error should certainly be less than or equal to the maximum error attained by E_I. Now the maximum value of $1/(1 + \eta)^{1/2}$ for $0 \le \eta \le \frac{1}{2}$ occurs when $\eta = 0$, and whereas η lies strictly within the interval $(0, \frac{1}{2})$ in our case, for practical purposes we shall use the value attained at $\eta = 0$ as our maximum error, namely

$$E_I \le \frac{1}{16} = 0.0625$$

In Example 2 we find that the computational error is $0.558 - 0.5$, i.e. 0.058 which is certainly less than the maximum predicted error. Similarly, for the trapezoidal rule, we obtain from Eq. (19)

$$E_I \le \max \left\{ -\frac{h^3}{12} f^{(2)}(\eta) \right\} \qquad 0 < \eta < \frac{1}{2}$$

where $h = \frac{1}{2}$, $f^{(2)}(\eta) = -\frac{1}{4}(1 + \eta)^{-3/2}$, so that

$$E_I \leq \frac{1}{32 \times 12} = 0.002\ 604$$

The computational error in Example 3 is easily seen to be $0.558 - 0.556 = 0.002$, again less than the predicted maximum error.

Exercises

15. Modify the program of Exercise 5 for the rectangular rule to integrate a general function $f(x)$ over a given interval $[a, b]$.
16. Use the program of Exercise 15 to evaluate the definite integrals

$$\int_0^1 \sin(x)\,dx, \quad \int_{-\pi/2}^{\pi/2} \cos(x)\,dx, \quad \int_0^3 \exp(x)\,dx.$$

17. By evaluating the definite integrals of Exercise 16, compare the computed errors with the estimates given by using Eq. (18).
18. Modify the program of Exercise 5 for the trapezoidal rule to integrate a general function $f(x)$ over a given interval $[a, b]$.
19. Use your program of Exercise 18 to evaluate the integrals of Exercise 16. Compare the computed results with the exact integrals.
20. The rectangular rule given by Eq. (5) is just one possible integrating formula using one tabular point. For example, choosing $x_0 = b$ would produce another such rectangular rule with *a different error but the same error bound*. Consider the *position* of the tabular point as an additional parameter. By the method of undetermined coefficients determine the value of this parameter (i.e. the position of the single tabular point) in order that the resulting rectangular rule is exact for linear polynomials as well as constants. Verify your result by calculating the error term E_I directly from Eq. (17).
21. To propose a first-order integrating formula involving just one tabular point, we may as an alternative to that derived in Exercise 20, specify the function value and its first derivative at the given point $x_0 = a + \frac{1}{2}h$, $b - a = h$. Therefore we consider representing the definite integral $\int_a^b f(x)\,dx$ as $\alpha_0 f_0 + \alpha_1 f_0^{(1)}$, where α_0 and α_1 are to be determined. Show that this method is identical to that of Exercise 20 and hence using the Hermite interpolation error term given by Eq. (53) of Chapter 4, determine the error term E_I for this first-order method.
22. In a similar manner to Exercise 21, we may propose, as an alternative method of deriving the Simpson's $\frac{1}{3}$-rule given in Exercise 13, the integrating formula

$$\int_a^b f(x)\,dx = \alpha_0 f(x) + \alpha_1 f(x_0 + h) + \alpha f^{(1)}(x_0 + h) + \alpha_2 f(x_0 + 2h)$$

where $x_0 = a$, $x_2 = b$, and $b - a = 2h$. α_0, α_1, α_2, and α are to be determined. Using the method of undetermined coefficients to attain maximal accuracy, verify that this does indeed give rise to the Simpson's $\frac{1}{3}$-rule. Hence using the error term given by Eq. (53) of Chapter 4 determine the error term for Simpson's $\frac{1}{3}$-rule.

2 The Coefficients in the Newton–Cotes Formulas

Let the equally spaced tabular points be denoted by x_i, $i = 0, 1, 2, \ldots, n$, with $a = x_0 + ph$, $b = x_0 + qh$, $p \geq 0$, $q \leq n$, and p and q not necessarily integers. The interpolating integrating formula based on these $n+1$ points is then given by

$$I_{n+1}(f) = \int_a^b \sum_{j=0}^n L_j(x)f_j \, dx \tag{20}$$

where $L_j(x)$ are the Lagrangian polynomials given in Chapter 4.

Interchanging the order of summation and integration in Eq. (20), we obtain

$$I_{n+1}(f) = \sum_{j=0}^n \int_a^b L_j(x)f_j \, dx \tag{21}$$

Writing $x = x_0 + sh$, then using s as the independent variable,

$$I_{n+1}(f) = \sum_{j=0}^n \int_p^q \prod_{\substack{k=0 \\ k \neq j}}^n \frac{(x_0+sh)-(x_0+kh)}{(x_0+jh)-(x_0+kh)} f_j h \, ds$$

$$= h \sum_{j=0}^n \int_p^q \prod_{\substack{k=0 \\ k \neq j}}^n \frac{(s-k)}{(j-k)} f_j \, ds \tag{22}$$

Equation (22) may be reorganized to produce

$$I_{n+1}(f) = h \sum_{j=0}^n f_j \prod_{\substack{k=0 \\ k \neq j}}^n \frac{1}{(j-k)} \int_p^q \prod_{k=0}^n \frac{(s-k)}{(s-j)} \, ds \tag{23}$$

Hence the jth coefficient in the integrating formulas is given by

$$\alpha_j = h \frac{1}{\displaystyle\prod_{\substack{k=0 \\ k \neq j}}^n (j-k)} \int_p^q \prod_{k=0}^n \frac{s-k}{s-j} \, ds \tag{24}$$

Equation (24) may be simplified by noting that

$$\frac{1}{\displaystyle\prod_{\substack{k=0 \\ k \neq j}}^n (j-k)} = \frac{(-1)^{n-j}}{j!(n-j)!} = \frac{(-1)^{n-j}}{n!}\binom{n}{j}$$

Therefore the jth coefficient is

$$\alpha_j = h \frac{(-1)^{n-j}}{n!}\binom{n}{j} \int_p^q \frac{\displaystyle\prod_{k=0}^n (s-k)}{(s-j)} \, ds \quad j = 0, 1, \ldots, n \tag{25}$$

For large n these coefficients clearly involve some considerable manipulative effort for their evaluation. Consequently, for the Newton–Cotes formulas of closed type we tabulate the coefficients for $n = 1, 2, \ldots, 8$ in Table 1. In addition the open Newton–Cotes formulas' coefficients are shown in Table 2 for values of $n = 2, 3, \ldots, 7$. For the closed formulas it can be seen that the

Table 1 Coefficients α_j for closed Newton–Cotes formulas using $(n+1)$ points; $b - a = nh$

n	$j = 0$	1	2	3	4	5	6	7	8
1	$\dfrac{h}{2}$	$\dfrac{h}{2}$							
2	$\dfrac{h}{3}$	$\dfrac{4h}{3}$	$\dfrac{h}{3}$						
3	$\dfrac{3h}{8}$	$\dfrac{9h}{8}$	$\dfrac{9h}{8}$	$\dfrac{3h}{8}$					
4	$\dfrac{14h}{45}$	$\dfrac{64h}{45}$	$\dfrac{24h}{45}$	$\dfrac{64h}{45}$	$\dfrac{14h}{45}$				
5	$\dfrac{95h}{288}$	$\dfrac{375h}{288}$	$\dfrac{250h}{288}$	$\dfrac{250h}{288}$	$\dfrac{375h}{288}$	$\dfrac{95h}{288}$			
6	$\dfrac{41h}{140}$	$\dfrac{216h}{140}$	$\dfrac{27h}{140}$	$\dfrac{272h}{140}$	$\dfrac{27h}{140}$	$\dfrac{216h}{140}$	$\dfrac{41h}{140}$		
7	$\dfrac{5257h}{17\,280}$	$\dfrac{25\,039h}{17\,280}$	$\dfrac{9261h}{17\,280}$	$\dfrac{20\,923h}{17\,280}$	$\dfrac{20\,923h}{17\,280}$	$\dfrac{9261h}{17\,280}$	$\dfrac{25\,039h}{17\,280}$	$\dfrac{5257h}{17\,280}$	
8	$\dfrac{3956h}{14\,175}$	$\dfrac{23\,552h}{14\,175}$	$\dfrac{-3712h}{14\,175}$	$\dfrac{41\,984h}{14\,175}$	$\dfrac{-18\,160h}{14\,175}$	$\dfrac{41\,984h}{14\,175}$	$\dfrac{-3712h}{14\,175}$	$\dfrac{23\,552h}{14\,175}$	$\dfrac{3956h}{14\,175}$

Table 2 Coefficients α_j for open Newton–Cotes formulas using $n+1$ points; $b-a=nh$

n	$j=1$	2	3	4	5	6
2	$2h$					
3	$\dfrac{3h}{2}$	$\dfrac{3h}{2}$				
4	$\dfrac{8h}{3}$	$\dfrac{-4h}{3}$	$\dfrac{8h}{3}$			
5	$\dfrac{55h}{24}$	$\dfrac{5h}{24}$	$\dfrac{5h}{24}$	$\dfrac{55h}{24}$		
6	$\dfrac{33h}{10}$	$\dfrac{-42h}{10}$	$\dfrac{78h}{10}$	$\dfrac{-42h}{10}$	$\dfrac{33h}{10}$	
7	$\dfrac{4277h}{1440}$	$\dfrac{-3171h}{1440}$	$\dfrac{3934h}{1440}$	$\dfrac{3934h}{1440}$	$\dfrac{-3171h}{1440}$	$\dfrac{4277h}{1440}$

Note: A quick check on the correctness of the coefficients is made by summing and comparing the coefficients with the length of interval $[a, b]$.

coefficients α_j in the integrating formulas are all positive for $n \leq 7$.† However, for $n = 8$ we now obtain some negative coefficients. For cases where the coefficients are not all positive, the integrating formulas suffer a build-up of round-off error. Therefore in addition to the coefficients given by Eq. (25) being complicated to evaluate for large n, we have poor round-off properties and consequently Newton–Cotes formulas are seldom used for n larger than 9. (We describe an alternative approach to high-degree Newton–Cotes formulas in sections 4 and 5.) Similarly, in the case of the open Newton–Cotes formulas, larger values of n produce integrating formulas with poor round-off properties. In practice there is seldom any advantage in using open Newton–Cotes formulas in preference to closed formulas although the former do produce useful numerical schemes for the numerical solution of ordinary differential equations discussed in Chapter 11.

3. The Error Term E_I‡

We have derived the simple error terms in Eqs. (18) and (19) and Exercise 22 in the cases either where the second mean value theorem is directly applicable or where it could be applied after some careful manipulation. In this section we require to give the error terms E_I for general n, although bear in mind the comments of the previous section on the application of the Newton–Cotes formulas for large n.

† In fact it can be shown that all of the coefficients are positive only in the cases $n \leq 7$ and $n = 9$ for the closed formulas and for only $n = 3$ and 5 for the open formulas.
‡ The reader may omit the details of this section at first reading. The essential results are contained in Eqs. (41) and (48). Also see Tables 3 and 4.

Consider n even. For convenience we renumber the tabular points so that x_0 becomes the middle tabular point. We therefore have tabular points x_i, $i = 0$, $\pm 1, \ldots, \pm r$, where $2r = n$, where we assume $a = -mh$ and $b = mh$, m a positive integer. With this notation the integrating formulas over the equal intervals are given by

$$\int_a^b f(x)\,dx = h \int_{-m}^m \sum_{j=-r}^r \prod_{\substack{k=-r \\ k \neq j}}^r \left(\frac{s-k}{j-k}\right) f_j \, ds$$

$$+ h^{n+2} \int_{-m}^m \frac{f^{(n+1)}(\xi)}{(n+1)!} \prod_{k=-r}^r (s-k)\,ds \qquad (26)$$

where $\xi = \xi(s)$. Hence

$$E_I = h^{n+2} \int_{-m}^m \frac{f^{(n+1)}(\xi)}{(n+1)!} \prod_{k=-r}^r (s-k)\,ds \qquad (27)$$

To simplify E_I we will integrate by parts. To do this, consider first the function

$$\phi(s) = \int_{-m}^s \prod_{k=-r}^r (t-k)\,dt$$

where m is a positive integer. Clearly $\phi(-m) = 0$. Also,

$$\phi(m) = \int_{-m}^0 \prod_{k=-r}^r (t-k)\,dt + \int_0^m \prod_{k=-r}^r (t-k)\,dt \qquad (28)$$

Since n is even we write Eq. (28) as

$$\phi(m) = \int_{-m}^0 t \prod_{k=1}^r (t^2 - k^2)\,dt - \int_0^m (-t) \prod_{k=1}^r ((-t)^2 - k^2)\,dt \qquad (29)$$

Introducing a new independent variable $t' = -t$ in the second integral of Eq. (29), we find

$$\phi(m) = \int_{-m}^0 t \prod_{k=1}^r (t^2 - k^2)\,dt + \int_0^{-m} t' \prod_{k=1}^r (t^2 - k^2)\,dt' \qquad (30)$$

$$= \int_{-m}^0 t \prod_{k=1}^r (t^2 - k^2)\,dt - \int_{-m}^0 t \prod_{k=1}^r (t^2 - k^2)\,dt = 0 \qquad (31)$$

i.e. $\phi(m) = \phi(-m) = 0$.

We next require to prove that $\phi(s)$ does not vanish in the interval $(-m, m)$. To do this, let

$$I_j = \int_j^{j+1} \prod_{k=-r}^r (t-k)\,dt \qquad j = -m, -m+1, \ldots, m-1$$

Now

$$I_{-j} = \int_{-j}^{-j+1} \prod_{k=-r}^r (t-k)\,dt = \int_{-j}^{-j+1} t \prod_{k=1}^r (t^2 - k^2)\,dt$$

Following similar steps carried out in Eqs. (30) and (31), we find

$$I_{-j} = -I_{j-1} \tag{32}$$

That is, symmetrically placed intervals on either side of x_0 have integrals associated with them of equal magnitude but of opposite sign. Furthermore,

$$I_{j-1} = \int_{j-1}^{j} \prod_{k=-r}^{r} (t-k)\, dt$$

Introduce the change of variable $t' = t+1$ so that

$$I_{j-1} = \int_{j}^{j+1} \prod_{k=-r}^{r} ((t'-1)-k)\, dt'$$

$$= \int_{j}^{j+1} \prod_{k=-r}^{r} (t-k-1)\, dt$$

$$= \int_{j}^{j+1} \prod_{k'=-r+1}^{r+1} (t-k')\, dt$$

where we have used the obvious change of notation $k' = k+1$. It then easily follows that

$$I_{j-1} = \int_{j}^{j+1} \frac{(t-r-1)}{(t+r)} \prod_{k=-r}^{r} (t-k)\, dt$$

and so we have, on using the second mean value theorem on the interval $[j, j+1]$,

$$I_{j-1} = \frac{(\xi-r-1)}{(\xi+r)} I_j \qquad \text{with } j < \xi < j+1$$

since

$$\prod_{k=-r}^{r} (t-k) \text{ is of one sign on } [j, j+1]$$

Therefore $|I_{j-1}| < |I_j|$ if $|(\xi-r-1)/(\xi+r)| < 1$, which is clearly the case. We have therefore proved:

Theorem 1 *For the tabular points x_i, $i = 0, \pm 1, \pm 2, \pm 3, \ldots, \pm r$, with $2r = n$, then $\int_{-m}^{m} \prod_{k=-r}^{r} (t-k)\, dt$ does not vanish in the interval $(-m, m)$ and takes the value zero at both end points.*

In the ensuing integration by parts we will require to evaluate $(d/ds)f^{(n+1)}(\xi(s))$. Hence we will determine this expression in the following theorem.

Theorem 2 *Let the error terms in the Lagrangian interpolating formulas on the $(n+1)$ equally spaced points x_i, $i = 0, \pm 1, \pm 2, \ldots, \pm r$, $2r = n$, be given by*

$$\frac{h^{n+1}}{(n+1)!} f^{(n+1)}(\xi(s)) \prod_{k=-r}^{r} (s-k)$$

where $x = x_0 + sh$; $s \in (-r, r)$, $\xi \in (-r, -r+1, \ldots, r, s)$. Then if $f^{(n+2)}(s)$ is continuous,

$$\frac{h^{n+1}}{(n+1)!} \frac{d}{ds} f^{(n+1)}(\xi) = \frac{h^{n+2}}{(n+2)!} f^{(n+2)}(\eta) \qquad \eta \in (-r, -r+1, \ldots, r, s)$$

Proof[†] Consider the Lagrangian interpolating formula on the $(n+1)$ equally spaced points x_i, $i = 0, \pm 1, \pm 2, \ldots, \pm r$, $n = 2r$, given by

$$f(x_0 + sh) = \sum_{j=-r}^{r} \prod_{\substack{k=-r \\ k \neq j}}^{r} \left(\frac{s-k}{j-k}\right) f_j$$

$$+ \frac{h^{n+1}}{(n+1)!} f^{(n+1)}(\xi) \prod_{k=-r}^{r} (s-k) \qquad \xi \in (-r, \ldots, r, s) \qquad (33)$$

where $s \in (-r, r)$ and we assume for the moment that $s \neq j$, $j = 0, \pm 1, \ldots, \pm r$. Denote $\prod_{k=-r}^{r}(s-k)$ as $\Pi_n(s)$. Divide both sides of (33) by $\Pi_n(s)$ so that

$$\frac{f(x_0 + sh)}{\Pi_n(s)} = \sum_{j=-r}^{r} \frac{f_j}{\Pi_{n,j}(j)(s-j)} + \frac{h^{n+1}}{(n+1)!} f^{(n+1)}(\xi) \qquad (34)$$

where we denote $\Pi_{n,j}(j) = \prod_{k=-r, k \neq j}^{r}(j-k)$. If we now differentiate both sides of Eq. (34) with respect to s, we obtain

$$\frac{d}{ds}\left(\frac{f(x_0 + sh)}{\Pi_n(s)}\right) = \sum_{j=-r}^{r} \frac{f_j}{-\Pi_{n,j}(j)(s-j)^2} + \frac{h^{n+1}}{(n+1)!} \frac{d}{ds} f^{(n+1)}(\xi) \qquad (35)$$

Now consider Eq. (33) with n replaced by $n+1$ where for convenience we assume the equally spaced points are x_i, $i = 0, \pm 1, \pm 2, \ldots, \pm r, r+1$. Then

$$f(x_0 + sh) = \sum_{j=-r}^{r+1} \prod_{\substack{k=-r \\ k \neq j}}^{r+1} \left(\frac{s-k}{j-k}\right) f_j + \frac{h^{n+2}}{(n+2)!} f^{(n+2)}(\xi) \prod_{k=-r}^{r+1}(s-k) \qquad (36)$$

$$\xi \in (-r, -r+1, \ldots, r+1, s)$$

Divide both sides of (36) by $\Pi_{n+1}(s)$ so that

$$\frac{f(x_0 + sh)}{\Pi_{n+1}(s)} = \prod_{\substack{k=-r \\ k \neq r+1}}^{r+1} \left(\frac{s-k}{r+1-k}\right) \Big/ \Pi_{n+1}(s) f_{r+1}$$

$$+ \sum_{j=-r}^{r} \prod_{\substack{k=-r \\ k \neq j}}^{r+1} \left(\frac{s-k}{j-k}\right) \Big/ \Pi_{n+1}(s) f_j + \frac{h^{n+2}}{(n+2)!} f^{(n+2)}(\xi)$$

$$\therefore \quad \frac{f(x_0 + sh)}{\Pi_{n+1}(s)} = \frac{f_{r+1}}{(s-(r+1))\Pi_{n+1}(r+1)}$$

$$+ \sum_{j=-r}^{r} \frac{f_i}{\Pi_{n+1,j}(j)(s-j)} + \frac{h^{n+2}}{(n+2)!} f^{(n+2)}(\xi) \qquad (37)$$

[†] The proof is essentially that given in Ralston [34].

It is easily shown (see Exercise 23) that

$$\Pi_{n+1}(s) = (s - (r+1)\Pi_n(s)$$

$$\Pi_{n+1}(r+1) = \Pi_n(r+1) \tag{38}$$

$$\Pi_{n+1,j}(j) = (j - (r+1))\Pi_{n,j}(j) \qquad j \neq r+1, j = -r, -r+1, \ldots, r$$

Therefore using relations (38) and (37) and rearranging, we obtain

$$\left(\frac{f(x_0 + sh)}{\Pi_n(s)} - \frac{f_{r+1}}{\Pi_n(r+1)}\right) \Big/ (s - (r+1))$$

$$= \sum_{j=-r}^{r} \frac{f_j}{(j - (r+1)(s-j)\Pi_{n,j}(j)} + \frac{h^{n+2}}{(n+2)!} f^{(n+2)}(\xi)$$

Now take the limit as $r + 1 \to s$. Hence

$$\frac{d}{ds}\left(\frac{f(x_0 + sh)}{\Pi_n(s)}\right) = \sum_{j=-r}^{r} \frac{f_j}{-(s-j)^2\Pi_{n,j}(j)} + \frac{h^{n+2}}{(n+2)!} f^{(n+2)}(\eta) \qquad \eta \in (-r, \ldots, r, s)$$

$$\tag{39}$$

On comparing Eqs. (39) and (35) we see that

$$\frac{h^{n+1}}{(n+1)!} \frac{d}{ds} f^{(n+1)}(\xi) = \frac{h^{n+2}}{(n+2)!} f^{(n+2)}(\eta)$$

as was required. By continuity, the result is also true if $s = j; j = 0, \pm 1, \ldots, \pm r$.

We may now proceed to obtain the required error bound using Theorems 1 and 2. Integrating Eq. (27) by parts, we obtain

$$E_I = \frac{h^{n+2}}{(n+1)!} f^{(n+1)}(\xi)\phi(s)\Big|_{-m}^{m} - \frac{h^{n+2}}{(n+1)!} \int_{-m}^{m} \phi(s) \frac{d}{ds} f^{(n+1)}(\xi) \, ds \tag{40}$$

The first term on the right-hand side of Eq. (40) is zero by Theorem 1. Moreover, since $\phi(s)$ is zero at $s = -m$, m, and never zero within the interval $(-m, m)$ i.e. it is of one sign in $(-m, m)$, we may now employ the second mean value theorem to the integral in Eq. (40) with the result

$$E_I = \frac{-h^{n+2}}{(n+1)!} \frac{d}{ds} f^{(n+1)}(\zeta) \int_{-m}^{m} \phi(s) \, ds \qquad \zeta \in (a, b)$$

Using Theorem 2, this is equivalent to

$$E_I = \frac{-h^{n+3}}{(n+2)!} f^{(n+2)}(\eta) \int_{-m}^{m} \phi(s) \, ds \qquad \eta \in (a, b)$$

Employing integration by parts once more, we obtain

$$E_I = \frac{-h^{n+3}}{(n+2)!} f^{(n+2)}(\eta)\left\{ s\phi(s)\Big|_{-m}^{m} - \int_{-m}^{m} s \frac{d}{ds}\left\{ \int_{-m}^{s} \prod_{k=-r}^{r} (t-k) \, dt \right\} ds \right\}$$

so that

$$E_I = \frac{h^{n+3}}{(n+2)!} f^{(n+2)}(\eta) \int_{-m}^{m} s \prod_{k=-r}^{r} (s-k) \, ds$$

This may be further simplified in an obvious manner so that

$$E_I = \frac{2h^{n+3}}{(n+2)!} f^{(n+2)}(\eta) \int_{0}^{m} \prod_{k=0}^{r} (s^2-k^2) \, ds \qquad (41)$$

is the required error bound for the Newton–Cotes formulas with an odd number of tabular points. We see straight away because of the $(n+2)$nd derivative in Eq. (41) that such integrating formulas are exact for a polynomial f of degree up to and including $(n+1)$.

Example 5 Consider the error term (41) for the closed Newton–Cotes formula with $n = 2$, namely Simpson's $\frac{1}{3}$-rule:

$$E_I = \frac{2h^5}{24} f^{(4)}(\eta) \int_{0}^{1} \prod_{k=0}^{1} (s^2-k^2) \, ds = \frac{-h^5}{90} f^{(4)}(\eta)$$

This should correspond to the readers' answer in Exercise 22.

The closed formula with $n = 4$ has an error term

$$E_I = \frac{2h^6}{6!} f^{(6)}(\eta) \int_{0}^{2} \prod_{k=0}^{2} (s^2-k^2) \, ds = \frac{2h^6}{6!} f^{(6)}(\eta) \left[\frac{s^7}{7} - s^5 + 4s^3 \right]_{0}^{2}$$

$$= -\tfrac{8}{945} h^7 f^{(6)}(\eta)$$

We now consider n odd (an even number of tabular points). For convenience we renumber the tabular points as $(2k-1)/2$, $k = -r+1, \ldots, r$, with $n = 2r-1$ (see Figure 3).

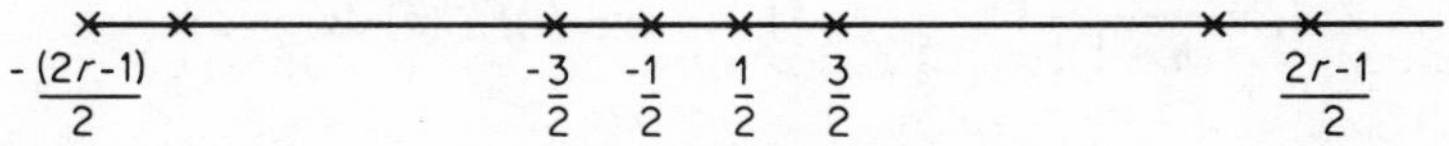

Figure 3

We assume the end points a and b are again coincident with the tabular points and let $a = -(m+\frac{1}{2})h$, $b = (m+\frac{1}{2})h$, where m is a positive integer. Using the Lagrangian interpolating formula of Chapter 4, our integrating formula becomes, using the present notation,

$$\int_{a}^{b} f(x) \, dx = h \sum_{j=-(2r-1)/2}^{(2r-1)/2} \int_{-m-\frac{1}{2}}^{m+\frac{1}{2}} \prod_{\substack{k=-(2r-1)/2 \\ k \neq j}}^{(2r-1)/2} \left(\frac{s-k}{j-k} \right) f_j \, ds$$

$$+ \frac{h^{n+2}}{(n+1)!} \int_{-m-\frac{1}{2}}^{m+\frac{1}{2}} f^{(n+1)}(\xi) \prod_{k=-(2r-1)/2}^{(2r-1)/2} (s-k) \, ds$$

where ξ is contained within the interval spanned by $-(2r-1)/2, \ldots,$ $(2r-1)/2$, and s. To obtain the error bound, we simplify the error term

$$E_I = \frac{h^{n+2}}{(n+1)!} \int_{-m-\frac{1}{2}}^{m-\frac{1}{2}} f^{(n+1)}(\xi) \prod_{k=-(2r-1)/2}^{(2r-1)/2} (s-k)\,ds$$

$$+ \frac{h^{n+2}}{(n+1)!} \int_{m-\frac{1}{2}}^{m+\frac{1}{2}} f^{(n+1)}(\xi) \prod_{k=-(2r-1)/2}^{(2r-1)/2} (s-k)\,ds$$

In the latter integral,

$$\prod_{k=-(2r-1)/2}^{(2r-1)/2} (s-k)$$

does not change sign in the interval $(m-\frac{1}{2}, m+\frac{1}{2})$ and consequently we may employ the second mean value theorem directly. Therefore

$$E_I = \frac{h^{n+2}}{(n+1)!} \int_{-m-\frac{1}{2}}^{m-\frac{1}{2}} f^{(n+1)}(\xi) \prod_{k=-(2r-1)/2}^{(2r-1)/2} (s-k)\,ds$$

$$+ \frac{h^{n+2}}{(n+1)!} f^{(n+1)}(\eta) \int_{m-\frac{1}{2}}^{m+\frac{1}{2}} \prod_{k=-(2r-1)/2}^{(2r-1)/2} (s-k)\,ds$$

which we denote by $I_1 + I_2$ using an obvious notation.

By rearrangement, we may write I_1 as

$$I_1 = \frac{h^{n+2}}{(n+1)!} \int_{-m-\frac{1}{2}}^{m-\frac{1}{2}} (s-r+\tfrac{1}{2}) f^{(n+1)}(\xi) \prod_{k=-(2r-3)/2}^{(2r-1)/2} (s-k)\,ds$$

But by the first mean value theorem we may write this as

$$I_1 = \frac{h^{n+2}}{(n+1)!} \int_{-m-\frac{1}{2}}^{m-\frac{1}{2}} \prod_{k=-(2r-3)/2}^{(2r-1)/2} (s-k)f^{(n)}(s)\,ds$$

$$- \frac{h^{n+2}}{(n+1)!} f^{(n)}(r-\tfrac{1}{2}) \int_{-m-\frac{1}{2}}^{m-\frac{1}{2}} \prod_{k=-(2r-3)/2}^{(2r-1)/2} (s-k)\,ds$$

Denote the second integral in this last equation by I_3. By introducing the obvious change of variable, I_3 may be written as

$$I_3 = \frac{h^{n+2}}{(n+1)!} f^{(n)}(r-\tfrac{1}{2}) \int_{-m}^{m} \prod_{k=-r+1}^{r-1} (s-k)\,ds$$

Consider the product in I_3 to be written as $s \prod_{k=1}^{r-1} (s^2-k^2)$. Therefore

$$I_3 = \frac{h^{n+2}}{(n+1)!} f^{(n)}(r-\tfrac{1}{2}) \left[\int_{-m}^{0} s \prod_{k=1}^{r-1} (s^2-k^2)\,ds + \int_{0}^{m} s \prod_{k=1}^{r-1} (s^2-k^2)\,ds \right]$$

$$= \frac{h^{n+2}}{(n+1)!} f^{(n)}(r-\tfrac{1}{2}) \left[\int_{0}^{-m} -s \prod_{k=1}^{r-1} (s^2-k^2)\,ds + \int_{0}^{m} s \prod_{k=1}^{r-1} (s^2-k^2)\,ds \right]$$

which is clearly zero. Hence

$$E_I = \frac{h^{n+2}}{(n+1)!} \int_{-m-\frac{1}{2}}^{m-\frac{1}{2}} \prod_{k=-(2r-3)/2}^{(2r-1)/2} (s-k) f^{(n)}(s)\,ds$$

$$+ \frac{h^{n+2}}{(n+1)!} f^{(n+1)}(\eta) \int_{m-\frac{1}{2}}^{m+\frac{1}{2}} \prod_{k=-(2r-1)/2}^{(2r-1)/2} (s-k)\,ds$$

A simple transformation in the first integral above allows us to write this as

$$E_I = \frac{h^{n+2}}{(n+1)!} \int_{-m}^{m} f^{(n)}(s-\tfrac{1}{2}) \prod_{k=-r+1}^{r-1} (s-k)\,ds$$

$$+ \frac{h^{n+2}}{(n+1)!} f^{(n+1)}(\eta) \int_{m-\frac{1}{2}}^{m+\frac{1}{2}} \prod_{k=-(2r-1)/2}^{(2r-1)/2} (s-k)\,ds$$

where the first integral is precisely of the form considered in the error term for the case n even, Eq. (27). We may therefore carry through precisely the steps described for n even (except we no longer require Theorem 2—why?). We will omit the details. We then find

$$E_I = \frac{2h^{n+2}}{(n+1)!} f^{(n+1)}(\xi) \int_0^m s \prod_{k=-r+1}^{r-1} (s-k)\,ds$$

$$+ \frac{h^{n+2}}{(n+1)!} f^{(n+1)}(\eta) \int_{m-\frac{1}{2}}^{m+\frac{1}{2}} \prod_{k=-(2r-1)/2}^{(2r-1)/2} (s-k)\,ds \tag{42}$$

We now endeavor to combine these two integrals.

First consider

$$\frac{d}{dt} \int_0^t s \prod_{k=-r+1}^{r-1} (s-k)\,ds = t \prod_{k=-r+1}^{r-1} (t-k) \tag{43}$$

Now the right-hand side of (43) can be written as

$$\frac{1}{(2r+1)} \prod_{k=-r+1}^{r-1} (2r+1)t(t-k)$$

$$= \frac{1}{(2r+1)} \prod_{k=-r+1}^{r-1} [(r+\tfrac{1}{2})t + (r+\tfrac{1}{2})t](t-k)$$

$$= \frac{1}{(2r+1)} \prod_{k=-r+1}^{r-1} [(t+\tfrac{1}{2})(t+r) - (t-\tfrac{1}{2})(t-r)](t-k)$$

$$= \frac{1}{(2r+1)} \left[(t+\tfrac{1}{2}) \prod_{k=-r}^{r-1} (t-k) - (t-\tfrac{1}{2}) \prod_{k=-r+1}^{r} (t-k) \right]$$

$$= \frac{1}{(2r+1)} \left[(t+\tfrac{1}{2}) \prod_{k=-(2r-1)/2}^{(2r-1)/2} \{t-(k-\tfrac{1}{2})\} - (t-\tfrac{1}{2}) \prod_{k=-(2r-1)/2}^{(2r-1)/2} \{t-(\tfrac{1}{2}+k)\} \right]$$

$$= \frac{1}{(2r+1)} \left[(t+\tfrac{1}{2}) \prod_{k=-(2r-1)/2}^{(2r-1)/2} \{(t+\tfrac{1}{2})-k\} - (t-\tfrac{1}{2}) \prod_{k=-(2r-1)/2}^{(2r-1)/2} \{(t-\tfrac{1}{2})-k\} \right]$$

$$= \frac{1}{(2r+1)} \frac{d}{dt} \int_{t-\frac{1}{2}}^{t+\frac{1}{2}} s \prod_{k=-(2r-1)/2}^{(2r-1)/2} (s-k)\,ds \tag{44}$$

Combining (43) and (44) we have

$$\int_0^t s \prod_{k=-r+1}^{r-1} (s-k)\,ds = \frac{1}{(2r+1)} \int_{t-\frac{1}{2}}^{t+\frac{1}{2}} s \prod_{k=-(2r-1)/2}^{(2r-1)/2} (s-k)\,ds + c$$

where c is a constant of integration. It is easy to see that in fact $c=0$ since both integrals reduce to zero at $t=0$. Therefore, we use the last result with $t=m$ in (42), with the result

$$E_I = \frac{2h^{n+2}}{(n+2)!} f^{(n+1)}(\xi) \int_{m-\frac{1}{2}}^{m+\frac{1}{2}} s \prod_{k=-(2r-1)/2}^{(2r-1)/2} (s-k)\,ds$$

$$+ \frac{h^{n+2}}{(n+1)!} f^{(n+1)}(\eta) \int_{m-\frac{1}{2}}^{m+\frac{1}{2}} \prod_{k=-(2r-1)/2}^{(2r-1)/2} (s-k)\,ds \qquad (45)$$

But the integrals of (45) are of the same sign so that we may write

$$E_I = \frac{2h^{n+2}}{(n+2)!} f^{(n+1)}(\xi) \left[\int_{m-\frac{1}{2}}^{m+\frac{1}{2}} \left\{ s \prod_{k=-(2r-1)/2}^{(2r-1)/2} (s-k) \right. \right.$$

$$\left. \left. + \frac{(n+2)}{2} \prod_{k=-(2r-1)/2}^{(2r-1)/2} (s-k) \right\} ds \right]$$

where

$$\xi \in \left(-\frac{(2r-1)}{2}, \ldots, \frac{(2r-1)}{2}, s \right)$$

Hence

$$E_I = \frac{2h^{n+2}}{(n+2)!} f^{(n+1)}(\xi) \left[\int_{m-\frac{1}{2}}^{m+\frac{1}{2}} \prod_{k=-(2r-1)/2}^{(2r-1)/2} (s-k)\left(s+\frac{(n+2)}{2}\right) ds \right]$$

$$= \frac{2h^{n+2}}{(n+2)!} f^{(n+1)}(\xi) \int_{m-\frac{1}{2}}^{m+\frac{1}{2}} \prod_{k=-(2r-1)/2}^{(2r-1)/2} (s-k)(s+r+\tfrac{1}{2})\,ds$$

$$= \frac{2h^{n+2}}{(n+2)!} f^{(n+1)}(\xi) \int_{m-\frac{1}{2}}^{m+\frac{1}{2}} \prod_{k=-(2r-1)/2}^{(2r+1)/2} (s-k)\,ds \qquad (46)$$

which on rearrangement may be written as

$$E_I = \frac{2h^{n+2}}{(n+2)!} f^{(n+1)}(\xi) \int_m^{m+1} \prod_{k=-r}^{r} (s-k)\,ds$$

It may easily be shown (see Exercise 24) that

$$\int_t^{t+1} \prod_{k=-r}^{r} (s-k)\,ds = (2r+1) \int_0^{t+\frac{1}{2}} \prod_{k=-(2r-1)/2}^{(2r-1)/2} (s-k)\,ds \qquad (47)$$

Therefore, substituting for this expression in (46) we obtain

$$E_I = \frac{2h^{n+2}}{(n+1)!} f^{(n+1)}(\xi) \int_0^{m+\frac{1}{2}} \prod_{k=-(2r-1)/2}^{(2r-1)/2} (s-k)\,ds \qquad (48)$$

which is the required form of the error bound for the Newton–Cotes formulas, n odd.

Example 6   The error bound for the trapezoidal rule is given by Eq. (48) with $n = 1$, $r = 1$, $m = 0$. Hence

$$E_I = \frac{2h^3}{2} f^{(2)}(\xi) \int_0^{\frac{1}{2}} (s^2 - \tfrac{1}{4}) \, ds = \frac{-h^3}{12} f^{(2)}(\xi) \qquad \xi \in (a, b)$$

(compare Eq. (46)).

The Simpson's $\frac{3}{8}$-rule is similarly found from Eq. (48) with $n = 3$, $r = 2$, $m = 1$. Hence

$$E_I = \frac{2h^5}{4!} f^{(4)}(\xi) \int_0^{3/2} \prod_{k=-3/2}^{3/2} (s - k) \, ds$$

$$= \frac{h^5}{12} f^{(4)}(\xi) \left[\frac{s^5}{5} - \frac{5s^3}{6} + \frac{9}{16} s \right]_0^{3/2} = -\frac{3h^5}{80} f^{(4)}(\xi)$$

For completeness we give tables of the error terms for the Newton–Cotes closed and open formulas—see Tables 3 and 4 respectively.

n	n odd	n even
	Table 3 Error terms	
1	$-\dfrac{1}{12} h^3 f^{(2)}(\xi)$	
2		$-\dfrac{1}{90} h^5 f^{(4)}(\xi)$
3	$-\dfrac{3}{80} h^5 f^{(4)}(\xi)$	
4		$-\dfrac{8}{945} h^7 f^{(6)}(\xi)$
5	$-\dfrac{275}{12\,096} h^7 f^{(6)}(\xi)$	
6		$-\dfrac{9}{1400} h^9 f^{(8)}(\xi)$
7	$-\dfrac{8183}{518\,400} h^9 f^{(8)}(\xi)$	
8		$-\dfrac{2368}{467\,775} h^{11} f^{(10)}(\xi)$

Table 4   Error terms

n	n odd	n even
2		$\dfrac{1}{3}h^3f^{(2)}(\xi)$
3	$\dfrac{3}{4}h^3f^{(2)}(\xi)$	
4		$\dfrac{14}{45}h^5f^{(4)}(\xi)$
5	$\dfrac{95}{144}h^5f^{(4)}(\xi)$	
6		$\dfrac{41}{140}h^7f^{(6)}(\xi)$
7	$\dfrac{5257}{8640}h^7f^{(6)}(\xi)$	

Exercises

23. Verify relations (38).
24. Verify Eq. (47).

4 Composite Rules

In previous sections we have proposed dividing our interval of integration into a certain number of equal intervals. In terms of this subdivision an integrating formula of a certain accuracy is obtained. If a higher order of accuracy is required then we would need to introduce new points upon which the new increased accuracy method is based. Clearly there are certain drawbacks which occur. First, the introduction of an additional single point would require the complete recalculation of the coefficients in our integrating method.† Second, to obtain a specified accuracy, a high-order integrating formula might be necessary which has its obvious drawbacks in terms of knowing the bound on the high-order derivatives as well as the round-off properties of the method. Consequently, we ask whether we may obtain the required precision without the necessity of introducing high-order methods. We see that this *can* be done in considering the error terms *if h* can be made small enough. This is therefore tantamount to suggesting subdividing the interval $[a, b]$ into several (many) subintervals, upon each of which piecewise integrating formulas are proposed. Therefore let us subdivide our interval $[a, b]$ into n subintervals $[a_i, a_{i+1}]$, $i = 0, 1, \ldots, n-1$, with $a_0 = a$, $a_n = b$. We

† Compare this with the addition of a single point in the Lagrangian interpolating formula— Chapter 3.

then propose considering the definite integral written in the form

$$\int_a^b f(x)\,dx = \sum_{i=0}^{n-1} \int_{a_i}^{a_{i+1}} f(x)\,dx \tag{49}$$

For each of these subintervals $[a_i, a_{i+1}]$ we can define our integrating formula with its corresponding error term defined in terms of the derivative of f evaluated within that subinterval, and the h appearing in the error term referring to the further subdivision of the intervals $[a_i, a_{i+1}]$. Thus consider the rectangular rule (5) defined on $[a_i, a_{i+1}]$:

$$\int_{a_i}^{a_{i+1}} f(x)\,dx = hf_i + \frac{h^2}{2} f^{(1)}(\eta_i) \qquad a_i < \eta_i < a_{i+1} \tag{50}$$

and where $a_{i+1} - a_i = h$. Substituting Eq. (50) into (49) and then summing over i, we find

$$\int_a^b f(x)\,dx = h \sum_{i=0}^{n-1} f_i + \frac{h^2}{2} \sum_{i=0}^{n-1} f^{(1)}(\eta_i) \tag{51}$$

Since we assume continuity of the derivatives of f in the interval $a < x < b$, there exists a number η satisfying $a < \eta < b$ such that

$$\sum_{i=0}^{n-1} f^{(1)}(\eta_i) = nf^{(1)}(\eta)$$

and hence Eq. (51) becomes

$$\int_a^b f(x)\,dx = h \sum_{i=0}^{n-1} f_i + (b-a)\frac{h}{2} f^{(1)}(\eta) \qquad a < \eta < b \tag{52}$$

Equation (52) is called the *composite rectangular rule*.

From its formation, the composite rule (52) can attain a high accuracy by taking h small enough, provided, of course, we know the corresponding function values f_i. Also, none of the drawbacks mentioned at the beginning of this section are present in Eq. (52).

Instead of using the rectangular rule, let us now substitute the trapezoidal rule (9) into equation (49); we then obtain

$$\int_a^b f(x)\,dx = \frac{h}{2} \sum_{i=0}^{n-1} (f_i + f_{i+1}) - \frac{h^3}{12} \sum_{i=0}^{n-1} f^{(2)}(\eta_i) \qquad a_i < \eta_i < a_{i+1} \tag{53}$$

Again using continuity of the derivatives, Eq. (53) can be simplified to

$$\int_a^b f(x)\,dx = h\left[\tfrac{1}{2}(f_0 + f_n) + \sum_{i=1}^{n-1} f_i\right] - \frac{(b-a)h^2}{12} f^{(2)}(\eta) \qquad a < \eta < b \tag{54}$$

Equation (54) is the *composite trapezoidal rule*. It is clear that this latter formula would be preferable to (52) when evaluating a given integral on a computer. Equation (54) is only slightly more complicated to program than Eq. (52) whilst the gain in accuracy is an order of h. Further, the number of subdivisions in the trapezoidal rule is the same as the rectangular rule.

In proposing a composite Simpson's $\frac{1}{3}$-rule, the interval $[a, b]$ is divided into n subintervals of length $2h$ where $2h = (b - a)/n$.

The integrating formula is

$$\int_a^b f(x)\,\mathrm{d}x = \frac{h}{3} \sum_{i=0}^{n-1} (f_{2i} + 4f_{2i+1} + f_{2i+2})$$

$$-\frac{h^5}{90} \sum_{i=0}^{n-1} f^{(4)}(\eta_i) \qquad a_{2i} < \eta_i < a_{2i+2}$$

where $f_j = f(a + jh)$

$$\int_a^b f(x)\,\mathrm{d}x = \frac{h}{3}\left[f_0 + f_{2n} + 2 \sum_{i=1}^{n-1} f_{2i} + 4 \sum_{i=0}^{n-1} f_{2i+1} \right]$$

$$-\frac{(b-a)h^4}{180} f^{(4)}(\eta) \qquad a < \eta < b \tag{55}$$

Finally, the composite Simpson's $\frac{3}{8}$-rule can be shown to be

$$\int_a^b f(x)\,\mathrm{d}x = \frac{3h}{8}\left[f_0 + f_{3n} + 2 \sum_{i=1}^{n-1} f_{3i} + 3 \sum_{i=0}^{n-1} f_{3i+1} + 3 \sum_{i=0}^{n-1} f_{3i+2} \right]$$

$$-\frac{3(b-a)h^4}{80} f^{(4)}(\eta) \tag{56}$$

where $h = (b - a)/3n$.

Clearly these formulas (55) and (56) have coefficients which are slightly more complicated than the composite rectangular and trapezoidal rules. However, they clearly gain in terms of the higher order of accuracy which they possess.

This process of composition clearly can be applied for any of the methods defined by the coefficients of section 2. Furthermore, if the need arises, the interval $[a, b]$ can be subdivided into intervals of differing lengths in each of which a further subdivision with equal subintervals is used. For example, the interval $[0, 1]$ could be divided into the subintervals $[0, \frac{1}{4}]$, $[\frac{1}{4}, \frac{3}{4}]$ and $[\frac{3}{4}, 1]$ and (say) Simpson's $\frac{1}{3}$-rule proposed on each subinterval thereby necessitating a subdivision of $h = \frac{1}{12}$ in the first and last intervals and $h = 1/6$ in the middle interval. Such an approach would clearly be of use where the integrand had a derivative $(f^{(4)}(\eta)$ in this case) which varied rapidly in some part of the region but was better behaved in the remainder (see Example 8).

Example 7 The effect of the composite rules on the accuracy of the results obtained from an integration formula can be easily seen by evaluating the integral of Example 2 using two applications of the rectangular rule and comparing the answer with just one application of the rectangular rule. The effect can also be seen by evaluating the integral with two applications of the trapezoidal rule and comparing the result with the previous answer derived using one application of the trapezoidal rule. In both the composite rectangular (two applications) and the composite trapezoidal (two applications) rules, a value of $(1+x)^{1/2}$ at $x = \frac{1}{4}$ is required.

For the composite rectangular rule we obtain

$$I_R = [(\tfrac{1}{2})/2][(1)^{1/2} + (1.25)^{1/2}] = 0.5295$$

which is considerably better than the previously obtained 0.5. For the composite trapezoidal rule we obtain

$$I_T = [(\tfrac{1}{2})/4][(1)^{1/2} + 2 \cdot (1.25)^{1/2} + (1.5)^{1/2}] = 0.5576$$

which is correct to three decimal places.

Example 8   Let us determine the value of the definite integral

$$\int_0^1 (x(-x+1.1))^{1/2}\, dx$$

using:

(i) the composite trapezoidal rule with a uniform spacing h;

(ii) the modified trapezoidal rule having a uniform spacing h in which the first and last subintervals are further divided into subintervals of length h^2.

We assume the interval $[0, 1]$ is first divided into equal intervals $(h = 1/n)$ with data points $x_i = ih$, $i = 0, 1, \ldots, n$. The composite trapezoidal rule is then given by Eq. (54), namely

$$\int_0^1 [x(-x+1.1)]^{1/2}\, dx \doteq \frac{h}{2}[nh(-nh+1.1)]^{1/2} + h\sum_{i=1}^{n-1} [ih(-ih+1.1)]^{1/2} \tag{57}$$

For the 'modified' composite trapezoidal rule, we further divide the interval $(0, h)$ and $((n-1)h, nh)$ so that the integrating formula is

$$\int_0^1 [x(-x+1.1)]^{1/2}\, dx = \frac{h^2}{2}[h(-h+1.1)]^{1/2} + h^2\sum_{i=1}^{n-1} [ih^2(-ih^2+1.1)]^{1/2}$$

$$+ \frac{h}{2}\{[h(-h+1.1)]^{1/2} + [(n-1)h(-(n-1)h+1.1)]^{1/2}\}$$

$$+ h\sum_{i=2}^{n-2} [ih(-ih+1.1)]^{1/2}$$

$$+ \frac{h^2}{2}\{[(n-1)h(-(n-1)h+1.1)]^{1/2} + (0.1)^{1/2}\}$$

$$+ h^2\sum_{i=1}^{n-1} [(n-1)h+ih^2][-\{(n-1)h+ih^2\}+1.1]^{1/2} \tag{58}$$

The results obtained using Eqs. (57) and (58) for a series of values of n are given in Table 5.

Table 5

	Composite trapezoidal method (57)	Modified composite trapezoidal method (58)
$n = 10$	0.4773	0.4516
100	0.4566	0.4546
1000	0.4560	0.4538
10000	0.4537	0.4537

As is easily verified, the theoretical value is 0.453715(5).

Exercises

25. Develop a program for the composite trapezoidal rule for n tabular points, the parameter value to be read in as data, for a general function f whose values are given at n points as data or are calculated from a given analytical form of f.
26. Develop a program for the composite Simpson's $\frac{1}{3}$-rule.
27. Develop a program for the composite Simpson's $\frac{3}{8}$-rule.
28. Use your program of Exercises 25–27 to evaluate

$$I = \int_0^1 \exp(-x) \sin(x) \, dx$$

for: (a) $n = 100$, (b) $n = 20$, (c) $n = 300$.
How do your answers compare for accuracy for the three cases when compared with the theoretical solution?
29. If the value of the integral

$$\int_{-1}^1 \exp(\pi x) \cos^2(x) \, dx$$

is required, how many subdivisions are necessary in order to guarantee an accuracy of 10^{-4} when using the composite: (a) trapezoidal rule, (b) Simpson's $\frac{1}{3}$-rule?
30. Evaluate

$$\int_0^6 \frac{\sin(x/4) \, dx}{\sqrt{(1+x^2)}}$$

using: (a) the trapezoidal rule with two points, (b) the composite trapezoidal rule guaranteeing an accuracy of 10^{-3}, (c) the composite Simpson's $\frac{1}{3}$-rule to guarantee an accuracy of 10^{-5}.
31. Use the program of Exercise 26 to evaluate

$$\int_0^1 \exp(-\sqrt{\pi} \, x^2) \, dx$$

for a given number of mesh points (100 say). Halve the mesh and evaluate the integral again. Is the increase in accuracy consistent with that predicted by the error term?

32. Determine the form of the composite integrating formula for $n = 4$. What is the error term E_I in this case?

33. Repeat Example 8 using the composite Simpson's $\frac{1}{3}$-rule and a modified Simpson's $\frac{1}{3}$-rule where the interval $[0, 1]$ is first divided into $2n$ equal intervals and the extreme intervals further divided into $2n$ equal subintervals. How do the numerical results compare, for the same number of subdivisions of $[0, 1]$, with those reported in Example 8?

34. If the form of the integrand of a given definite integral suggests that it has a large (singular) derivative at one end point of the interval of integration, what strategy would you suggest for modifying the composite integrating formula?

5 Romberg Integration

In this section we will consider an alternative method to that of the previous section for attaining an increasing order of accuracy from our integrating formulas with their associated disadvantages. The basis of our algorithm, the method of Romberg integration, is the composite trapezoidal rule. In addition we will require the following result.

Lemma 1 *The error term E_I can be written as an asymptotic series*

$$E_I = \sum_{i=1}^{\infty} \beta_i h^{2i} \tag{59}$$

where $h = (b - a)/n$ and the β_i, $i = 1, 2, \ldots$, do not depend upon h if h is small enough (i.e. n large enough).

The interested reader is referred to the proof given in Ralston [34]. The significance of the asymptotic form of E_I rather than the closed form given previously appears in the following way. Let us first consider just the first term of E_I, namely

$$E_I = \beta_1 h^2 + \text{terms depending on higher powers of } h$$

We will rewrite this in a convenient form as

$$E_I = \beta_1 h^2 + O(h^4)$$

For a particular n we will obtain a numerical value from the application of the trapezoidal rule when applied to evaluate the definite integral $I = \int_a^b f(x)\, dx$. For convenience let $a = 0$, $b = 1$, and $n = 1$. Thus

$$h = (b - a)/n = 1$$

Then

$$I = \tfrac{1}{2} h_1 (f_1 + f_0) + \beta_1 h_1^2 + O(h^4) \tag{60}$$

where $h_1 = h = 1$.

The numerical value we could obtain when applying the trapezoidal rule with $n = 1$ would be $\tfrac{1}{2} h_1 (f_1 + f_0)$. If we could determine the value of β_1 (which in general we cannot), we would be able to state that by adding $\beta_1 h_1^2$ to

$\frac{1}{2}h_1(f_1+f_0)$ the resulting numerical value would be more accurate than $\frac{1}{2}h_1(f_1+f_0)$ alone.

The simple idea which allows us to obtain more accurate results from a simple modification of Eq. (60) arises from remembering that β_1 is independent of h. Thus if we were to halve h, namely apply the trapezoidal rule in a composite manner (two applications), we would obtain the integrating formula

$$I = \frac{h_2}{2}(f_1+2f_{1/2}+f_0)+\beta_1 h_2^2+O(h^4) \tag{61}$$

where $h_2 = h = \frac{1}{2}$.

We have added an extra tabular point in Eq. (61), namely $f_{1/2}$. This point is the value of f at $\frac{1}{2}(b+a)$.

Once again the actual numerical value we obtain is $\frac{1}{2}h_2(f_1+2f_{1/2}+f_0)$, where to know β_1 would mean obtaining a more accurate result by adding $\beta_1 h_2^2$ to $\frac{1}{2}h_2(f_1+2f_{1/2}+f_0)$. If we assume, however, that we have obtained two numerical results, one using $\frac{1}{2}h_1(f_1+f_0)$ and another using $\frac{1}{2}h_2(f_1+2f_{1/2}+f_0)$ (call these I_1 and I_2 respectively), then by multiplying Eq. (60) by h_2^2 and Eq. (61) by h_1^2 we will be able to eliminate the unknown term involving β_1. We then obtain a scheme which has the following form:

$$h_2^2 I - h_1^2 I = \frac{1}{2}h_2^2 h_1(f_1+f_0)-\frac{1}{2}h_1^2, h_2(f_1+2f_{1/2}+f_0)$$
$$+ h_1^2 h_2^2 \beta_1 - h_1^2 h_2^2 \beta_1 + O(h^4) \tag{62}$$

i.e. $(h_2^2-h_1^2)I = h_2^2 I_1 - h_1^2 I_2 + O(h^4)$ or

$$I = \frac{h_2^2 I_1 - h_1^2 I_2}{(h_2^2-h_1^2)}+O(h^4) \tag{63}$$

Therefore the combination of I_1 and I_2 given by Eq. (63) is a more accurate approximation of I than either I_1 or I_2. This equation holds for any h_1 and h_2 ($h_1 \neq h_2$). However, it is advantageous to choose $h_1 = 2h_2 = h$. In this case,

$$I = \frac{4I_2-I_1}{3}+O(h^4) \tag{64}$$

Consequently, by applying the trapezoidal rule to an integral for n points followed by an application with $2n$ points, a formula like (64) results which provides a more accurate approximation than either of the previous results. We may therefore proceed and eliminate successive terms from E_I in (59) by taking linear combinations of successive results obtained by applying the composite trapezoidal rule on differing numbers of points.

In order to describe the method of Romberg integration, we require to introduce a new notation to describe the two-tier system of formulas. Define

$$I_{0,k} = \frac{(b-a)}{2^k}[\frac{1}{2}f_0+f_1+\cdots+f_{2^k-1}+\frac{1}{2}f_{2^k}] \qquad k = 0, 1, \ldots \tag{65}$$

as the composite trapezoidal rule defined on $2^k + 1$ points in the interval $[a, b]$.

Define, recursively, the steps representing the successive elimination of the terms involving $\beta_1, \beta_2, \beta_3, \ldots$ in the error term E_I. As

$$I_{m,k} = \frac{4^m I_{m-1,k+1} - I_{m-1,k}}{4^m - 1} \qquad k = 0, 1, \ldots; m = 1, 2, \ldots \qquad (66)$$

where $I_{0,k}$ is given by Eq. (65). Successive values of m represent the elimination of successive β's in E_I. For example, $m = 1$ produces (64) the elimination of β_1; $m = 2$ represents the elimination of β_2 from E_I; etc.

The implementation of Romberg integration is therefore to evaluate approximations to the required integral I for various $k = 0, 1, 2, \ldots$ using Eq. (65). Having tabulated the results, define successive $I_{m,k}$ for $m = 1$, $k = 0, 1, 2, \ldots$; $m = 2$, $k = 0, 1, 2, \ldots$; etc., using Eq. (66). The $I_{m,k}$ so defined have the property that successive $I_{m,k}$, $m = 1, 2, \ldots$, are more accurate with increasing m.

Example 9 Determine the value of the integral

$$I = \int_0^1 x(1+x^2)^{1/2} \, dx$$

using Romberg integration.

We will use just three values of k for the composite trapezoidal rule to illustrate the procedure.

Step (i): one application ($k = 0$ in Eq. (65))

$$I_{0,0} = \frac{1}{1} \left(\tfrac{1}{2}f(0) + \tfrac{1}{2}f(1) \right) = \tfrac{1}{2}(0(1+0)^{1/2} + 1(1+1)^{1/2}) = 0.7071$$

Step (ii): two applications ($k = 1$ in Eq. (65))

$$\begin{aligned}
I_{0,1} &= \tfrac{1}{2}(\tfrac{1}{2}f(0) + f(\tfrac{1}{2}) + \tfrac{1}{2}f(1)) \\
&= \tfrac{1}{2}(0(1+0)^{1/2} + \tfrac{1}{2}(1+\tfrac{1}{4})^{1/2} + \tfrac{1}{2}(1+1)^{1/2}) \\
&= 0.6331
\end{aligned}$$

Step (iii): four applications ($k = 2$ in Eq. (65))

$$\begin{aligned}
I_{0,2} &= \frac{1}{4}\left(\frac{1}{2}f(0) + f\left(\frac{1}{4}\right) + f\left(\frac{1}{2}\right) + f\left(\frac{3}{4}\right) + \frac{1}{2}f(1) \right) \\
&= \frac{1}{4}\left(0(1+0)^{1/2} + \frac{1}{4}\left(1+\frac{1}{16}\right)^{1/2} + \frac{1}{2}\left(1+\frac{1}{4}\right)^{1/2} + \frac{3}{4}\left(1+\frac{9}{16}\right)^{1/2} + \frac{1}{2}(1+1)^{1/2} \right) \\
&= 0.6153
\end{aligned}$$

Step (iv): use Eq. (66) for $m = 1$, $k = 0, 1$.

$$I_{1,0} = \frac{4I_{0,1} - I_{0,0}}{3} = \frac{4 \times 0.6331 - 0.7071}{3} = 0.6084$$

$$I_{1,1} = \frac{4I_{0,2} - I_{0,1}}{3} = \frac{4 \times 0.6153 - 0.6331}{3} = 0.6094$$

Step (v): use Eq. (66) with $m = 2$, $k = 0$

$$I_{2,0} = \frac{16 I_{1,1} - I_{1,0}}{15} = \frac{16 \times 0.6094 - 0.6084}{15} = 0.6094(6)$$

As is easily shown, the exact result is $0.6094(7)$. Our result is therefore correct to four decimals.

It will be noted that the strategy of dividing h successively by 2 is a sound one since each application of the composite rule (Eq. (65)) utilizes all of the function values previously used for smaller values of k and $m = 0$. A division of h by some other number would require introducing a complete set of new function values. Consequently, the Romberg integration method requires fewer function values than might at first appear. In addition, the elimination process represented by (66) is a simple manipulation requiring minimal computer time. The most time-consuming part of the algorithm is the calculation of the $I_{0,k}$, $k = 0, 1, 2, \ldots$, the original composite trapezoidal rule.

Exercises

35. Show that the term $I_{2,0}$ eliminates the term involving β_2 in the expansion (59) of E_I for the trapezoidal rule.
36. Develop a subprogram (procedure in ALGOL, subroutine in FORTRAN) which will use the trapezoidal rule to evaluate $\int_a^b f(x)\, dx$ over an arbitrary number of subintervals $(b - a)/2^k$ in length, where k is a parameter of the subprogram. Using the subprogram, write a program to evaluate $I_{m,k}$ of Romberg integration, Eq. (66).
37. Use your program of Exercise 36 to evaluate:

$$\text{(a)} \quad \int_0^1 \exp(-x) \sin(x)\, dx$$

$$\text{(b)} \quad \int_{-\pi/2}^{\pi/2} \sin^3(x)\, dx$$

where five decimal places of accuracy are required.

6 Gaussian Integration

In the previous sections we have assumed that the function values $f(x_i)$, $i = 0, 1, \ldots, n$, were given at abscissas x_i, $i = 0, 1, \ldots, n$, which were equally spaced. This can clearly be of advantage when we require to integrate a function which is in tabulated form. In the case that the function $f(x)$ is known analytically and is sufficiently cheap to evaluate, then there is no obvious need to require the integrating formula to employ data at equally spaced abscissas. Let us consider the definite integral $\int_a^b f(x)\, dx$, using an integration scheme employing two distinct points. To facilitate the determination of the parameters in the associated two-point integrating formula, let us introduce the change of variable

$$y = (2x - (a + b))/(b - a) \tag{67}$$

so that the interval $[a, b]$ is transformed into the interval $[-1, 1]$. Let the function $f(x)$ be thereby transformed to a new function $F(y)$ and note that $dy = 2\,dx/(b-a)$, so that

$$\int_a^b f(x)\,dx \to \int_{-1}^1 (b-a)\tfrac{1}{2}F(y)\,dy \tag{68}$$

where $F(y) = f(\tfrac{1}{2}(b-a)y + \tfrac{1}{2}(a+b))$.

Our integrating formula is

$$\int_{-1}^1 \tfrac{1}{2}(b-a)F(y)\,dy = \alpha_0 F(y_0) + \alpha_1 F(y_1) \tag{69}$$

We now have four parameters α_0, α_1, y_0, and y_1 at our disposal and consequently we can expect our integrating formula (69) to be exact for polynomials up to and including those of degree three. Hence we can determine the parameters by the method of undetermined coefficients. Substitute $F(y) = y^k$, $k = 0, 1, 2, 3$, into Eq. (69). As a result we obtain the four nonlinear equations in α_0, α_1, y_0, and y_1, namely

$$b - a = \alpha_0 + \alpha$$

$$0 = \alpha_0 y_0 + \alpha_1 y_1$$

$$(b-a)\tfrac{1}{3} = \alpha_0 y_0^2 + \alpha_1 y_1^2$$

$$0 = \alpha_0 y_0^3 + \alpha_1 y_1^3$$

Solving these equations we find

$$\alpha_0 = \alpha_1 = \tfrac{1}{2}(b-a) \qquad y_0 = -y_1 = -1/\sqrt{3}$$

and our integrating formula is then

$$\int_{-1}^1 \tfrac{1}{2}(b-a)F(y)\,dy = \frac{(b-a)}{2}[F(-1/\sqrt{3}) + F(1/\sqrt{3})]$$

i.e.

$$\int_{-1}^1 F(y)\,dy = [F(-1/\sqrt{3}) + F(1/\sqrt{3})]$$

Clearly the algebra involved in the derivation of these integrating formulas is far more tedious than in the derivation of the integrating formulas employing equally spaced abscissas. To obtain the general formula we do not recommend the reader solve the integrating formula from first principles. Instead we proceed as follows. To retain a consistent notation consider $\int_{-1}^1 f(x)\,dx$. Then consider the Hermite interpolating formula of Chapter 4 given by

$$f(x) = \sum_{i=0}^n h_i(x)f(x_i) + \sum_{i=0}^r g_i(x)f^{(1)}(x_i) + E$$

where

$$h_i(x) = [1 - 2(x - x_i)]L_i^{(1)}(x_i)L_i^2(x) \qquad i = 0, 1, \ldots, n$$

and   (70)

$$g_i(x) = (x - x_i)L_i^2(x) \qquad i = 0, 1, \ldots, n$$

where

$$E = \frac{\pi_{n+1}(x)\pi_{r+1}(x)f^{(n+r+2)}(\zeta)}{(n+r+2)!}$$

In this section we will take $r = n$ and hence the interpolating polynomial is exact for polynomials of degree $2n + 1$ or less. To obtain our integrating formula we integrate Eq. (70) between -1 and 1 to obtain

$$\int_{-1}^{1} f(x)\, dx = \int_{-1}^{1} \sum_{i=0}^{n} h_i(x)f(x_i)\, dx$$

$$+ \int_{-1}^{1} \sum_{i=0}^{n} g_i(x)f^{(1)}(x_i)\, dx + \int_{-1}^{1} E\, dx$$

Assuming the continuity of the functions $h_i(x)$ and $g_i(x)$, we interchange the integration and summation signs to obtain

$$\int_{-1}^{1} f(x)\, dx = \sum_{i=0}^{n} \int_{-1}^{1} h_i(x)f(x_i)\, dx$$

$$+ \sum_{i=0}^{n} \int_{-1}^{1} g_i(x)f^{(1)}(x_i)\, dx + E_I \qquad (71)$$

where

$$E_I = \int_{-1}^{1} \frac{\pi_{n+1}^2(x)}{(2n+2)!} f^{(2n+2)}(\zeta)\, dx \qquad (72)$$

Now, our aim is to choose an integrating formula of the form

$$\int_{-1}^{1} f(x)\, dx = \sum_{i=0}^{n} \alpha_i f(x_i) + E_I$$

and so Eq. (71) will give us such a formula if we choose the abscissas x_i such that

$$\int_{-1}^{1} g_i(x)f^{(1)}(x_i)\, dx = 0 \qquad i = 0, 1, \ldots, n$$

i.e. such that

$$\int_{-1}^{1} g_i(x)\, dx = 0$$

since $f^{(1)}(x_i) \neq 0$ in general. But from Eq. (70), $g_i(x) = (x - x_i)L_i^2(x)$ and so we require to choose the x_i so that

$$\int_{-1}^{1} g_i(x)\,dx = \int_{-1}^{1} (x - x_i)L_i^2(x)\,dx = 0 \tag{73}$$

for $i = 0, 1, \ldots, n$. But it is easy to show that $L_i(x)$ can be written as

$$L_i(x) = \frac{\pi_{n+1}(x)}{(x - x_i)\pi_{n+1}^{(1)}(x_i)} \tag{74}$$

where $\pi_{n+1}^{(1)}(x_i) = (d/dx)\pi_{n+1}(x)|_{x=x_i}$. The right-hand side of Eq. (73) can then be written in the form

$$\int_{-1}^{1} \frac{\pi_{n+1}(x)L_i(x)\,dx}{\pi_{n+1}^{(1)}(x_i)} = 0 \tag{75}$$

But $\pi_{n+1}(x)$ is a polynomial of degree $n+1$ and $L_i(x)$, the Lagrangian polynomials, are of degree n. Thus a condition which will ensure that Eq. (25) is always satisfied is that the x_i be chosen so that $\pi_{n+1}(x)$ should be orthogonal to all polynomials of degree n or less on the interval $[-1, 1]$. The function $\pi_{n+1}(x)$ which satisfies this requirement is the appropriate polynomial defined by

$$P_0(x) = 1$$

$$P_1(x) = x$$

$$P_i(x) = \frac{1}{i}[(2i - 1)xP_{i-1}(x) - (i - 1)P_{i-2}(x)] \qquad i = 2, 3, \ldots$$

namely the Legendre polynomials multiplied by the constant $2^{n+1}((n+1)!)^2/(2(n+1))!$. The zeros of these polynomials are the required abscissas x_i for our integrating formula. It is well known that the zeros of the Legendre polynomials are all real so that the question of "complex abscissas" does not arise. The zeros of the Legendre polynomials have been tabulated for all n of practical interest (see Krylov, [29, p. 337]) and these can be stored on a computer for direct use when demanded. For conciseness we have listed in Table 6 the zeros of the first six Legendre polynomials.

Thus, knowing the number of points we want to use in the integrating formula, we merely have to look up the zeros of the Legendre polynomials of that degree and thus determine the abscissas of the integrating formula. The coefficients α_i, $i = 0, 1, \ldots, n$, are found by integrating the polynomials $h_i(x)$ given by Eq. (70). It is seen that

$$\alpha_i = \int_{-1}^{1} h_i(x)\,dx = \int_{-1}^{1} L_i^2\,dx$$

Table 6

Degree of Legendre polynomials	Zeros of Legendre polynomials	Weights α_i
1	0	2
2	$\pm 1/\sqrt{3}$	1, 1
3	0	$\frac{8}{9}$
	$\pm\sqrt{3/5}$	$\frac{5}{9}, \frac{5}{9}$
4	$\pm 0.339\,981\,043\,6$	$0.652\,145\,154\,9$
	$\pm 0.861\,136\,311\,6$	$0.347\,854\,845\,1$
5	0	$0.568\,888\,888\,9$
	$\pm 0.538\,469\,310\,1$	$0.478\,628\,670\,5$
	$\pm 0.906\,179\,845\,9$	$0.236\,926\,885\,1$
6	$\pm 0.238\,619\,186\,1$	$0.467\,913\,934\,6$
	$\pm 0.661\,209\,386\,5$	$0.360\,761\,573\,0$
	$\pm 0.932\,469\,514\,2$	$0.171\,324\,492\,4$

Now if we consider the integrating formula (71) (with $g_i(x) = 0$) applied to one of the Lagrangian polynomials $L_j(x)$, $j = 0, 1, \ldots, n$, we have

$$\int_{-1}^{1} L_j(x)\,dx = \sum_{k=0}^{n} \left(\int_{-1}^{1} h_k(x)\,dx \right) L_j(x_k)$$

$$= \sum_{k=0}^{n} \left\{ \int_{-1}^{1} L_k^2(x)\,dx \right\} L_j(x_k)$$

since the error term E_I is zero (why?) Thus

$$\int_{-1}^{1} L_j(x) = \int_{-1}^{1} L_j^2(x)\,dx$$

since

$$L_j(x_k) = \delta_{jk}$$

Hence

$$\alpha_i = \int_{-1}^{1} L_i(x)\,dx \qquad i = 0, 1, \ldots, n \qquad (76)$$

Values of the coefficients α_i are tabulated in Table 6, for $n = 0, 1, \ldots, 5$. A comprehensive list of coefficients α_i can again be obtained by referring to [29].

Integrating formulas in which the abscissas x_i and parameters α_i are chosen to produce maximal accuracy are called *Gaussian* integrating formulas. When the abscissas are the zeros of the Legendre polynomials, as described above, the formulas are called *Legendre–Gauss* integrating formulas.

The error term E_I of the Legendre–Gauss formulas can be simplified at once by the second mean value theorem since, in (72), $\pi_{n+1}^2(x)$ is of one sign in $[-1, 1]$. Hence

$$E_I = \frac{f^{(2n+2)}(\xi)}{(2n+2)!} \int_{-1}^{1} \pi_{n+1}^2(x)\, dx \qquad -1 < \xi < 1$$

The use of Gaussian integration is clearly an excellent example of when to use a computer and not to use paper and pencil, since in general the zeros of the Legendre polynomials are irrational numbers and to retain sufficient figures in a hand calculation makes the calculation prohibitive.

Example 10 Determine the value of the definite integral $\int_{-1}^{1} x \sin(x)\, dx$ with $n = 2$ in the Legendre–Gauss formula.

We have

$$\int_{-1}^{1} x \sin(x)\, dx = \tfrac{8}{9}(0.\sin(0)) + \tfrac{5}{9}(\sqrt{3/5} \sin(\sqrt{3/5})) + \tfrac{5}{9}(-\sqrt{3/5} \sin(-\sqrt{3/5}))$$

$$= \tfrac{10}{9}\sqrt{3/5}\,(0.699\,45) = 0.601\,99$$

If we evaluate the error bound we find

$$E_I \leq \max_{-1 \leq \zeta \leq 1} \frac{f^{(6)}(\zeta)}{6!} \int_{-1}^{1} x^2(x^2 - \sqrt{3/5})(x^2 + \sqrt{3/5})\, dx$$

$$= \max_{-1 \leq \zeta \leq 1} f^{(6)}(\zeta)/6300$$

Now $f^{(6)}(x) = 6 \cos(x) - x \sin(x)$, which takes on its maximum at $x = 0$. Hence $E_I \leq 6/6300 = 0.00095$.

The computed result clearly differs from the theoretical solution (0.60234) by less than this amount.

Exercises

38. Develop a program for the Gauss–Legendre integrating formula for an arbitrary function $f(x)$ for $n + 1$ points, the abscissas given by Table 6.
39. Use the Legendre–Gauss integrating formula based on two points to solve:

$$\text{(a)} \quad \int_{0}^{1} \exp(-x) \sin(x)\, dx$$

$$\text{(b)} \quad \int_{-\pi/2}^{\pi/2} \sin^3(x)\, dx$$

Compare your results with those obtained using the trapezoidal rule, Simpson's $\tfrac{1}{3}$-rule, and Romberg integration (cf. Exercise 37).
40. A satellite having unequal principal moments of intertia orbits the earth. The period of the satellite is given by

$$p = 4/(c\sqrt{2}) \int_{0}^{\pi/2} \frac{dx}{(1 - \sin^2(\alpha) \sin^2(x))^{1/2}}$$

where

$$c = \sqrt{\frac{R^2 g}{2 r_0^3}\left(\frac{I_y - I_x}{I_z}\right)}$$

where the constants R = radius of the earth = 20.9×10^6 ft, g = acceleration due to the gravity at the earth's surface = 32.2 ft/sec^2, r_0 = radius of the circular orbital path = $1.2R$, I_x, I_y, and I_z are the principal moments of interia of the satellite where $I_y = I_z$; $I_x = 0.45 I_y$. α is the angle of inclination of the satellite's main axis to the radius vector from the earth to the satellite.

Calculate the integral to obtain p using:
(i) the composite trapezoidal rule with $n = 10$ (Ex. 25);
(ii) the composite Simpson's $\frac{1}{3}$-rule with $n = 20$ (Ex. 26);
(iii) the Gauss–Legendre integrating formula with $n = 5$ (Ex. 38) for $\alpha = \pi/15$, $\pi/10$, $\pi/5$, $\pi/4$, $\pi/3$, $7\pi/16$.

The use of weight functions

In previous sections we have considered the definite integral $\int_a^b f(x)\,dx$ represented as a linear combination of the function values of f evaluated at given (or calculated) tabular points. However, in certain circumstances, we find that it becomes more convenient to write $f(x)$ as $w(x)\gamma(x)$ where we consider $w(x)$ to be a *weight function* which is of constant sign in $[a, b]$. It is then proposed to construct an integrating formula

$$\int_a^b w(x)\gamma(x)\,dx = \sum_{i=0}^n \alpha_i \gamma(x_i) + E_I$$

where E_I will depend upon the derivatives of γ alone, i.e., a derivative of $w(x)$ will *not* appear in E_I. This would be of particular value in those problems where our original $f(x)$ has badly behaved derivatives in some portion of the interval $[a, b]$. For example $f = x^{1/2}(1.1 - x)^{1/2}$ in Example 8 has a singular derivative at $x = 0$. In such a case we would associate with $w(x)$ the portion of f which is badly behaved, and with γ the remaining part of f. Consider the Gaussian integrating formula

$$I = \int_a^b w(x)\gamma(x)\,dx$$

$$= \sum_{i=0}^n \int_a^b w(x) h_i(x)\gamma(x_i)\,dx$$

$$+ \sum_{i=0}^n \int_a^b w(x) g_i(x)\gamma^{(1)}(x_i)\,dx + E_I \tag{77}$$

To obtain an integrating formula of the form

$$I = \sum_{i=0}^n \alpha_i \gamma(x_i) + E_I$$

of maximal accuracy, we clearly require to choose the abscissas x_i such that

$$\int_a^b w(x)g_i(x)\,\mathrm{d}x = 0$$

Namely, we require $\pi_n(x)$ to be orthogonal on $[a, b]$ with respect to the weight function $w(x)$, to all polynomials of degree n or less, i.e. we require to choose x_i, $i = 0, 1, \ldots, n$, such that

$$\int_a^b w(x)\pi_{n+1}(x)p_n(x)\,\mathrm{d}x = 0$$

where $p_n(x)$ is any polynomial of degree n or less.

Depending upon the particular weight function $w(x)$ and the interval $[a, b]$, we will obtain a particular variant of Gaussian integration.

The coefficients α_i of the integrating formula can be determined as before, namely

$$\alpha_i = \int_a^b w(x)L_i(x)\,\mathrm{d}x$$

the error term is clearly

$$E_I = \frac{1}{(2n)!}\int_a^b w(x)\pi_{n+1}^2(x)\gamma^{(2n)}(\zeta)\,\mathrm{d}x$$

which can be simplified using the second mean value theorem to give

$$E_I = \frac{1}{(2n)!}\gamma^{(2n)}(\zeta)\int_a^b w(x)\pi_{n+1}^2(x)\,\mathrm{d}x \qquad \zeta \in (a, b) \tag{78}$$

since π_{n+1}^2, and by assumption, $w(x)$, are of constant sign in (a, b).

Our motivation for the use of weight functions now becomes clearer, since if $f(x)$ has a singular derivative, the singular portion of $w(x)$ does not appear differentiated in the error term.

Example 11 Evaluate the definite integral

$$I = \int_0^1 f(x)\,\mathrm{d}x = \int_0^1 (1-t^2)^{1/2}\cos(t)\,\mathrm{d}t$$

by means of:
  (i) the Legendre–Gauss formula for $f(x)$ with $n = 1$
  (ii) the Legendre–Gauss formula for $\cos(t)$ using $w(t) = (1-t^2)^{1/2}$ as a weight function with $n = 1$.
The theoretical value of I is $\frac{1}{2}\pi J_1(1)$, where $J_1(1)$ is the Bessel function of order 1 with argument equal to 1. This has a value of 0.440 05, so that $I = 0.691\,19$.

  (i) To employ the Legendre–Gauss formula to the interval $[0, 1]$ we must transform the independent variable using the transformation on page 228 with

$a = 0$, $b = 1$, namely $t = (x+1)/2$, and hence $dx = 2dt$. Then

$$I = \int_{-1}^{1} \tfrac{1}{2}[1-((x+1)/2)]^2 \cos((x+1)/2)\,dt$$

Now, using the Legendre–Gauss formula with abscissas $\pm 1/\sqrt{3}$ we obtain

$$I \doteq \tfrac{1}{2}\{[1-((1/\sqrt{3}+1)/2)^2]^{1/2}\cos((1/\sqrt{3}+1)/2)$$
$$+[1-((1-1/\sqrt{3})/2)^2]^{1/2}\cos((1-1/\sqrt{3})/2)\}$$
$$= 0.694\,49$$

(ii) Once again we transform the given integral into one defined on $[-1, 1]$. We define

$$w(x) = [1-(x+1)/2]^{1/2} \qquad \gamma(x) = \tfrac{1}{2}\cos((x+1)/2)$$

Hence the Legendre–Gauss formula on two points is

$$\alpha_0\gamma(x_0)+\alpha_1\gamma(x_1) \quad \text{where } x_0 = 1/\sqrt{3},\ x_1 = -1/\sqrt{3}$$

and

$$\alpha_0 = \int_{-1}^{1} [1-((x+1)/2)^2]^{1/2}\{(x-x_1)/(x_0-x_1)\}\,dx$$

$$\alpha_1 = \int_{-1}^{1} [1-((x+1)/2)^2]^{1/2}\{x-x_0)/(x_1-x_0)\}\,dx$$

Performing the integration we obtain

$$\alpha_0 = 0.579\,75, \qquad \alpha_1 = 0.991\,04$$

and then

$$I \doteq \tfrac{1}{2}(0.408\,60+0.968\,80) = 0.688\,80$$

Hence we have obtained an approximation to the integral which is slightly more accurate than that obtained in (i). The important point, of course, is that the error bound for part (i) involves $f^{(4)}(\eta)$ which, as $\eta \to 1$, is unbounded. We can consequently say little about E_I. However, for (ii) we have

$$E_I = \frac{1}{(2n!)}\gamma^{(2n)}(\zeta)\int_{-1}^{1}(x-x_0)^2(x-x_1)^2(1-((x+1)/2)^2)^{1/2}\,dx \qquad \zeta \in [-1, 1]$$

and $n = 2$. $|\gamma^{(2n)}(\zeta)|$ is now clearly bounded and we find

$$|E_I| \le \frac{1}{4!}\cos(1)/2\int_{-1}^{1}(x-1/\sqrt{3})^2(x+1/\sqrt{3})^2(1-((x+1)/2)^2)^{1/2}\,dx \le 0.004.$$

The actual error is $0.691\,19 - 0.688\,80 = 0.002\,39$ which is certainly less than the bound.

A special weight function

In the remainder of this section we will consider the more important examples of weight functions $w(x)$.

If we define

$$w(x) = (1-x^2)^{-1/2} \quad \text{and} \quad a = -1, b = 1 \tag{79}$$

then the definite integral under consideration is

$$\int_{-1}^{1} f(x)\,dx = \int_{-1}^{1} (1-x^2)^{-1/2} g(x)\,dx$$

The abscissas in the integrating formula

$$\int_{-1}^{1} f(x)\,dx = \sum_{i=0}^{n} \alpha_i \gamma(x_i)$$

are chosen so that $\pi_{n+1}(x)$ is orthogonal to all polynomials of degree n or less on the interval $[-1, 1]$ with respect to the weight function $w(x)$. But for $w(x)$ defined by (79) we know from Chapter 4, Theorem 1, that the resulting $\pi_{n+1}(x)$ are the Chebyshev polynomials whose zeros are at $x_i = \cos\{(2i-1)\pi/(2n+2)\}$, $i = 0, 1, \ldots, n$. That is, the abscissas in the integrating formula are the zeros of the Chebyshev polynomials of degree $n+1$. The coefficients are once more given by

$$\alpha_i = \int_{-1}^{1} w(x) L_i(x)\,dx \qquad i = 0, 1, \ldots, n$$

But by Eq. (74) this can be rewritten as

$$\alpha_i = \int_{-1}^{1} \frac{w(x)\pi_{n+1}(x)\,dx}{(x-x_i)\pi_{n+1}^{(1)}(x_i)} \qquad i = 0, 1, \ldots, n \tag{80}$$

and hence using property (P3), Chapter 4, we find

$$\alpha_i = \int_{-1}^{1} \frac{(1-x^2)^{-1/2} T_{n+1}(x)\,dx}{(x-x_i)T_{n+1}^{(1)}(x)} \dagger \tag{81}$$

To proceed further with the evaluation of these coefficients we introduce the *Christoffel–Darboux* relation. Let polynomials $\phi_k(x)$ of degree k, $k = 0, 1, \ldots,$ be orthogonal with respect to a weight function $w(x)$ over an interval $[a, b]$. Let the leading coefficients of $\phi_k(x)$ be a_k (i.e. the coefficient of x^k in $\phi_k(x)$ is a_k). Then the orthogonal polynomials satisfy the Christoffel–Darboux relation:

$$\frac{a_k}{a_{k+1}b_k}(\phi_{k+1}(x)\phi_k(y) - \phi_{k+1}(y)\phi_k(x)) = (x-y)\sum_{j=0}^{k} \frac{\phi_j(x)\phi_j(y))}{b_j}$$

$$k = 0, 1, 2, \ldots \tag{82}$$

† The coefficients α_i are known as the *Christoffel numbers*.

where

$$b_j = \int_a^b w(x)\phi_j^2(x)\,dx$$

(For a proof of this identity see Isaacson and Keller [24, p. 205]). Substituting $y = x_i$, a zero of $\phi_k(x)$, into Eq. (82), we obtain

$$\frac{-a_k}{a_{k+1}b_k}(\phi_{k+1}(x_i)\phi_k(x)) = (x - x_i)\sum_{j=0}^{k-1}\frac{\phi_j(x)\phi_j(x_i)}{b_j} \tag{83}$$

since $\phi_k(x_i) = 0$. Multiply both sides of Eq. (83) by $w(x)\phi_0(x)/(x - x_i)$ and integrate with respect to x over $[a, b]$; then

$$\frac{-a_k}{b_k a_{k+1}}\int_a^b \frac{\phi_{k+1}(x_i)\phi_k(x)w(x)\phi_0(x)\,dx}{(x - x_i)} = \int_a^b \sum_{j=0}^{k-1}\frac{\phi_j(x)\phi_j(x_i)w(x)\phi_0(x)\,dx}{b_j}$$

Using the orthogonality of the $\phi_k(x)$ we obtain

$$\frac{-a_k}{b_k a_{k+1}}\phi_{k+1}(x_i)\int_a^b \frac{w(x)\phi_k(x)\phi_0(x)\,dx}{(x - x_i)} = \phi_0(x_i) \tag{84}$$

Now $\phi_0(x)$ is a polynomial of degree zero, i.e. a constant, and so $\phi_0(x) = \phi_0(x_i)$ may be divided from both sides of Eq. (84). We thus obtain

$$\frac{-a_k}{b_k a_{k+1}}\phi_{k+1}(x_i)\int_a^b \frac{w(x)\phi_k(x)\,dx}{(x - x_i)} = 1$$

or

$$\int_a^b \frac{w(x)\phi_k(x)\,dx}{(x - x_i)\phi_k^{(1)}(x_i)} = \frac{-b_k a_{k+1}}{a_k \phi_{k+1}(x_i)\phi_k^{(1)}(x_i)} \tag{85}$$

We can now use Eq. (85) in Eq. (81) by writing

$$a = -1 \qquad b = 1 \qquad \phi_k(x) = T_{n+1}(x) \qquad w(x) = (1 - x^2)^{-1/2}$$

and

$$b_k = \int_{-1}^1 (1 - x^2)^{-1/2}T_{n+1}^2(x) \qquad a_{k+1} = 2^k$$

Thus

$$\alpha_i = -\int_{-1}^1 \frac{2(1 - x^2)^{-1/2}T_{n+1}^2(x)\,dx}{T_{n+2}(x_i)T_{n+1}^{(1)}(x_i)} \qquad i = 0, 1, \ldots, n \tag{86}$$

where we stress the fact that x_i is a zero of $T_{n+1}(x)$. We note that

$$T_{n+2}(x_i) = \cos((n+1)\arccos(x_i) + \arccos(x_i))$$

$$= \cos((n+1)\arccos(x_i))x_i$$

$$- \sin((n+1)\arccos(x_i))\sin(\arccos(x_i))$$

using the elementary relation for the sum of two cosines. Since x_i is a zero of T_{n+1}, we see therefore that

$$T_{n+2}(x_i) = -\sin\left((n+1)\ \text{arc cos}\ \{\cos\left((2i+1)\pi\right)/(2(n+1))\}\right.$$
$$\times (1 - \cos^2\left(\text{arc cos}\ (x_i)\right))^{1/2}$$
$$= (-1)^{i+1}(1 - x_i^2)^{1/2} \tag{87}$$

Also,

$$T_{n+1}^{(1)}(x_i) = \frac{d}{dx} \cos\left((n+1)\ \text{arc cos}\ (x)\right)\big|_{x=x_i}$$

$$= (n+1)\sin\left((n+1)\ \text{arc cos}\ (x_i)\right)/\sqrt{1-x_i^2}$$

$$= \frac{(-1)^i(n+1)}{\sqrt{1-x_i^2}} \tag{88}$$

Combining Eqs. (87) and (88) in Eq. (86), we find

$$\alpha_i = \frac{1}{n+1} \int_{-1}^{1} (1-x^2)^{-1/2} T_{n+2}^2(x)\ dx$$

which is easily seen to produce

$$\alpha_i = \frac{\pi}{n+1} \qquad \text{for all}\ i = 0, 1, 2, \ldots, n$$

Thus our integrating formula has the particularly simple form

$$\int_{-1}^{1} f(x)\ dx = \int_{-1}^{1} (1-x^2)^{-1/2} \gamma(x)\ dx \doteq \frac{\pi}{(n+1)} \sum_{i=0}^{n} \gamma(x_i) \tag{89}$$

and x_i are the $(n+1)$ zeros of the Chebyshev polynomials of degree $(n+1)$. Formula (89) is known as the *Chebyshev–Gauss* integrating formula.

Exercises

41. Program the Chebyshev–Gauss formula for an arbitrary function $\gamma(x)$ and an arbitrary number $(n+1)$ of tabular points.
42. Use the program of Exercise 41 to evaluate the integrals:

   (i) $\displaystyle\int_0^{\pi} \exp\left(\cos\left(\theta\right)\right)\ d\theta \qquad n = 2;\ n = 3;\ n = 4;\ n = 5.$

   (ii) $\displaystyle\int_{-1}^{1} \frac{x\ dx}{(1-x^2)^{1/2}} \qquad n = 3;\ n = 8;\ n = 10$

   (iii) $\displaystyle\int_{-1}^{1} \frac{\cos^{-1}(x)\ dx}{(1-x^2)^{1/2}} \qquad n = 5;\ n = 20$

43. Determine the particular form of the error term for Chebyshev–Gauss integration.
44. For the integrand of part (ii) of Exercise 42, do the computed results satisfy the error bounds suggested by Exercise 43?

45. Compare the numerical results obtained for (iii) of Exercise 42. In the light of the form of the error bound in Exercise 43 for this integrand, comment on your numerical results.

46. A particle of mass m is attracted towards a fixed point O by a force which varies inversely as the square of the distance between them. It is required to find the time $t = T$ taken for the particle to reach O when it is released from a point distance X from the point O with a zero velocity at time $t = 0$. The required time is given by the integral

$$T = \int_X^0 \frac{dx}{\left\{ \dfrac{2\alpha}{m} \left(\dfrac{1}{x} - \dfrac{1}{X} \right) \right\}^{1/2}}$$

where α is a constant. For $m = 5$, $\alpha = 2$, $X = 1$, evaluate the time taken for the particle to reach O ($x = 0$) using Chebyshev–Gauss integrating formula for $n = 10$ and $n = 20$.

There are many other special weight functions and corresponding orthogonal polynomials for particular finite intervals $[a, b]$. For example, the Chebyshev–Gauss integrating formula is itself a special case of a more general integrating formula obtained as a result of specifying $w(x) = (1-x)^\alpha (1+x)^\beta$, α, $\beta > -1$, namely the *Jacobi–Gauss* integrating formula. However, a discussion of these more general weight functions is beyond the scope of this text and the interested reader is referred to Krylov [29], Davis and Rabinowitz [8], and Stroud and Secrest [41].

Gaussian integration formulas for infinite intervals

It is common in practice to require to evaluate integrals defined on intervals $[a, b]$ where either a or b or both are infinite. In this case a weight function $w(x)$ would be used which guarantees that the integral converges. We will consider two cases:

(i) $a = 0$ $b = \infty$ $w(x) = \exp(-x)$

(ii) $a = -\infty$, $b = \infty$ $w(x) = \exp(-x^2)$

Once again we will position the tabular points x_i so that $\pi_{n+1}(x)$ is orthogonal on the given interval to polynomials of degree n or less with respect to $w(x)$. For (i) it is found that $\pi_{n+1}(x)$ are the Laguerre polynomials $\mathscr{L}_{n+1}(x)$ where

$$\mathscr{L}_n(x) = \exp(x) \frac{d^n}{dx^n} (x^n \exp(-x))$$

from which

$$\mathscr{L}_n(x) = (-1)^n \left(x^n - \frac{n^2}{1!} x^{n-1} + n^2 \frac{(n-1)^2}{2!} x^{n-2} + \cdots \right) \qquad n = 0, 1, 2, \ldots$$

The zeros of the Laguerre polynomials have been tabulated for all practical values of n (see Krylov [29]). For convenience we have provided the zeros of the Laguerre polynomials up to degree 6 in Table 7. The coefficients α_i

can once again be calculated using Eq. (80) and the Christoffel–Darboux relation (82) to give

$$\alpha_i = \frac{[(n+1)!]^2}{x_i \mathscr{L}_{n+1}^{(1)}(x_i)}$$

from which

$$\alpha_i = \frac{[(n+1)!]^2}{\mathscr{L}_{n+1}^{(1)}(x_i)\mathscr{L}_{n+2}(x_i)}$$

Table 7

Zeros of Laguerre polynomials (degree $n+1$)	α_i	$\alpha_i \times$ exponential of zeros
$n=0$	$n=0$	$n=0$
1.000 000 0	1.000 000 0	2.718 281 8
$n=1$	$n=1$	$n=1$
.585 786 4	.853 553 4	1.533 326 0
3.414 213 6	.146 446 6	4.450 957 3
$n=2$	$n=2$	$n=2$
.415 774 6	.711 093 0	1.077 692 8
2.294 280 4	.278 517 7	2.762 143 0
6.289 945 1	†.(1)103 892 5	5.601 094 6
$n=3$	$n=3$	$n=3$
.322 547 7	.603 154 1	.832 739 1
1.745 761 1	.357 418 6	2.048 102 4
4.536 620 3	.(1)388 879 1	3.631 146 3
9.395 070 9	.(3)539 294 7	6.487 145 1
$n=4$	$n=4$	$n=4$
.263 560 3	.521 755 6	.679 094 0
1.413 403 1	.398 666 8	1.638 487 9
3.596 425 8	.(1)759 424 5	2.769 443 2
7.085 810 0	.(2)361 175 9	4.315 656 9
12.640 800 8	.(4)233 699 7	7.219 186 4
$n=5$	$n=5$	$n=5$
.222 846 6	.458 964 7	.573 535 5
1.188 932 1	.417 000 8	1.369 252 5
2.992 736 3	.113 373 4	2.260 684 6
5.775 143 6	.(1)103 992 0	3.350 524 6
9.837 467 4	.(3)261 017 2	4.886 826 8
15.982 874 0	.(6)898 547 9	7.849 015 9

† The number in the parentheses stands for the number of zeros between the decimal point and the first significant figure.

where x_i is a zero of the Laguerre polynomial of degree $(n+1)$. For convenience, values of α_i for $n = 1, 2, \ldots, 5$ are also given in Table 7. A fuller list can be found in [29]. The error term for the *Laguerre–Gauss* integrating formula

$$\int_0^\infty \exp(-x)\gamma(x)\,dx = \sum_{i=0}^n \alpha_i\gamma(x_i) \tag{90}$$

is

$$E_I = \frac{[(n+1)!]^2}{(2(n+1))!}\,\gamma^{(2n+2)}(\eta) \qquad \eta \in (0, \infty)$$

Clearly, since η belongs to the infinite interval, the error bound E_I will only be meaningful for the function γ which has $2(n+1)$th derivatives bounded on all $(0, \infty)$. That is, the integrating formula should be avoided for functions where this is not true.

The Laguerre–Gauss integrating formula can also be used for definite integrals on $[0, \infty]$ in which the weight function $\exp(-x)$ does not explicitly occur. This is simply achieved by writing

$$\int_0^\infty f(x)\,dx \qquad \text{as} \qquad \int_0^\infty \exp(-x)\exp(x)f(x)\,dx$$

The effect of multiplication of $f(x)$ by $\exp(x)$ is simply to alter the weights α_i to $\alpha_i\exp(x_i)$ in the Laguerre–Gauss formula (90) with $\gamma(x)$ replaced by $f(x)$. It is a simple matter to obtain the new coefficients; however, we have also tabulated these coefficients in Table 7 for $n = 0, 1, \ldots, 5$.

Example 12 Evaluate the definite integral $I = \int_0^\infty \exp(-x)\cos(bx)\,dx$ for $b = 1, 2, \pi$, using the Laguerre–Gauss integration formula with $n = 0, 1, 2, 3, 4$.

The formula (9), using the appropriate entries from Table 7, produced the results given in Table 8. The exact value of the integral I is $1/(1+b^2)$, the value of which is given in the last row of Table 8.

Table 8

n	$I\,(b=1)$	$I\,(b=2)$	$I\,(b=\pi)$	$|E|$
0	0.540 302	−0.416 147	−1	$0.5b^2$
1	0.570 209	0.456 991	−0.266 255	$0.1667b^4$
2	0.476 521	0.455 074	0.360 068	$0.05b^5$
3	0.502 494	0.110 216	0.563 807	$0.0143b^8$
4	0.500 539	0.118 383	0.264 929	$0.0039b^{10}$
Theoretical value of I	0.5	0.2	0.091 999 7	

The effect of b in this example is quite apparent. The larger is b, the more rapidly oscillatory becomes the integrand $\cos{(bx)}$. This, as we would expect, is mirrored in the numerical results by correspondingly less accurate approximations. Therefore, to obtain an acceptable degree of accuracy for larger b an increased number of tabular points would have to be taken.

Exercises

47. Program the Laguerre–Gauss integrating formula for an arbitrary function $\gamma(x)$ and a number of tabular points $n+1 < 6$ using the zeros and coefficients given in Table 7.

48. Use your program from Exercise 47 to evaluate the integral $\int_0^\infty \exp{(-x)} \sin{(bx)}\,dx$ whose theoretical value is $b/(1+b^2)$. How large would n have to be to guarantee an accuracy of (i) 10^{-4}, (ii) 10^{-8}?
    Do your numerical results approach this accuracy for the values of n used in the program of Exercise 47?

49. Using the modified coefficients $\alpha_i \exp{(x_i)}$ contained in column 3 of Table 7, determine, using a modification to the program of Exercise 47, the value of the integrals:

$$\text{(i)} \quad \int_0^\infty \exp{(-2x)}\,dx$$

$$\text{(ii)} \quad \int_0^\infty \frac{dx}{\exp{(x)} + \exp{(-x)} - 1}$$

50. Evaluate $\int_0^\infty dx/(\exp{(x)} + \exp{(-x)})]$ using the Gauss–Laguerre formulas for $n = 1, 2, \ldots, 5$.

51. Consider the following alternative approach to evaluate the integral of Exercise 50. Since $\exp{(-x)} \to 0$ "quickly" with increasing x, an approximation to the required value of the integral may be obtained by considering the integral $\int_0^\infty \exp{(-x)}\,dx$, which can be calculated analytically, and adding to it the area between the curves $y = 1/(\exp{(x)} + \exp{(-x)})$ and $y = \exp{(-x)}$ for $0 \le x \le X$, where $\exp{(-X)} < 10^{-p}$, p a chosen integer. Use the program of Exercise 25 for the composite trapezoidal rule to evaluate the integrals $\int_0^X \exp{(-x)}\,dx$ and

$$\int_0^X \frac{dx}{\exp{(x)} + \exp{(-x)}}$$

for n, the parameter of the program, taking the values $n = 10, 20, 40$ for values of $p = 5, 10$. How do the results compare with the value obtained in Exercise 50?

We now consider the infinite interval $(-\infty, \infty)$. In this case we look at integrals of the form $I = \int_{-\infty}^\infty \exp{(-x^2)}\gamma(x)\,dx$. The polynomials which are orthogonal on the entire real line $-\infty < x < \infty$ with respect to the weight function $w(x) = \exp{(-x^2)}$ are the *Hermite* polynomials† defined by

$$H_m(x) = (-1)^m \exp{(x^2)} \frac{d^m}{dx^m} \exp{(-x^2)} \qquad m = 0, 1, 2, \ldots$$

† These polynomials should not be confused with the polynomials $h_i(x)$ used in the definition of Hermite interpolation in Chapter 4.

or

$$H_m(x) = \sum_{j=0}^{[m/2]} \frac{(-1)^j m!(2x)^{m-2j}}{j!(m-2j)!}$$

where $[m/2]$ is the integer part of $m/2$.

The zeros of these polynomials have been tabulated for a large number of values of n. Table 9 contains values of the zeros for the Hermite polynomials of up to degree 5. The coefficients in the resulting $(n+1)$-point Gaussian integrating formula

$$I = \int_{-\infty}^{\infty} \exp(-x^2)\gamma(x)\,dx \doteq \sum_{i=0}^{n} \alpha_i \gamma(x_i) \tag{90}$$

with x_i, $i = 0, 1, \ldots, n$, the zeros of the Hermite polynomial of degree $n+1$, are given by

$$\alpha_i = \frac{\pi^{1/2} 2^n n!}{(n+1)\{H_n(x_i)\}^2}$$

These coefficients are tabulated in Table 9 for $n = 0, 1, \ldots, 5$. A fuller list may be found once again in Krylov [29]. The error term for the Hermite–Gauss integrating formula (90) is given by

$$E_I = \frac{(n+1)!\sqrt{\pi}\gamma^{(2n+2)}(\eta)}{2^{n+1}(2(n+1))!} \tag{91}$$

Table 9   Zeros of Hermite polynomials of degree $n+1$ and associated coefficients

	Zeros x_i	Coefficients
$n = 0$		
	0	1.772 453 85
$n = 1$		
	±0.707 106 78	0.886 226 93
$n = 2$		
	0	1.181 635 90
	±1.224 744 87	0.295 408 98
$n = 3$		
	±0.524 647 62	0.804 914 09
	±1.650 680 12	0.081 312 84
$n = 4$		
	0	0.945 308 72
	±0.958 572 46	0.393 619 32
	±2.020 182 87	0.019 953 24
$n = 5$		
	±0.436 077 41	0.724 629 59
	±1.335 849 07	0.157 067 32
	±2.350 604 97	0.004 530 01

Exercises

52. Program the Hermite–Gauss integrating formula for an arbitrary function $\gamma(x)$ and a number of tabular points $(n+1) < 6$ using the zeros and coefficients given in Table 9 which are to be read as data.

53. Using the program of Exercise 52, evaluate the definite integrals $\int_{-\infty}^{\infty} |x| \exp(-mx^2) \, dx$, $m = 1, 2, 10$, for $n = 0, 1, \ldots, 5$. How do the computed solutions compare with the theoretical solution? Are the results consistent with the error bound (91)?

7 Composite Gaussian Integration

In the final section of this chapter we will discuss briefly the natural application of composition, as was applied to the Newton–Cotes formulas in section 4, to the Gaussian integrating formulas. The motivation for considering such formulas is similar to that outlined for the Newton–Cotes formulas: the higher the order of the integrating formulas, the higher order is the derivative occurring in the error term, and the consequent greater difficulty in obtaining its form and a resulting bound. Further, with the higher-order Gaussian formulas it is clear that a greater amount of information is needed with regard to the zeros of the associated polynomials and the coefficients in the integrating formula. Therefore, to attain a prescribed accuracy, an alternative to choosing a higher-order integrating formula is to make the size of the interval sufficiently small so that the error bound is reduced by the required amount.

Consider $\int_a^b f(x) \, dx$. In determining the Legendre–Gauss formula, we transformed the interval $x \in [a, b]$ to the interval $y \in [-1, 1]$ by the linear transformation $y = (2x - (a + b))/(b - a)$. Then

$$\int_a^b f(x) \, dx = \tfrac{1}{2}(b-a) \int_{-1}^{1} f(\tfrac{1}{2}(b-a)y + \tfrac{1}{2}(a+b)) \, dy \tag{92}$$

In proposing the Legendre–Gauss formula for the right-hand side of Eq. (92) we clearly now incur the error given by

$$E_I = \frac{f^{(2n+2)}(\zeta)}{(2n+2)!} \left(\frac{b-a}{2}\right)^{2n+2} \int_{-1}^{1} \pi_{n+1}^2(x) \, dx \tag{93}$$

from Eq. (72) (why?). Let us now consider dividing the interval into two equal intervals such that $a = x_0$, $x_1 = (a+b)/2$, $x_2 = b$. Then propose an $(n+1)$-point Legendre–Gauss integrating formula on each interval. Then the errors in each interval will be given by substituting x_0, x_1, and x_2 into Eq. (93). For $x \in [x_0, x_1]$,

$$E_{I_1} = \frac{f^{(2n+2)}(\zeta_1)}{(2n+2)!} \left(\frac{x_1 - x_0}{2}\right)^{2n+2} \int_{-1}^{1} \pi_{n+1}^2(x) \, dx \qquad \zeta \in (x_0, x_1)$$

and for $x \in [x_1, x_2]$,

$$E_{I_2} = \frac{f^{(2n+2)}(\zeta_2)}{(2n+2)!} \left(\frac{x_2 - x_1}{2}\right)^{2n+2} \int_{-1}^{1} \pi_{n+1}^2(x) \, dx \qquad \zeta_2 \in (x_1, x_2)$$

If the $(2n+2)$nd derivative is continuous in (a, b) we find that the error over (a, b) obtained by adding E_{I_1} and E_{I_2} and substituting for x_0, x_1 and x_2 is given by

$$E_I = \left\{ f^{(2n+2)}(\zeta_1)\left(\frac{b-a}{4}\right)^{2n+2} \right.$$

$$\left. + f^{(2n+2)}(\zeta_2)\left(\frac{b-a}{4}\right)^{2n+2} \right\} \frac{1}{(2n+2)!} \int_{-1}^{1} \pi_{n+1}^2(x)\, dx$$

$$= \frac{1}{2^{2n+1}} \left\{ f^{(2n+2)}(\zeta)\left(\frac{b-a}{2}\right)^{2n+2} \right\} \frac{1}{(2n+2)!} \int_{-1}^{1} \pi_{n+1}^2(x) \qquad \zeta \in (a, b)$$

(94)

Thus the effect of introducing two intervals on $[a, b]$ and using $(n+1)$-point integrating formulas in each interval is to introduce the factor $1/(2^{2n+1})$ into the error in (94) compared with the error term (93). By dividing the interval into m subintervals, it is a simple matter to see that the effect is to reduce the error term by a factor $1/m^{2n+1}$. Therefore, by choosing m sufficiently large we can make the error term arbitrarily small. This, of course, can be done even for quite modest n. To obtain the composite Legendre–Gauss integrating formula, let us divide the interval $[a, b]$ into $m+1$ equal subintervals such that

$$a = x_0 < x_1 < \cdots < x_m = b \quad \text{and} \quad x_{i+1} - x_i = h \quad i = 0, 1, \ldots, m-1$$

On one subinterval we have

$$\int_{x_i}^{x_{i+1}} f(x)\, dx = \frac{h}{2} \int_{-1}^{1} f\left(\frac{hy}{2} + \tfrac{1}{2}(x_i + x_{i+1})\right) dy$$

using (92), and hence the integral on the ith subinterval,

$$I_i \doteq \frac{h}{2} \sum_{j=0}^{n} \alpha_j f\left(\frac{hy_j}{2} + \tfrac{1}{2}(x_i + x_{i+1})\right) \qquad i = 0, 1, \ldots, m-1 \qquad (95)$$

where y_j, $j = 0, 1, \ldots, n$, are the zeros of the $(n+1)$th-degree Legendre polynomials and α_j the associated coefficients (see Table 6). Hence the integral over $[a, b]$ is represented as

$$I \doteq \sum_{i=0}^{n-1} I_i = \frac{h}{2} \sum_{i=0}^{m-1} \sum_{j=0}^{n} \alpha_j f\left(\frac{hy_j}{2} + \tfrac{1}{2}(x_i + x_{i+1})\right) \qquad (96)$$

The composite Legendre–Gauss integrating formula therefore can be proposed which requires a minimal number of zeros and coefficients to be stored. However, with increasing m the number of function values will increase and if these are expensive computationally, the cost of storage will have to be weighed against the cost of computation. Further, note that unlike the composite Newton–Cotes formulas no function values are shared with the subintervals $[x_i, x_{i+1}]$ (other than a single tabular point per interval) since all the zeros y_j are internal to these subintervals.

Example 13   Evaluate $\int_0^1 x^{3/2} \exp(x^{5/2})\,dx$ by the Legendre–Gauss integrating formula using $n+1$ tabular points where n takes the values 1, 2, 3, 4, 5, and by the composite Legendre–Gauss formula (96) using two points in each of four subintervals $h = 0.25$ in length.

To employ the Legendre–Gauss formula to the integral we employ once more the linear transformation (67) so that

$$\int_0^1 x^{3/2} \exp(x^{5/2})\,dx = \tfrac{1}{2}\int_{-1}^1 f(y/2 + 1/2)\,dy \tag{97}$$

Using the Legendre–Gauss formula as indicated, the results shown in Table 10 were obtained using the zeros and coefficients given in Table 6. In comparison, the value obtained using the composite formula (96) with $m = 4$, $n = 1$, as shown in column 3 of Table 10. The correct value of the integral is given in the final column.

Table 10

n	Legendre–Gauss	Composite Legendre–Gauss $m = 4,\ n = 1$	Theoretical
1	0.608 438 82	0.687 052 1	0.687 312 73
2	0.685 465 48		
3	0.687 194 21		
4	0.687 292 93		
5	0.687 305 21		

As can be seen from Table 10, the composite formula produces a more accurate result than the corresponding results obtained from the Legendre–Gauss formula with $n = 1$ and 2 for this example, and for the chosen values of m and n the remaining results in column 2 are more accurate than those in column 3. However, to obtain more accurate results using the composite formula is a simple task since we require no other zeros or coefficients than those already used in (96), namely $\alpha_0 = \alpha_1 = 1$ and $y_0 = -y_1 = 1/\sqrt{3}$. Thus choosing m larger, i.e., taking a smaller h, we can obtain more accurate approximations to the required integral. We leave this to Exercise 55.

Exercises

54. Program the composite Legendre–Gauss formula for a parameter value $n+1$ representing the number of points used in, an additional, parameter m subintervals. The required zeros and coefficients should be read as data.
55. Use your program of Exercise 54 for the integral of Example 13 with $n = 1$ for values of $m = 6, 8, 10$, and 12. Compare your results with those given in Table 10.

56. Compute the integral $\int_0^1 x^{11/2} \exp(x^{13/2}) \, dx$ using:
   (a) the Legendre–Gauss formula for $n = 1, 2, \ldots, 5$;
   (b) the composite formula using the program of Exercise 54, for $n = 1$, $m = 4, 6$, 8, 10, and 12;
   (c) the composite formula using the program of Exercise 54 for $n = 2$, $m = 4, 6$, 8, 10, and 12.
   Compare your results with the theoretical solution.

CHAPTER 8

Numerical Differentiation

1 Introduction

In the previous chapter we discussed one particular application of the Lagrangian interpolation polynomial. In the present chapter we consider briefly an alternative application of the Lagrangian interpolation polynomial, namely to obtain numerical approximations to a specified derivative of a function $f(x)$ which may be either known analytically but is too complicated to differentiate, or is known empirically at a finite number of points. Consider a function $f(x)$ of the single real variable x defined on an interval I containing the tabular points $(x_0, x_1, \ldots, x_n)$. We will assume $f(x)$ to be continuously differentiable on I a sufficient number of times to ensure what follows is valid.

Let the polynomial interpolant of f be represented by $y(x)$ (see Chapter 4, Eqs. (11), (13), (15)). Then we will approximate the derivative $f^{(p)}(x)$, p a given (specified) positive integer, by $y^{(p)}(x)$ obtained by differentiating $y(x)$, p times. In theory any p may be considered. However, as we will see, for practical reasons p will be restricted to small, just 1 or 2.

From Chapter 4, Eq. (17), we have an expression for the error term:

$$E(x) = f(x) - y(x) = \frac{f^{(n+1)}(\zeta)}{(n+1)!} \pi_{n+1}(x) \tag{1}$$

where $\zeta(x)$ is some point contained in the interval spanned by $x_0, x_1, \ldots, x_n$ and x. In particular, the error term of the first derivative is given by

$$E^{(1)}(x) = f^{(1)}(x) - y^{(1)}(x) = \frac{d}{dx} \frac{[f^{(n+1)}(\zeta)\pi_{n+1}(x)]}{(n+1)!}$$

$$= \frac{\pi_{n+1}^{(1)}(x)f^{(n+1)}(\zeta) + (d/dx)f^{(n+1)}(\zeta)\pi_{n+1}(x)}{(n+1)!} \tag{2}$$

If we require the derivative $f^{(1)}$ at a tabular point x_i, $i = 0, 1, \ldots, n$, then $\pi_{n+1}(x_i) = 0$ (why?) and we are left with the result

$$E^{(1)}(x_i) = f^{(1)}(x_i) - y^{(1)}(x_i) = \frac{\pi_{n+1}^{(1)}(x_i) f^{(n+1)}(\zeta)}{(n+1)!} \qquad i = 0, 1, \ldots, n \qquad (3)$$

To obtain the error in the approximation at nontabular points we have to obtain the derivative of $f^{(n+1)}(\zeta)$, remembering that $\zeta = \zeta(x)$ is an unknown function of x. We have met this problem in Chapter 7, namely the result contained in Theorem 2.

Analogous to this result we have

$$\frac{1}{(n+1)!} \frac{d}{dx} f^{(n+1)}(\zeta) = \frac{1}{(n+2)!} f^{(n+2)}(\eta) \qquad \eta \in (x_0, x_1, \ldots, x_n, x) \qquad (4)$$

(see Exercise 1 for proof). Using Eq. (4) in Eq. (2), we obtain

$$E^{(1)}(x) = f^{(1)}(x) - y^{(1)}(x) = \frac{\pi_{n+1}^{(1)}(x) f^{(n+1)}(\eta)}{(n+1)!} + \frac{\pi_{n+1}(x) f^{(n+2)}(\zeta)}{(n+2)!} \qquad (5)$$

To derive the error $E^{(p)}(x) = f^{(p)}(x) - y^{(p)}(x)$, we need the natural generalization of Eq. (4) given in the following result.

Theorem 1 *Let*

$$\frac{\pi_{n+1}(x) f^{(n+1)}(\zeta)}{(n+1)!}$$

be the error term in the Lagrangian interpolation formula. Then if $f^{(n+j)}(x)$ is continuous for $j = 1, 2, \ldots$,

$$\frac{1}{(n+1)!} \frac{d^j}{dx^j} f^{(n+1)}(\zeta) = \frac{j!}{(n+1+j)!} f^{(n+j+1)}(\eta_j) \qquad j = 1, 2, \ldots$$

where η_j is contained within the interval spanned by $x_0, x_1, \ldots, x_n$, and x.

The proof of this result follows similarly to that given for Theorem 2 of Chapter 7; the details are left to the reader in Exercise 2.

Now using Theorem 1 and an application of Leibnitz' rule (see Jones and Jordan [27]) we can obtain the form of the error in the approximation for the pth derivative of $f(x)$, namely

$$E^{(p)}(x) = f^{(p)}(x) - y^{(p)} = \sum_{j=0}^{p} \frac{p!}{j!} \frac{\pi_{n+1}^{(j)}(x)}{(n+1+p-j)!} f^{(n+p+1-j)}(\eta_j) \qquad p \leq n \qquad (6)$$

where $y^{(p)}(x)$ is the pth derivative of the Lagrangian interpolating polynomial given by

$$y^{(p)}(x) = \frac{d^p}{dx^p} \left[\sum_{j=0}^{n} \prod_{\substack{i=0 \\ i \neq j}}^{n} \frac{(x - x_i)}{(x_j - x_i)} f_j \right] = \sum_{j=0}^{n} \left[\frac{d^p}{dx^p} \prod_{\substack{i=0 \\ i \neq j}}^{n} \frac{(x - x_i)}{(x_j - x_i)} \right] f_j \qquad (7)$$

Clearly the coefficients contained in the square brackets in Eq. (7) are independent of the function values of f. Consequently, these could be calculated once and for all and applied to the pth derivative of any given function f. However, as mentioned above we refrain from deriving expressions for these coefficients for general p since we will confine our attention to values of $p = 1$ and 2. The reason for this is given in the next section.

Exercises

1. Using an analysis analogous to that carried out in Theorem 2 of Chapter 7, prove relation (4).
2. Prove Theorem 1.
3. Verify Eq. (6).
4. Assume the number of data points $n+1$ is even. Show that, by *choosing* the abscissas x_i, $i = 0, 1, \ldots, n$, symmetrically about the point a (say) at which the first derivative of $f(x)$ is required, the error term is given by

$$E(x) = f^{(1)}(x) - y^{(1)}(x) = \frac{1}{(n+2)!} f^{(n+2)}(\zeta) \prod_{j=0}^{(n-1)/2} [-(a-x_j)^2]$$

2 The Error Term; Equally Spaced Data

Hence forth for convenience we will assume the tabular points are equally spaced; it is a simple matter to extend our discussion to the case of unequally spaced points. As before, we will assume the mesh spacing between the tabular points is h. The error term for the formulas for numerical differentiation on equally spaced tabular points is then given by an obvious modification of Eq. (6):

$$E^{(p)}(m) = f^{(p)}(m) - y^{(p)}(m)$$

$$= h^{n+1} \sum_{j=0}^{p} h^{-j} \frac{p!}{j!} \frac{\pi_{n+1}^{(j)}(m)}{(n+1+p-j)!} f^{(n+1+p-j)}(\eta_j) \qquad p \le n \qquad (8)$$

where we have used the obvious notation $x = x_0 + mh$, and $\pi_{n+1}^{(j)}(m)$ is the jth derivative with respect to m of the product polynomial $\prod_{i=0}^{n}(m-i)$. From the right-hand side of Eq. (8) we would surmise that for given fixed n and p that as $h \to 0$, the error would tend to zero. For exact computation (no round-off) this will be the case. However, we are not able to compute without making round-off, and whether the total error (the error in Eq. (8) plus the round-off error) tends to zero will depend upon the way the round-off error behaves as h tends to zero. To discover this behavior, let us assume the function values f are subject to a round-off error ε_j in the rth decimal place. Then instead of the exact computation $y^{(p)}(m)$ we will calculate $y_{\text{comp}}^{(p)}$ (say). Now

$$y^{(p)}(m) = h^{-p} \sum_{j=0}^{n} \left[\frac{d^p}{dm^p} \prod_{\substack{i=0 \\ i \ne j}}^{n} \frac{(m-i)}{(j-i)} \right] f_j \qquad (9)$$

and so

$$y^{(p)}_{\text{comp}}(m) = h^{-p} \sum_{j=0}^{n} \left[\frac{d^p}{dm^p} \prod_{\substack{i=0 \\ i \neq j}}^{n} \frac{(m-i)}{(j-i)} \right] (f_j + \varepsilon_j) \qquad (10)$$

Subtract Eq. (9) from Eq. (10) and denote the round-off error by $R_p(m)$; then we obtain

$$R_p(m) = h^{-p} \sum_{j=0}^{n} \left[\frac{d^p}{dm^p} \prod_{\substack{i=0 \\ i \neq j}}^{n} \frac{(m-i)}{(j-i)} \right] \varepsilon_j$$

and hence

$$|R_p(m)| \leq h^{-p} \sum_{j=0}^{n} \left| \frac{d^p}{dm^p} \prod_{\substack{i=0 \\ i \neq j}}^{n} \frac{(m-i)}{(j-i)} \right| |\varepsilon_j| \qquad (11)$$

If we let the maximum round-off error in the function values f_i be denoted by ε, then[†] $|\varepsilon| \leq 5 \cdot 10^{-r-1}$, and hence Eq. (11) gives

$$|R_p(m)| \leq h^{-p} 5 \cdot 10^{-r-1} \sum_{j=0}^{n} \left| \frac{d^p}{dm^p} \prod_{\substack{i=0 \\ i \neq j}}^{n} \frac{(m-i)}{(j-i)} \right|$$

We therefore see that the round-off error is inversely proportional to a power of h, and consequently if h is small the round-off error can become large. We therefore have the two errors working against one another. For large h the truncation error given by Eq. (6) is large and the round-off error is small. For small h the truncation error is small whereas the round-off error given by Eq. (11) is large. There must therefore be an optimum value of h, i.e. a value of h which will minimize both the errors of truncation and round-off. We omit a discussion of the derivation of the expression for such an h; the interested reader is referred to Ralston [34]. The motivation for restricting p to be small now becomes clear. To allow p large will produce numerical differentiation formulas which have disastrous round-off properties; to obtain any significance in the computation would require an unrealistic number of figures in the computation. The calculation of derivatives numerically is therefore an unstable process, especially when the function values f_i are subject to round-off errors themselves. A better approach therefore, in general, would be to use least squares approximation to the given tabular data and then differentiate the resulting formula. The effect of the least squares approach is to smooth out the round-off errors.

Alternatively, spline approximation of the given data may be used as described in Chapter 4, section 10. Differentiating the resulting spline will then give an approximation to the required derivative. In either of these two cases, if p becomes large then the degree of the associated polynomials must

---

[†] If chopped arithmetic is used then $|\varepsilon| \leq 10^{-r}$.

also increase. For $p = 1$, the convergence of the derivative of the interpolating cubic spline to the required derivative of the function $f(x)$ is proved in Johnson and Riess [25].

3 Numerical Differentiation Formulas for $p = 1$

Consider $p = 1$. To obtain the required approximation to $f^{(1)}(x)$, differentiate the Lagrangian interpolating polynomial $y(m)$ on equally spaced points, namely

$$y(mh) = \sum_{j=0}^{n} \prod_{\substack{i=0 \\ i \neq j}}^{n} \frac{(m-i)}{(j-i)} f_j$$

We then obtain

$$y^{(1)}(mh) = h^{-1} \sum_{j=0}^{n} \left[\frac{d}{dm} \sum_{\substack{i=0 \\ i \neq j}}^{n} \frac{(m-i)}{(j-i)} \right] f_j$$

$$= h^{-1} \sum_{j=0}^{n} \left[\sum_{\substack{k=0 \\ k \neq j}}^{n} \frac{1}{(j-k)} \prod_{\substack{i=0 \\ i \neq j \\ i \neq k}}^{n} \frac{(m-i)}{(j-i)} \right] f_j \tag{12}$$

Example 1   Consider $n = 1$ (two points)

$$y^{(1)}(mh) = h^{-1} \sum_{j=0}^{1} \left[\sum_{\substack{k=0 \\ k \neq j}}^{1} \frac{1}{(j-k)} \prod_{\substack{i=0 \\ i \neq j \\ i \neq k}}^{1} \frac{(m-i)}{(j-i)} \right] f_j$$

$$= h^{-1} \left\{ \sum_{\substack{k=0 \\ k \neq 0}}^{1} \left[\frac{1}{-k} \prod_{\substack{i=1 \\ i \neq k}}^{1} \frac{(m-i)}{-i} \right] f_0 + \sum_{\substack{k=0 \\ k \neq 1}}^{1} \left[\frac{1}{(1-k)} \prod_{\substack{i=1 \\ i \neq k}}^{1} \frac{(m-i)}{(1-i)} \right] f_1 \right.$$

so

$$y^{(1)}(mh) = \frac{f_1 - f_0}{h} \tag{13}$$

From Example 1 we see that using two points x_0 and x_1 the approximation to the first derivative of f is the gradient of the line passing through the points (x_0, f_0) and (x_1, f_1) (see Figure 1).

If the function f is a straight line then the approximation and the actual derivative of f will coincide; the error term will be zero. Otherwise the gradient of the straight line will differ at some (all?) points in $[x_0, x_1]$ from the first derivative of f; the error term will be nonzero.

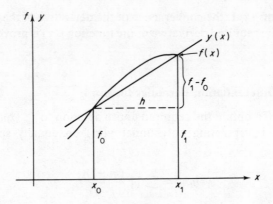

Figure 1

Example 2   Consider $n = 2$ (three points)

$$y^{(1)}(mh) = h^{-1}\left[\sum_{j=0}^{2}\left[\sum_{\substack{k=0\\k\neq j}}^{2}\frac{1}{(j-k)}\prod_{\substack{i=0\\i\neq j\\i\neq k}}^{2}\frac{(m-i)}{(j-i)}\right]f_j\right]$$

$$= h^{-1}\left[\sum_{\substack{k=0\\k\neq 2}}^{2}\left[\frac{1}{(2-k)}\prod_{\substack{i=0\\i\neq j\\i\neq k}}^{2}\frac{(m-i)}{(2-i)}\right]f_2 + \sum_{\substack{k=0\\k\neq 1}}^{2}\left[\frac{1}{(1-k)}\prod_{\substack{i=0\\i\neq 1\\i\neq k}}^{2}\frac{(m-i)}{(1-i)}\right]f_1\right.$$

$$\left.+ \sum_{\substack{k=0\\k\neq 0}}^{2}\left[\frac{1}{-k}\prod_{\substack{i=0\\i\neq 0\\i\neq k}}^{2}\frac{(m-i)}{-i}\right]f_0\right]$$

$$= h^{-1}\left[\frac{1}{2}\frac{(m-1)}{(2-1)}f_2 + \frac{1}{(2-1)}\frac{mf_2}{2} + \frac{1}{1}\frac{(m-2)}{(1-2)}f_1 + \frac{1}{(1-2)}\frac{mf_1}{1}\right.$$

$$\left.+ \frac{1}{-1}\frac{(m-2)}{(-2)}f_0 + \frac{1}{-2}\frac{(m-1)}{(-2)}f_0\right]$$

so

$$y^{(1)}(mh) = \frac{(2m-1)f_2 - 4(m-1)f_1 + (2m-3)f_0}{2h} \tag{14}$$

In Example 2 the function f has been interpolated by the quadratic polynomial passing through the three points (x_0, f_0), (x_1, f_1), and (x_2, f_2) (see Figure 2). In contrast to Example 1, the gradient of the approximating quadratic y will depend upon the position mh in a linear manner. If f is a quadratic then the two functions f and y will coincide and once again at any point in $[x_0, x_2]$ the gradient of y and the first derivative $f^{(1)}(mh)$ will coincide; the error term is zero. If f is not a quadratic then the approximation $y^{(1)}(mh)$ to $f^{(1)}(mh)$ will

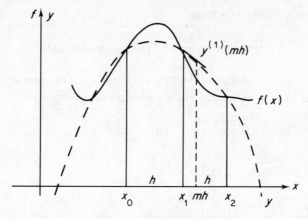

Figure 2

be in error. In particular, we note the characteristic behavior of the derivative of the interpolating polynomial differing considerably at the data points even though the functions $y(x)$ and $f(x)$ agree there (see Figure 2).

The particular form of the error term for $p = 1$ is obtained simply from Eq. (8), namely

$$E_1^{(1)}(mh) = h^{n+1} \sum_{j=0}^{1} h^{-j} \frac{1}{j!} \frac{\pi_{n+1}^{(j)}(m)}{(n+2-j)!} f^{(n+2-j)}(\eta_j) \tag{15}$$

Example 3 Considering Eq. (15) with $n = 1$.

$$E^{(1)}(mh) = h^2 \left[\frac{\pi_2}{3!}(m) f^{(3)}(\eta) + h^{-1} \frac{\pi_2^{(1)}(m)}{2!} f^{(2)}(\eta_1) \right] \tag{16}$$

If we choose m so as to coincide with a tabular point ($m = 0$ or 1), $\pi_2(m) = 0$ and the error term reduces to

$$E_1^{(1)}(mh) = h \frac{\pi_2^{(1)}(m)}{2!} f^{(2)}(\eta_1) \qquad m = 0, 1$$

It is easily seen that

$$\pi_2^{(1)}(0) = -1 \qquad \pi_2^{(1)}(1) = 1$$

and so

$$E_1^{(1)}(0) = -\tfrac{1}{2} h f^{(2)}(\eta); \quad E_1^{(1)}(1) = \tfrac{1}{2} h f^{(2)}(\eta) \qquad \eta \in (x_0, x_1)$$

and Eq. (8) becomes

$$f^{(1)}(mh) = \frac{f_1 - f_0}{h} - (-1)^m \tfrac{1}{2} h f^{(2)}(\eta) \qquad \eta \in (x_0, x_1); m = 0, 1$$

For nontabular m, the error term is given in general by Eq. (16). However, if we choose m so that $\pi_2^{(1)}(m) = 0$ (i.e. we require the first derivative of f at the point where $\pi_2^{(1)}(m) = 0$), then

$$E_1^{(1)}(mh) = h^2 \frac{\pi_2(m)}{3!} f^{(3)}(\eta) \qquad \eta \in (x_0, x_1) \qquad (17)$$

Since $\pi_2^{(1)}(m) = (d/dm)(m(m-1)) = 2m - 1$ it is easy to see that approximating $f^{(1)}(mh)$ at $\frac{1}{2}h$ (the midpoint) attains an increase in accuracy. In this case,

$$f^{(1)}(mh) = \frac{f_1 - f_0}{h} \frac{-h^2}{24} f^{(3)}(\eta) \qquad \eta \in (x_0, x_1); \; m = \frac{1}{2} \qquad (18)$$

Example 4 Consider $n = 2$ in Eq. (15).

$$E_2^{(1)}(mh) = h^3 \left[\frac{\pi_3(m)}{4!} f^{(4)}(\eta_0) + \frac{h^{-1} \pi_3^{(1)}(m)}{3!} f^{(3)}(\eta_1) \right] \qquad (19)$$

Choosing to approximate the derivative of f at a tabular point ($m = 0, 1, 2$) produces an error term

$$E_2^{(1)}(mh) = \frac{h^2 \pi_3^{(1)}(m)}{3!} f^{(3)}(\eta) \qquad \eta \in (x_0, x_2)$$

and hence

$$f^{(1)}(mh) = \frac{(2m-1)f_2 - 4(m-1)f_1 + (2m-3)f_0}{2h} + \frac{h^2 \pi_3^{(1)}(m)}{3!} f^{(3)}(\eta) \quad (20)$$

Now $\pi_3^{(1)}(m) = 3m^2 - 6m + 2$. Hence if $m = 0$, Eq. (20) becomes

$$f^{(1)}(0) = \frac{-f_2 + 4f_1 - 3f_0}{2h} + \frac{2h^2}{3!} f^{(3)}(\eta)$$

if $m = 1$ then

$$f^{(1)}(h) = \frac{f_2 - f_0}{2h} \frac{-h^2}{3!} f^{(3)}(\eta) \qquad (21)$$

if $m = 2$ then

$$f^{(1)}(2h) = \frac{3f_2 - 4f_1 + f_0}{2h} + \frac{2h^2}{3!} f^{(3)}(\eta)$$

We note the (obvious) fact that approximating the derivative $f^{(1)}(x)$ at the midpoint produces a formula comprising two function values but retaining the accuracy of the three-point formula. Equation (21) is of course just Eq. (18) expressed on an interval of length $2h$. This "bonus" accuracy is not an isolated occurrence and in fact it may be shown that for an odd number $(n+1)$ of tabular points, the numerical differentiation formula is exact for polynomial f of degree $n+1$ if the tabular points are placed symmetrically with respect to the point at which the approximation to the derivative of f is required. This is exemplified by Eq. (18) for $n = 1$.

Exercises

5. Derive the numerical differentiation formula for the first derivative of $f(x)$ using four tabular points.
6. Obtain values of the first derivative of the function whose tabulated values are given in Table 1 for $x = 0, 0.2, 0.3, 0.4$, using the tabular points for $n = 1, 2$, and 3 as indicated. Given that the function is $\sin(x)$, how does the error behave as a function of the position of x relative to the end points of the interval $[0.0, 0.6]$ for each value of n?

Table 1

x	0.0	0.1	0.2	0.3	0.4	0.5	0.6
n							
1	0.0000						0.5646
2		0.0998		0.2955		0.4794	
3	0.0000						0.5646

$$f(x)$$

7. Derive the error term for $n = 3$ for m coincident with a tabular point.
8. Using your knowledge of the derivative of $\sin(x)$, are the error bounds pessimistic for $n = 1, 2$, and 3 in the light of the numerical results of Exercise 6?

4 Numerical Differentiation Formulas for $p = 2$

To obtain the formula for general n for $p = 2$, we use as a starting point the derivative of the Lagrangian interpolating polynomial obtained in the previous section, Eq. (12). Differentiating this formula, we find in a straightforward manner

$$y^{(2)}(mh) = h^{-2} \sum_{j=0}^{n} \left\{ \sum_{\substack{k=0 \\ k \neq j}}^{n} \frac{1}{(j-k)} \left[\sum_{\substack{l=0 \\ l \neq j \\ l \neq k}}^{n} \frac{1}{(j-l)} \prod_{\substack{i=0 \\ i \neq j \\ i \neq k \\ i \neq l}}^{n} \frac{(m-i)}{(j-i)} \right] \right\} f_j \tag{22}$$

The error term is given by substituting $p = 2$ in Eq. (8). By implication we must therefore have at least three points ($n = 2$) in order to propose an approximation to the second derivative of $f(x)$.

Example 5 Consider $n = 2$ in Eq. (22). Then

$$y^{(2)}(mh) = h^{-2} \left\{ \frac{(1)}{(-1)} \frac{(1)}{(2)} f_0 + \frac{(1)}{(-2)} \frac{(1)}{(1)} f_0 + \frac{(1)}{(1)} \frac{(1)}{(-1)} f_1 + \frac{(1)}{(-1)} \frac{(1)}{(1)} f_2 \right.$$
$$\left. + \frac{(1)}{(1)} \frac{(1)}{(2)} f_2 \right\}$$
$$= h^{-2}(f_0 - 2f_1 + f_2)$$

Thus as we would expect, the second derivative of $f(x)$ is given as a constant (the function is independent of m) valid at each point within the interval $[x_0, x_2]$.

Example 6 Consider $n = 3$ in Eq. (22). We obtain

$$y^{(2)}(mh) = h^{-2} \sum_{\substack{j=0}}^{3} \sum_{\substack{k=0 \\ k \neq j}}^{3} \frac{1}{(j-k)} \sum_{\substack{l=0 \\ l \neq j \\ l \neq k}}^{3} \frac{1}{(j-l)} \prod_{\substack{i=0 \\ i \neq j \\ i \neq k \\ i \neq l}}^{3} \frac{(m-i)}{(j-i)} f_j$$

$$= h^{-2} \left\{ \sum_{\substack{k=0 \\ k \neq 0}}^{3} \frac{1}{-k} \left[\sum_{\substack{l=0 \\ l \neq 0 \\ l \neq k}}^{3} \frac{1}{-l} \prod_{\substack{i=0 \\ i \neq 0 \\ i \neq k \\ i \neq l}}^{3} \frac{(m-i)}{-i} \right] f_0 \right.$$

$$+ \sum_{\substack{k=0 \\ k \neq 1}}^{3} \frac{1}{(l-k)} \left[\sum_{\substack{l=0 \\ l \neq 1 \\ l \neq k}}^{3} \frac{1}{(1-l)} \prod_{\substack{i=0 \\ i \neq 1 \\ i \neq k \\ i \neq l}}^{3} \frac{(m-i)}{(1-i)} \right] f_1$$

$$+ \sum_{\substack{k=0 \\ k \neq 2}}^{3} \frac{1}{(2-k)} \left[\sum_{\substack{l=0 \\ l \neq 2 \\ l \neq k}}^{3} \frac{1}{(2-l)} \prod_{\substack{i=0 \\ i \neq 2 \\ i \neq k \\ i \neq l}}^{3} \frac{(m-i)}{(2-i)} \right] f_2$$

$$+ \sum_{\substack{k=0 \\ k \neq 3}}^{3} \frac{1}{(3-k)} \left[\sum_{\substack{l=0 \\ l \neq k \\ l \neq 3}}^{3} \frac{1}{(3-l)} \prod_{\substack{i=0 \\ i \neq 3 \\ i \neq k \\ i \neq l}}^{3} \frac{(m-i)}{(3-i)} \right] f_3 \right\}$$

$$= h^{-2} \left\{ \frac{(6m-12)}{-6} f_0 + \frac{(6m-10)}{2} f_1 + \frac{(6m-8)}{-2} f_2 + \frac{(6m-6)}{6} f_3 \right\}$$

$$= h^{-2} \{ (2-m)f_0 + (3m-5)f_1 + (4-3m)f_2 + (m-1)f_3 \}$$

a linear function of the variable m.

The particular form of the error term for the numerical differentiation formula for the second derivative is given by

$$h^{n+1} \sum_{j=0}^{2} h^{-j} \frac{2!}{j!} \frac{\pi_{n+1}^{(j)}(m)}{(n+3-j)!} f^{(n+3-j)}(\eta_j)$$

Substituting the expressions for the derivatives of the polynomials, we find

$$E^{(2)}(mh) = h^{n+1} \left[\frac{2! \pi_{n+1}(m)}{(n+3)!} f^{(n+3)}(\eta_0) + h^{-1} \frac{2!}{1} \frac{\pi_{n+1}^{(1)}(m)}{(n+2)!} f^{(n+2)}(\eta_1) \right.$$

$$\left. + h^{-2} \frac{2!}{2!} \frac{\pi_{n+1}^{(2)}(m)}{(n+1)!} f^{(n+1)}(\eta_2) \right] \qquad (23)$$

where

$$\pi_{n+1}^{(1)}(m) = \sum_{j=0}^{n} \prod_{\substack{i=0 \\ i \neq j}}^{n} (m-i); \quad \pi_{n+1}^{(2)}(m) = \sum_{j=0}^{n} \sum_{\substack{k=0 \\ k \neq j}}^{n} \prod_{\substack{i=0 \\ i \neq j \\ i \neq k}}^{n} (m-i)$$

Example 7   Consider the error term for $n = 2$. Using Eq. (23),

$$E_2^{(2)}(mh) = h^3 \left[\frac{2\pi_3(m)}{5!} f^{(5)}(\eta_0) + h^{-1} \frac{2\pi_3^{(1)}(m)}{4!} f^{(4)}(\eta_1) \right.$$
$$\left. + h^{-2} \frac{\pi_3^{(2)}(m)}{3!} f^{(3)}(\eta_2) \right]$$

If we consider the error at a tabular point, $\pi_3(m) = 0$ and we remain with

$$E_2^{(2)}(mh) = h^3 \left[\frac{2^{-1}2}{4!} (3m^2 - 6m + 2) f^{(4)}(\eta_1) + \frac{h^{-2}}{3!} (6m - 6) f^{(3)}(\eta_2) \right]$$

For $m = 0$,

$$E_2^{(2)}(0) = \frac{h^2}{6} f^{(4)}(\eta_1) - h f^{(3)}(\eta_2)$$

For $m = 1$,

$$E_2^{(2)}(h) = \frac{-h^2}{12} f^{(4)}(\eta_1)$$

For $m = 2$,

$$E_2^{(2)}(2h) = \frac{h^2}{6} f^{(4)}(\eta_1) + h f^{(3)}(\eta_2)$$

Once again we note the increased accuracy when the point of approximation is symmetrically placed with respect to the tabular points. Combining the error terms above, we see

$$f^{(2)}(0) = h^{-2}(f_0 - 2f_1 + f_2) + \frac{h^2}{6} f^{(4)}(\eta_1) - h f^{(3)}(\eta_2)$$

$$f^{(2)}(h) = h^{-2}(f_0 - 2f_1 + f_2) - \frac{h^2}{12} f^{(4)}(\eta_1)$$

$$f^{(2)}(2h) = h^{-2}(f_0 - 2f_1 + f_2) + \frac{h^2}{6} f^{(4)}(\eta_1) + h f^{(3)}(\eta_2)$$

We leave the explicit formulation of the numerical differentiation formula for $n = 3$ to Exercise 9. The error term is obtained in Exercise 10.

Exercises

9. Derive the explicit form of the numerical differentiation formula for the second derivative of $f(x)$ using four equally spaced points.
10. What form does the error term take for the formula of Exercise 9 for $m = 0, 1, 2,$ and 3?
11. Using the tabulated values in Table 1 for $n = 2$ and 3, obtain approximations to the second derivative of the tabulated function at the tabular points. How accurate are the results? Is this accuracy superior or inferior to the approximations for the first derivative?

CHAPTER 9

The Solution of Nonlinear Algebraic Equations

1 Iteration

In this chapter we will consider solving the nonlinear algebraic equation $F(x) = 0$. Throughout this chapter we assume F is a reasonably well-behaved function of the single variable x. That is, we assume $F(x)$ to be a real single-valued function of the variable x which possesses a certain number of continuous derivatives; the actual number depending upon the particular application. Our main tool for solving such an equation is that of iteration, i.e. an application of a formula in a repetitive manner whereby a sequence of numbers is generated from a starting (initial) value or "guess".

Example 1 For the equation $x = f(x)$ an iterative process can be set up and denoted by

$$x_{(n+1)} = f(x_{(n)}) \qquad n = 0, 1, 2, \ldots \tag{1}$$

$x_{(0)}$ a given starting value. This formula generates the sequence of iterates $x_{(1)}, x_{(2)}, \ldots$ starting from a given number $x_{(0)}$, by using the most recent number in the sequence. Iterative methods which use just one point, $x_{(n)}$, at a time to define the new iterate, $x_{(n+1)}$, are called *one-point* iteration methods (see Traub [43]).

For an iterative process to be used, the formula must be valid under all conditions, or in circumstances where this is not so an additional rule must exist to indicate the procedure to be followed in such a case. That is, an iterative process must be *unambiguously* defined.

Example 2 The iterative procedure

$$x_{(n+1)} = \sqrt{x_{(n)} - 1} \qquad n = 0, 1, 2, \ldots, x_{(0)} \text{ given}$$

can break down if we restrict the iterates $x_{(n+1)}$ to the real numbers, if an $x_{(n)}$ becomes less than 1.

In our study of iterative methods we will be concerned in the main with three aspects. First, we will consider the development of a workable algorithm. Second, we will establish conditions under which the sequence of numbers calculated by the algorithm converges to the required solution (if it is to be of any practical use). Third, we will establish the rate at which such iterates converge.

The question of convergence of an algorithm can essentially be divided into two problems: the so-called "local" convergence properties where under given conditions the algorithm will be convergent if the starting point ($x_{(0)}$) of the algorithm is "sufficiently close" to the required root; and the so-called "global" convergence where the algorithm will converge under given conditions for any starting value in the interval (which may be infinite) specified by the implied conditions. In general, the question of local convergence can be answered in a simple manner. However, the problem of global convergence is more difficult and in many cases cannot be answered (and the corresponding algorithms are hence of questionable use). Consequently, it is a common feature of iterative processes to combine a globally convergent scheme, which produces a slowly convergent sequence, with a method with a higher rate of convergence which is locally convergent. The starting value of the latter method is that value produced by the globally convergent scheme which satisfies the definition of "sufficiently close" implied in the convergence condition for the locally convergent method.

2 A General First-order Method (the Method of Simple Iteration)†

In this section we will generate an iterative algorithm by rewriting the given function F in terms of an *iterating function f* whereby x can be obtained from F in an explicit manner. Thus we write $F(x) = x - f(x)$. Clearly there are many functions $f(x)$ for which $F(x) = x - f(x) = 0$.

Example 3 $F(x) = x^3 - 3x^2 - 2x + 1.0 = 0$ can be written as $x - f(x) = 0$ in any one of the following ways:

(a) $x - \frac{1}{2}(x^3 - 3x^2 + 1) = 0$ $\qquad$ i.e. $f = \frac{1}{2}(x^3 - 3x^2 + 1)$

(b) $x - [(x^3 - 2x + 1)/3]^{1/2} = 0$ $\qquad$ $f = [(x^3 - 2x + 1)/3]^{1/2}$

(c) $x - (x^3 + 1)/(2 + 3x) = 0$ $\qquad$ $f = (x^3 + 1)/(2 + 3x)$

(d) $x + [-(x^3 + 1)/x + 2]/3 = 0$ $\qquad$ $f = -[-(x^3 + 1)/x + 2]/3$

(e) $x - (3x^2 - 1)/(x^2 - 2) = 0$ $\qquad$ $f = (3x^2 - 1)/(x^2 - 2)$

(f) $x - (3x^2 + 2x - 1)^{1/3} = 0$ $\qquad$ $f = (3x^2 + 2x - 1)^{1/3}$

(g) $x - (3x^2 + 2x - 1)/x^2 = 0$ $\qquad$ $f = (3x^2 + 2x - 1)/x^2$

---

† The method is also known as *Picard iteration* or as *functional iteration*.

To avoid the ambiguity in the algorithm mentioned in section 1, the simplest condition which can be imposed is that for $x \in [a, b]$, $f(x) \in [a, b]$ for given constants a, b. In this manner the algorithm

$$x_{(n+1)} = f(x_{(n)})$$

can produce only values of x belonging to the given interval when $x_{(0)} \in [a, b]$. However, this condition will not guarantee that a value of x exists for which $x = f(x)$. Clearly, if f is discontinuous in $[a, b]$ the possibility may arise whereby the equation $x - f(x)$ has no root in $[a, b]$. Hence to ensure at least one root exists we assume further that $f(x)$ is continuous for $x \in [a, b]$. For our present consideration we require to ensure that there is at most one root. This will be guaranteed if the conditions of the following theorem are satisfied. We assume the existence of $f^{(1)}(x)$, the first derivative of f with respect to x, for $x \in (a, b)$.

Theorem 1 *Assume* $|f^{(1)}(x)| \leq L < 1$ *for* $x \in (a, b)$; *then the equation* $x - f(x) = 0$ *has at most one root in* $[a, b]$.

Proof By the mean value theorem, for $x_1, x_2 \in [a, b]$,

$$f(x_1) - f(x_2) = (x_1 - x_2) f^{(1)}(\zeta) \qquad \zeta \in (x_1, x_2) \qquad (2)$$

Assume there are two distinct roots $X_1, X_2 \in [a, b]$. Hence

$$X_1 - f(X_1) = 0 \qquad (3)$$

$$X_2 - f(X_2) = 0 \qquad (4)$$

Subtract Eq. (4) from Eq. (3) so that we obtain

$$(X_1 - X_2) - [f(X_1) - f(X_2)] = 0$$

Using Eq. (2) and taking absolute values, we find

$$|X_1 - X_2| = |(X_1 - X_2) f^{(1)}(\zeta)| = |X_1 - X_2| \, |f^{(1)}(\zeta)| \qquad \zeta \in (X_1, X_2)$$

On our assumption $X_1 - X_2 \neq 0$, we may divide by $|X_1 - X_2|$ so that

$$1 = f^{(1)}(\zeta)$$

But $|f^{(1)}(\zeta)| \leq L < 1$ and so we have a contradiction. Hence our assumption $X_1 \neq X_2$ must be false and thus we have at most one root.

Combining the above results we have:

Corollary 1 *The equation* $x - f(x) = 0$ *has a unique solution* $X \in [a, b]$ *if*
  (i) $f(x) \in [a, b]$
  (ii) $f(x)$ *is continuous*
  (iii) *there exists a constant* $L < 1$ *so that* $|f^{(1)}(x)| \leq L$ *for all* $x \in [a, b]$

In the proof of Theorem 1 we assumed the existence of the first derivative of $f(x)$ in (a, b). If $f^{(1)}(x)$ does not exist everywhere in (a, b) then the proof of Theorem 1 is no longer valid. To cover such a situation, the following theorem may be applied.

Theorem 2 *Assume* $|f(x_1) - f(x_2)| \leq L|x_1 - x_2|$ *with* $0 < L < 1$ *for* $x_1, x_2 \in [a, b]$; *then the equation* $x - f(x)$ *has at most one root in* $[a, b]$.

Exercises

1. Prove Theorem 2 following similar steps to those described for Theorem 1.
2. State the corresponding corollary to Corollary 1 under the weaker assumptions of Theorem 2.

Example 4 Find the root of $x^2 - 4x + 2.3 = 0$ contained in the interval $[0, 1]$ by an iterative method.

We may derive an iterative process by writing the given function in one of the following ways:

   (i) $x - (x^2 + 2.3)/4 = 0$
   (ii) $x - (4x - 2.3)^{1/2} = 0$

If we start with an initial guess $x_{(0)} = 0.6$ obtained by graphing the function for $x \in [0, 1]$, we find that applying the iterative process based on (i),

$$x_{(n+1)} = (x_{(n)}^2 + 2.3)/4$$

we obtain the following sequence:

$$x_{(1)} = ((0.6)^2 + 2.3)/4 = 0.665$$

$$x_{(2)} = ((0.665)^2 + 2.3)/4 = 0.686$$

$$x_{(3)} = ((0.686)^2 + 2.3)/4 = 0.693$$

$$x_{(4)} = ((0.693)^2 + 2.3)/4 = 0.695$$

$$x_{(5)} = ((0.695)^2 + 2.3)/4 = 0.696$$

Thus if we require the root correct to two decimal places we have two iterates in the sequence which agree to two decimal places and we conclude the required root is 0.70.

Alternatively, using the iterative process based on (ii), namely

$$x_{(n+1)} = (4x_{(n)} - 2.3)^{1/2}$$

using the same initial guess $x_{(0)} = 0.6$, we obtain the following sequence:

$$x_{(0)} = 0.6$$

$$x_{(1)} = 0.316$$

$$x_{(2)} = (1.264 - 2.3)^{1/2}$$

The last expression cannot be evaluated in terms of real arithmetic and so we conclude something has gone wrong with the process. These two sequences exemplify the situation mentioned in section 1, namely convergence and nonconvergence (divergence) respectively.

An iterative method which produces a convergent sequence of iterates is said to be a *convergent process*.

The functions in Example 4 indicate very clearly that starting with the same F written as $x - f(x)$, we are not necessarily bound to get a convergent process. The question therefore naturally arises: "under what conditions will this process converge?" In this case the conditions which guarantee convergence coincide with our earlier conditions guaranteeing a unique solution of $x - f(x) = 0$. We therefore have the following result.

Theorem 3 *Let $f(x)$ satisfy the conditions of Corollary 1. Then the iterative method*

$$x_{(n+1)} = f(x_{(n)}) \tag{1}$$

converges to the unique solution $x \in [a, b]$ of $x - f(x)$ for any $x_{(0)} \in [a, b]$.

Proof Under the hypothesis of Corollary 1 we know there exists a unique solution to $x - f(x) = 0$ for $x \in [a, b]$. Consider $\xi_{(n)} = x_{(n)} - X$ to be the error at the nth stage of the iteration. Then by definition,

$$x_{(n)} - X = f(x_{(n-1)}) - X = f(x_{(n-1)}) - f(X)$$

From Corollary 1, we have

$$|f^{(1)}(x)| \le L < 1 \quad \text{for all } x \in [a, b] \tag{5}$$

Hence

$$|x_{(n)} - X| = |f^{(1)}(\zeta)| \, |x_{(n-1)} - X| \le L |x_{(n-1)} - X| \qquad \zeta \in (a, b)$$

where we have used the mean value theorem and the inequality (5). A repeated application of this inequality produces

$$|x_{(n)} - X| \le L^n |x_{(0)} - X|$$

Since $L < 1$,

$$\lim_{n \to \infty} |x_{(n)} - X| = 0$$

which is equivalent to

$$\lim_{n \to \infty} x_{(n)} = X$$

That is, the method (1) is convergent to the unique solution $X \in [a, b]$.

Exercise

3. Prove that Eq. (1) is also convergent under the weaker conditions of the corollary of Exercise 2.

Example 5. To illustrate the convergence properties of the method (1), let us consider once again the equation given in Example 4 namely $x^2 - 4x + 2.3 = 0$. In this example we found that two ways of writing the algorithm for this equation produced the functions

$$f_1(x) = (x^2 + 2.3)/4$$

and

$$f_2(x) = (4x - 2.3)^{1/2}$$

For $f_1(x)$ we see that $f_1^{(1)}(x) = x/2$. Thus for $0 \le x \le 1$, in which the solution is required, $|f_1^{(1)}(x)| < 1$. Hence $f_1(x)$ satisfies the conditions of Corollary 1 and the iterative method is convergent, as is borne out by the numerical results. In contrast however, the function $f_2(x)$ produces a derivative $|f_2^{(1)}(x)| > 1$ for $0.575 \le x \le 1.595$, and for $x < 0.575$, $f_s^{(1)}$ is complex. Hence the method cannot converge, as is borne out by the numerical results.

Exercises

4. Which of the algorithms defined using the functions in (a)–(g) of Example 3 are convergent for a root in $[0, 1]$?
5. By rewriting the equation $0.1x^2 + 2x - 1 = 0$ in the form $x - f(x) = 0$, derive a convergent scheme. Evaluate the root nearer to zero correct to two decimal places using your algorithm. Does the sequence of iterates defined by the algorithm converge when used to find the second root?
6. Derive an alternative algorithm for the equation of Exercise 5. Is this algorithm convergent for both roots?
7. Calculate correct to two decimal places the root nearer zero of the equation $x^2 - 4x + 1.1 = 0$ using the method

$$x_{(n+1)} = 1.1/(4 - x_{(n)})$$

8. Calculate correct to three decimal places the root of $x^2 + 2.1x + 1.09 = 0$ nearer zero using the iterative procedure $x_{(n+1)} = -(1.09 + x_{(n)}^2)/2.1$.

3 The Graphical Interpretation of the Condition for Convergence

It will prove instructive if we consider graphically the stages of the simple iterative method (1) and discover why the procedure diverges for $|f^{(1)}(x)| > 1$ in the vicinity of the root X. Graphically this can be represented by drawing the line $y = x$ and the curve $y = f(x)$. Figure 1 shows a case where the gradient of $f(x)$ has modulus less than 1 in the interval $[a, b]$. Choose x as the starting value. Evaluate $f(x_{(0)})$, namely A. Hence evaluate $x_{(1)}$ from the equation $x_{(1)} = f(x_{(0)})$. Evaluate $f(x_{(1)})$; represent this by B. Then evaluate $x_{(2)}$ and hence $f(x_{(2)})$, i.e. C. Repeat the process and, as one can easily verify, we follow the route A, B, C, D, E, etc., ultimately arriving at the root.

Contrast this case with what happens when $|f^{(1)}(x)| > 1$ for x near X, as shown in Figure 2. Here, if the reader follows similar steps to those carried out in the first case, he will find a sequence of approximations A, B, C, ... all tending away (quite rapidly) from the root, i.e. the process is divergent.

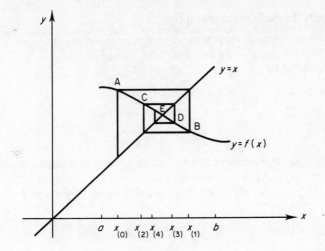

Figure 1

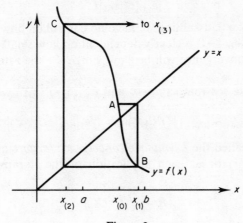

Figure 2

For the case $|f^{(1)}(x)| < 1$ for $x \in [a, b]$, the following analysis will indicate the effect upon the convergence of the *size* of $f^{(1)}(x)$. In this analysis we assume that the function f satisfies the properties of Corollary 1 and furthermore that $f^{(1)}(x) \neq 0$ for $x \in [a, b]$. Define the error at the nth stage of the iterative method by

$$\varepsilon_{(n)} = X - x_{(n)}$$

and similarly

$$\varepsilon_{(n-1)} = X - x_{(n-1)}$$

as the error at the previous stage. Then by Eq. (1) we have

$$X \overline{+} \varepsilon_{(n)} = f(X \overline{+} \varepsilon_{(n-1)}) \tag{6}$$

we now apply Taylor's theorem so that

$$X + \varepsilon_{(n)} = f(X) + \varepsilon_{(n-1)} f^{(1)}(\zeta) \qquad \zeta \in [a, b]$$

This is equivalent to

$$\varepsilon_{(n)} = \varepsilon_{(n-1)} f^{(1)}(\zeta) \tag{7}$$

so that

$$|\varepsilon_{(n)}| = |\varepsilon_{(n-1)}| |f^{(1)}(\zeta)|$$

Thus we see that the error at the nth stage is related linearly to the error at the $(n-1)$th stage. Such a method is said to have *linear convergence*, or is said to be a *first-order* method. Our assumption in Corollary 1 allows us to replace Eq. (7) by

$$|\varepsilon_{(n)}| \leq |\varepsilon_{(n-1)}| L$$

which applied recursively gives

$$|\varepsilon_{(n)}| \leq |\varepsilon_{(0)}| L^n$$

If we require $|\varepsilon_{(n)}| \to 0$, assuming we have no control on how big $\varepsilon_{(0)}$ can be,[†] the rate at which $|\varepsilon_{(n)}| \to 0$ is clearly dependent on how small is L. For L close to 1 the rate of convergence will be slow. For $L \ll 1$ the rate of convergence would be fast.

From Eq. (7) we see that as $n \to \infty$ and $f^{(1)}(\zeta) \to f^{(1)}(X)$ we have, approximately, that

$$|\varepsilon_{(n)}| \simeq |\varepsilon_{(n-1)}| |f^{(1)}(X)| \simeq \cdots \simeq |\varepsilon_{(0)}| |f^{(1)}(X)|^n$$

$|f^{(1)}(X)|$ is often called the *asymptotic convergence factor* since it dictates the reduction of the error at each application of the iterative method. The expression

$$R = 1/|f^{(1)}(X)|$$

is often called the *rate of convergence* since to reduce the error by a factor of 10^p, say, the number ν of iterations required is given approximately by

$$10^{-p} |\varepsilon_{(n)}| \simeq |f^{(1)}(X)| \nu |\varepsilon_{(n)}|$$

Hence $10^{-p} \simeq |f^{(1)}(X)| \nu$. Taking logarithms of both sides, we see that

$$\nu = p/\log\left(1/|f^{(1)}(X)|\right) = p/\log R$$

Thus the larger is R, the fewer will be the iterations required.

4 The Newton–Raphson Method

In the event that the simple iteration method (1) converges slowly, it is often preferable to apply a method which has a faster rate of convergence. Clearly

[†] I.e. that we cannot make $x_{(0)}$ arbitrarily close to X.

this can be achieved by obtaining a method for which the error at the nth stage is related quadratically to the error at the previous stage. That is, we look for a second-order method. To derive such a method, let us ask whether we can choose $f(x)$ in writing $F(x) = 0$ as $x - f(x) = 0$ so that the resulting algorithm has second-order convergence. In Eq. (2) we may expand $f(X + \varepsilon_{(n-1)})$ to second-order terms using Taylor series so that

$$X + \varepsilon_{(n)} = f(X) + \varepsilon_{(n-1)} f^{(1)}(X) + \tfrac{1}{2} \varepsilon^2_{(n-1)} f^{(2)}(\zeta)$$

so that

$$\varepsilon_{(n)} = \varepsilon_{(n-1)} f^{(1)}(X) + \tfrac{1}{2} \varepsilon^2_{(n-1)} f^{(2)}(\zeta)$$

The method is therefore second-order if $f^{(1)}(X)$ is zero. So we endeavor to *choose* $f(x)$ so that $f^{(1)}(X)$ is zero. To do this, let us write $F(x) = (x - x)/h(x) + F(x)$, where $h(x)$ is to be chosen to produce the required zero derivative of our f. This equation therefore suggests the iterative method

$$x_{(n+1)} = x_{(n)} - h(x_{(n)})F(x_{(n)}), \text{ or } f(x_{(n)}) = x_{(n)} - h(x_{(n)})F(x_{(n)})$$

Thus we see that in this way in writing $F(x) = x - f(x)$ we have chosen

$$f(x) = x - h(x)F(x) \tag{8}$$

Differentiating Eq. (8) we obtain

$$f^{(1)}(x) = 1 - F^{(1)}(x)h(x) - F(x)h^{(1)}(x)$$

But we are considering $F(X) = 0$, so that substituting X for x we obtain

$$f^{(1)}(X) = 1 - F^{(1)}(X)h(X) = 0$$

Hence

$$h(X) = [F^{(1)}(X)]^{-1} \tag{9}$$

Since we do not know X, we cannot choose h as $[F^{(1)}(X)]^{-1}$. However, we need only choose $h(x)$ so that substitution of X for x satisfies Eq. (9). This is simply achieved by writing

$$h(x) = [F^{(1)}(x)]^{-1}$$

and

$$f(x) = x - F(x)/F^{(1)}(x)$$

so that the iterative method (1) now becomes

$$x_{(n+1)} = x_{(n)} - F(x_{(n)})/F^{(1)}(x_{(n)}) \qquad n = 0, 1, \ldots \tag{10}$$

and has quadratic convergence. The algorithm defined by Eq. (10) is known as the *Newton–Raphson method*.

Exercises

9. Repeat Exercise 5 using the Newton–Raphson method. Compare your results with those obtained using the first-order method.

10. By applying the Newton–Raphson method to the quadratic equation $x^2 - a = 0$, a a positive constant, devise an algorithm for determining the square root of a positive constant.

11. Generalize the algorithm of Exercise 10 to determine the pth root of a positive constant.

12. Devise an algorithm for determining the reciprocal of a number without using division. Use your algorithm to calculate the reciprocal of 99, $\log 2$, $\sin \pi/3$, correct to four decimal places. (Be careful about starting values—see Theorem 4.)

13. The Newton–Raphson method has been derived by choosing a function f appropriately in the simple iteration method to produce an algorithm with second-order convergence. Confirm that the Newton–Raphson method is second-order convergent by carrying out an analysis similar to that given in section 2.
    Hint: You will need to use the binomial theorem to expand the denominator $F^{(1)}(x_{(n)})$.

Example 6 We will illustrate the faster rate of convergence of the Newton–Raphson method by calculating a root of the equation

$$F(x) = x^3 + x^{1/2} - 1 = 0$$

contained in $(0.3, 0.7)$ and comparing the sequence of iterates with that obtained using the first-order method.

For this equation the Newton–Raphson process is defined by

$$x_{(n+1)} = x_{(n)} - (x_{(n)}^3 + x_{(n)}^{1/2} - 1)/(3x_{(n)}^2 + \tfrac{1}{2}x_{(n)}^{-1/2})$$

Starting with $x_{(0)} = 0.3$ the sequence of iterates is:

$$x_{(1)} = 0.6595$$
$$x_{(2)} = 0.6080$$
$$x_{(3)} = 0.6054$$
$$x_{(4)} = 0.6054$$

so that after four iterations we have agreement to four decimal places. In contrast, the sequence of iterates for the first order method is:

$$x_{(1)} = 0.5329$$
$$x_{(2)} = 0.6195$$
$$x_{(3)} = 0.5999$$
$$x_{(4)} = 0.6073$$
$$x_{(5)} = 0.6047$$
$$x_{(6)} = 0.6057$$
$$x_{(7)} = 0.6053$$
$$x_{(8)} = 0.6054$$
$$x_{(9)} = 0.6054$$

Thus in contrast to the first-order method, the second-order Newton–Raphson process reduces the error at successive stages by the square of the error at the previous stage. As a result, we see that we obtain accuracy to the same number of decimal places in approximately half the number of iterations as were needed by the first-order method.

It is a simple extension to generalize the concept of second-order convergence to pth-order convergence when the error at one stage depends on the pth power of the error at the previous stage, i.e.

$$\varepsilon_{(n)} = \alpha \left(\varepsilon_{(n-1)} \right)^p$$

α (nonzero) is the *asymptotic error constant* for the pth-order method given by $1/p!\, f^{(p)}(X)$ where by construction the iterating function f has the property $f^{(j)}(X) = 0$, $j = 1, 2, \ldots, p-1$.

The consequence of the order of convergence can be seen from the following. Assume a method of order p produces a sequence of iterates which converges to the solution X. After n iterations, say, let the number of correct digits in the iterate $x_{(n)}$ be d_n so that the error $l_{(n)}$ satisfies

$$l_{(n)} \doteq 10^{-d_n}$$

If we denote the asymptotic error constant by C then

$$|l_{(n+1)}| = C |l_{(n)}|^p$$

so, assuming d_{n+1} correct digits at the $(1+n)$th iteration

$$10^{-d_{n+1}} \doteq C \times 10^{-p d_n}$$

i.e.

$$d_{n+1} \doteq p \ln (1/C) d_n$$

The number of correct digits d_{n+1} is therefore dependent on the order p and the asymptotic error constant C. For $C = 1$ (the worst case) we see that the number of correct digits at each iteration will increase p-fold; for a second order method the number of correct digits will approximately double. For $p = 1$, the first order method, the number of correct digits will only increase slowly for $C \doteq 1$ (but less than 1; why?).

In order that the Newton–Raphson method be useful we require to know conditions under which the algorithm is convergent. Many sets of conditions under which the algorithm is convergent have been derived in the literature which ensure this convergence. Here we will give just two convergence theorems. First we consider a local convergence criterion.

Theorem 4 (*local convergence*) *Consider* $F(x) = x - f(x) = 0$ *with* $f^{(1)}(X) = 0$. *Assume* $f(x)$ *and* $f^{(1)}(x)$ *are continuous for* $x \in I = (a, b)$. *Further assume for* $x \in I$, $f(x) \in I$. *Then there exists a number* $z_0 > 0$ *such that the iterative method* (10) *converges to the solution* X *for any* $x_{(0)}$ *satisfying* $|x_{(0)} - X| \le z_0$.

Proof Under our assumption that F has been rewritten so that $f(x)$ has been chosen to ensure that $f^{(1)}(X) = 0$ and that $f^{(1)}(x)$ is continuous for $x \in I$, it is clear that in a small interval, say $I_0 = (X - z_0, X + z_0)$, $|f^{(1)}(x)| \leq L < 1$ for $x \in I_0$. It is now elementary to apply the results of Theorem 3 and the Newton–Raphson method is convergent for $x_{(0)} \in I_0$.

The shortcoming of this theorem, of course, is that the interval I_0 may be extremely small for certain functions F. This would suggest, therefore, as advocated on page 262, that the first-order method (1) could be used to produce an $x_{(0)} \in I$ and which could be used as the starting value for the Newton–Raphson method.

An alternative to using the above approach is to ascertain global convergence conditions under which the Newton–Raphson method will converge. The reader should note that these conditions will again be given in terms of an interval which may again prove to be small. The proof of this theorem is essentially that contained in Henrici [20].

Theorem 5 *Let the function F be defined on the closed interval $[a, b]$. Further let F be continuous and have continuous derivatives $F^{(1)}(x), F^{(2)}(x)$ for $a \leq x \leq b$. If the following conditions are satisfied, then the Newton–Raphson method converges to the only solution in $[a, b]$ for any $x_{(0)} \in [a, b]$.*
(1) There is a change of sign of F in $[a, b]$, i.e. $F(a)F(b) < 0$.
(2) The first derivative of F does not vanish for $a \leq x \leq b$, i.e. $F^{(1)}(x) \neq 0$, $x \in [a, b]$.
(3) Either $F^{(2)}(x) > 0$ or $F^{(2)}(x) < 0$, for all $x \in [a, b]$.
(4) c denotes the end point a or b for which $|F^{(1)}|$ is smaller and

$$\left| \frac{F(c)}{F^{(1)}(c)} \right| \leq b - a$$

Proof The conditions in the statement of the theorem represent several possible cases. These may be represented by

(i) $F(a) > 0, F(b) < 0, F^{(2)}(x) \geq 0; c = b$

(ii) $F(a) < 0, F(b) > 0, F^{(2)}(x) \leq 0; c = b$

(iii) $F(a) > 0, F(b) < 0, F^{(2)}(x) \geq 0; c = a$

(iv) $F(a) < 0, F(b) > 0, F^{(2)}(x) \leq 0; c = a$

However, we do not have to prove the theorem separately for all cases. Case (i) can be derived from case (ii) by considering $-F$ instead of F. Since $F(x) = 0$ and $-F(x) = 0$ are the same, nothing is changed. Case (iv) can be derived from case (iii) by a similar approach. If X is a solution of (iii), then $-X$ is a solution of (i) and hence the sequence $\{-x_n\}$ is generated instead of $\{x_{(n)}\}$; again nothing is changed. Similarly, we can deal with cases (iv) and (ii).

Thus to prove the theorem we need only consider one of the cases (i)–(iv). The proof will immediately apply to the other cases. We will choose case (ii), for which the function is exemplified by Figure 3.

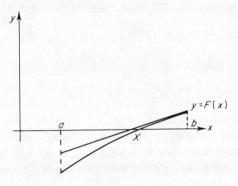

Figure 3

Let X be the required root of $F(x) = 0$. Let us assume that the starting value $x_{(0)}$ is taken in the range $a \le x_{(0)} \le X$. We would then like to show that for such an $x_{(0)}$, the successive iterates defined by Eq. (10) will only get larger than $x_{(0)}$ and will not get larger than X, i.e. they will tend to X. The proof will be by induction on n.

From Figure 3 we see that the value of $F(x)$ for x in the range $[a, X]$ is nonpositive. Hence by condition (2) of the theorem,

$$x_{(1)} = x_{(0)} - F(x_{(0)})/F^{(1)}(x_{(0)}) \ge x_{(0)} \tag{11}$$

We now want to show that this $x_{(1)}$ is not larger than X. By the mean value theorem,

$$-F(x_{(0)}) = F(X) - F(x_{(0)}) = (X - x_{(0)})F^{(1)}(\zeta) \qquad \zeta \in [x_{(0)}, X] \tag{12}$$

For the case (ii) we have assumed $F^{(1)}(x) > 0$ and $F^{(2)}(x) < 0$. Thus, $F^{(1)}(\theta) \ge F^{(1)}(\phi)$ for $\theta \le \phi$. Hence for $\zeta \ge x_{(0)}$, we can write Eq. (12) as

$$-F(x_{(0)}) = (X - x_{(0)})F^{(1)}(\zeta) \le (X - x_{(0)})F^{(1)}(x_{(0)})$$

Hence $-F(x_{(0)})/F^{(1)}(x_{(0)}) \le (X - x_{(0)})$ and so

$$x_{(1)} = x_{(0)} - F(x_{(0)})/F^{(1)}(x_{(0)}) \le x_{(0)} + X - x_{(0)} = X \tag{13}$$

Thus inequalities (11) and (13) prove the result for $n = 0$. Let us assume the result is true for $n = k - 1$, i.e. $x_{(k-1)} \le X$ and $x_{(k-1)} \ge x_{(k-2)}$. We will show that this result is true for $n = k$. We proceed as before. Since $F^{(2)}(x) \le 0$, and $F^{(1)}(x) > 0$, we have

$$-F(x_{(k-1)}) = F(X) - F(x_{(k-1)}) = (X - x_{(k-1)})F^{(1)}(\zeta) \qquad \zeta \in [x_{(k-1)}, X].$$

Therefore

$$-F(x_{(k-1)}) < (X - x_{(k-1)})F^{(1)}(x_{(k-1)})$$

so that

$$-F(x_{(k-1)})/F^{(1)}(x_{(k-1)}) \leq X - x_{(k-1)}$$

Hence applying the Newton–Raphson method for $k = n$ we have

$$x_{(k)} = x_{(k-1)} - F(x_{(k-1)})/F^{(1)}(x_{(k-1)}) \leq x_{(k-1)} + X - x_{(k-1)} = X$$

Thus $x_{(k)} \leq X$ for all k. Hence $F(x_{(k)}) \leq 0$ and so

$$x_{(k)} = x_{(k-1)} - F(x_{(k-1)}/F^{(1)}(x_{(k-1)}) \geq x_{(k-1)} \qquad \text{for all } k$$

If we choose a starting value $x_{(0)} \in [a, X]$ then the iterates $\{x_{(n)}\}$ define a monotonic increasing sequence. It is a well-known property that when such a sequence is bounded it tends to a limit (see Jones and Jordan [27]). If we let this limit be α and substitute the limit into Eq. (10), we find that

$$x_{(n+1)} = \alpha = x_{(n)} - F(x_{(n)})/F^{(1)}(x_{(n)}) = \alpha - F(\alpha)/F^{(1)}(\alpha)$$

Therefore $F(\alpha) = 0$, which implies that α is the required (unique) root X.

To complete the proof of convergence, we must also consider the possibility of $x_{(0)}$ being in the interval $[X, b]$. In this case, using the mean value theorem we find

$$F(x_{(0)}) = F(x_{(0)}) - F(X) = (x_{(0)} - X)F^{(1)}(\eta) \qquad X \leq \eta \leq x_{(0)}$$

Since $F^{(1)}(x)$ decreases as x increases, $F^{(1)}(\eta) \geq F^{(1)}(x_{(0)})$. Thus

$$F(x_{(0)}) \leq (x_{(0)} - X)F^{(1)}(x_{(0)})$$

and hence

$$F(x_{(0)})/F^{(1)}(x_{(0)}) \leq (x_{(0)} - X)$$

so that

$$x_{(1)} = x_{(0)} - F(x_{(0)})/F^{(1)}(x_{(0)}) \leq x_{(0)} - (x_{(0)} - X) = X$$

Thus the first iterate $x_{(1)}$ is situated on the lower side of X. All we have to do is show that $x_{(1)}$ is greater than or equal to a, for then we can apply the first part of the proof. To do this we note that

$$F(x_{(0)}) = F(b) - (b - x_{(0)})F^{(1)}(\theta) \qquad x_{(0)} \leq \theta \leq b$$

Thus

$$F(x_{(0)}) \leq F(b) - (b - x_{(0)})F^{(1)}(b) \tag{14}$$

since $F^{(1)}(x)$ is decreasing with increasing x. Thus

$$x_{(1)} = x_{(0)} - F(x_{(0)})/F^{(1)}(x_{(0)})$$

$$\geq x_{(0)} - F(x_{(0)})/F^{(1)}(b)$$

$$\geq x_{(0)} - F(b)/F^{(1)}(b) + (b - x_{(0)})$$

where we have used Eq. (14). Now using condition (4) of the theorem we see that

$$x_{(1)} \geq b - (b - a) = a$$

and so the first part of the proof can be applied.

Convergence is therefore proved.

Exercises

14. Develop a subprogram for the Newton–Raphson method for an arbitrary function $F(x)$ and its first derivative $F^{(1)}(x)$ where these two functions are given as parameters (representing function procedures (in ALGOL) or functions (in FORTRAN)). Your subprogram should also have parameters for the initial guess $x_{(0)}$, the tolerance required before convergence is deemed to have occurred, and the maximum number of iterations to be allowed. If convergence does not occur this should be indicated by a printed message output from within the subprogram. On exit from the subprogram the last calculated iterate should be contained in the parameter which originally contained $x_{(0)}$.
15. Use your subprogram of the previous exercise in a main program to calculate the roots indicated for the following equations where ε = tolerance; m = maximum number of iterations:
    (i) $x^{1/2} \ln(x) - 1.0 = 0$ (real root); $x_{(0)} = 1.0$, $\varepsilon = 10^{-5}$, $m = 20$.
    (ii) $-2\pi/3 + \sqrt{3} + 2 \sin(x) + x^{1/2} = 0$ (real root); $x_{(0)} = 0.5$, $\varepsilon = 10^{-6}$, $m = 30$.
    (iii) $x \exp(-x) - (\ln(x))^2 = 0$ (real root); $x_{(0)} = 0.5$, $\varepsilon = 10^{-6}$, $m = 30$.
16. For the cubic equation $ax^3 + bx^2 + cx + d = 0$, show that if there exist three real roots X_1, X_2, X_3, then if a starting value $x_{(0)}$ is chosen as $(X_1 + X_3)/2$ then the Newton–Raphson method converges in one step to the third root X_2.

5 The Graphical Representation of the Newton–Raphson Method

Let us consider the steps in the Newton–Raphson method applied to the function illustrated in Figure 4. Let the initial guess be $x_{(0)}$. The value of F is then $F(x_{(0)})$ and the value of $F^{(1)}$ is $F^{(1)}(x_{(0)})$. From $F(x_{(0)})$ we obtain the point A on the curve. The point x is then calculated using the Newton–Raphson method, namely

$$x_{(1)} = x_{(0)} - F(x_{(0)})/F^{(1)}(x_{(0)}) \tag{15}$$

Rearranging (15), we find that

$$-F(x_{(0)})/(x_{(1)} - x_{(0)}) = F^{(1)}(x_{(0)})$$

But $F^{(1)}(x_{(0)})$ is the tangent to the curve $y = F(x)$ at A, and hence x must be the point of intersection of this tangent with the x-axis. Hence we obtain $x_{(1)}$. Repeating the process, we obtain B as the point on the curve obtained by substituting $x_{(1)}$ into $F(x)$. We therefore obtain the next iterate $x_{(2)}$ as the tangent intersection with the x-axis. Clearly in this example the iterates $x_{(0)}$, $x_{(1)}$, $x_{(2)}$, ... converge rapidly to the required solution X. Consider in contrast the function defined in Figure 5. In this case, starting with $x_{(0)}$ as the initial guess for the root X we see that the Newton–Raphson method produces the

276

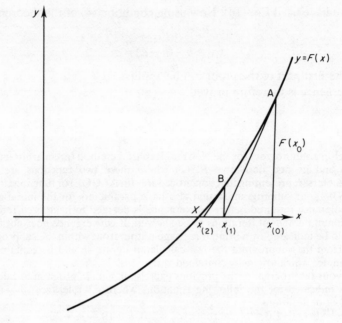

Figure 4

sequence of iterates $x_{(1)}, x_{(2)}, \ldots$ converging not to X but to another root of $F(x)$, which we have denoted by X_1. However, note that by a more careful starting value this example could also produce a convergent sequence for the root X.

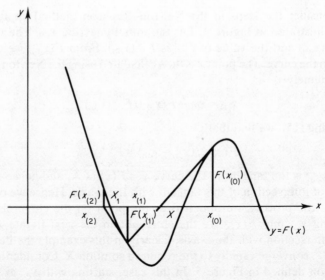

Figure 5

If the reader considers the Newton–Raphson method graphically, or otherwise, certain features of the algorithm will make themselves apparent. For example, in order to implement the method we must be able to evaluate the derivative $F^{(1)}(x)$ at each stage. (By assumption $F(x)$ exists everywhere in $[a, b]$.) In certain cases this may be time-consuming. In addition, in order to evaluate the right-hand side of Eq. (10) we must ensure that F is never zero in the interval. If the derivative $F^{(1)}$ does become zero close to the required root then special procedures will have to be implemented in order to obtain the root. Further, if $F^{(1)}$ is actually zero at the root then $F(x)$ has more than one root at this point, i.e. $F(x)$ has multiple roots. Such functions are depicted in Figures 6 and 7.

For $F(x) = (x-1)^2$ depicted in Figure 6, the function has a double root at $x = 1$. An alternative way of describing this situation is to say that $F(x)$ has a root at $x = 1$ of *multiplicity* 2. For $F(x) = (x-1)^3$ depicted in Figure 7, the function has a triple root at $x = 1$, i.e. $x = 1$ has multiplicity 3.

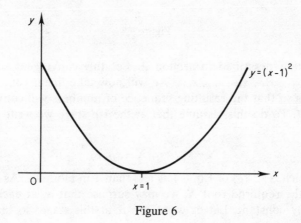

Figure 6

6 Aitken's Process, Steffensen's Method, Secant Method, Regula Falsi

We will deal first with the problem of evaluating the derivative $F^{(1)}(x)$. Let us assume we cannot (or will not) evaluate this function anywhere in the interval $[a, b]$. Then, clearly there are two things we can do. First, if we are to demand a rate of convergence in excess of unity we can endeavor to modify the first-order method (1) to increase the rate of convergence but without evaluating, or approximating to, $F^{(1)}(x)$ in so doing. Second, and alternatively, we can take an approximation to $F^{(1)}(x)$ without actually needing to calculate $F^{(1)}(x)$. From our discussion on divided differences in Chapter 4 we know that a divided difference of $F(x)$ can produce an approximation to $F^{(1)}(x)$ involving only values of $F(x)$. Both of these approaches produce algorithms which possess a rate of convergence in excess of unity. The first algorithm we will describe is known as *Aitken's δ^2-process*. To describe this technique let us assume we have calculated $n + 1$ iterates using a convergent method of

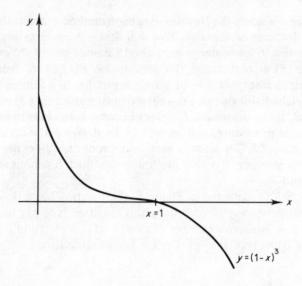

Figure 7

simple iteration described in section 2. Let this convergent sequence be denoted by $x_{(0)}, x_{(1)}, x_{(2)}, \ldots, x_{(n)}$. We will now take linear combinations of these iterates so that the resulting sequence of numbers will converge faster than the first. To do this, assume that at the ith stage we write Eq. (7) for the error as

$$X - x_{(i)} = \alpha_i (X - x_{(i-1)}) \tag{16}$$

where α_i, which is $f^{(1)}(\zeta)$ of Eq. (7), is less than 1 in modulus. As the iterates $x_{(i)}$ tend to the required root X, we may surmise that α_i at each stage will remain nearly constant. Let us write $\alpha_i = \alpha$ at this stage, so that Eq. (16) becomes

$$X - x_{(i)} = \alpha (X - x_{(i-1)}) \tag{17}$$

Also,

$$X - x_{(i+1)} = \alpha (X - x_{(i)}) \tag{18}$$

at the next stage. Under the assumption that α is common to the two equations (17), (18), we may eliminate α. We then obtain

$$\frac{X - x_{(i+1)}}{X - x_{(i)}} = \frac{X - x_{(i-1)}}{X - x_{(i)}} \tag{19}$$

approximately (why?). Rearranging Eq. (19), we find the unknown X is given by

$$X = \frac{x_{(i-1)} x_{(i+1)} - x_{(i)}^2}{x_{(i+1)} - 2x_{(i)} + x_{(i-1)}}$$

approximately. Introducing $\Delta x_{(i)} \equiv x_{(i+1)} - x_{(i)}$ and $\delta^2 x_{(i)} \equiv x_{(i+1)} - 2x_{(i)} + x_{(i-1)}$, we find that

$$X \simeq x_{(i+1)} - (\Delta x_{(i)})^2 / \delta^2 x_{(i)} \qquad i = 1, 2, \dots \tag{20}$$

The values calculated by the formula (20) are new approximations to the required root. That is, for successive i we have defined a new sequence of iterates, say $x^1_{(i)}$, which by derivation will converge faster than the convergent sequence $x_{(i)}$, $i = 0, 1, \dots, n$, in the sense that

$$(x^1_{(i)} - X)/(x_{(i)} - X) \to 0 \quad \text{as } i \to \infty$$

(see Henrici [20, p. 73]; Traub [43, p. 265]). However, since $\alpha_i = \alpha$ is only an approximation, the first-order term in the error may not be completely eliminated by the Aitken δ^2-process (20). Thus we suspect that the convergence is better than linear but not as good as quadratic. It is noted that the term

$$\frac{(\Delta x_{(i)})^2}{\delta^2 x_{(i)}}$$

in (20) often involves the quotient of very small numbers (particularly if $|f^{(1)}(X)|$ is close to 1) and consequently the algorithm will require the use of multiple-precision arithmetic when used on a computer.

We describe the effect of the δ^2-process in the following example where the root of the cubic equation $x^3 - 3x^2 + 2x - 0.375 = 0$ closest to 1 is determined. For comparison, we quote the results obtained by the method of simple iteration (1), the Aitken δ^2-process (20) with double precision (about 20 decimal places), and also the Newton–Raphson method (13).

Example 7 For $x^3 - 3x^2 + 2x - 0.375 = 0$, we define $f(x) = ((x^3 + 2x - 0.375)/3)^{1/2}$ for the simple iteration method. Thus the iterative method (1) in this case becomes $x_{(n+1)} = [(x^3_{(n)} + 2x_{(n)} - 0.375)/3]^{1/2}$. The Newton–Raphson method for this $F(x)$ is

$$x_{(n+1)} = x_{(n)} - \frac{(x^3_{(n)} - 3x^2_{(n)} + 2x_{(n)} - 0.375)}{(3x^2_{(n)} - 6x_{(n)} + 2)}$$

The results of calculating the sequence of iterates starting with the initial guess $x_{(0)} = 0.7$ are given in Table 1. Successive iterates whose difference was less than 0.5×10^{-4} were used to indicate that convergence had occurred. Alternatively, the iteration was terminated if the number of iterations exceeded 50. The slow rate of convergence for the simple iteration is apparent, as is the faster rate of convergence for the Newton–Raphson method.† As predicted, the rate of convergence of the δ^2-process lies between the two.

---

† An analysis given by Isaacson and Keller [24, p. 95] indicates that under certain given conditions the number of iterations required for the method of simple iteration to converge to a given accuracy compared with that for the Newton–Raphson method is in the ratio 130:7. This is substantiated by the numerical results in Table 1.

Table 1

First order	Aitken	Newton–Raphson	Steffensen	Secant	Regula falsi
.700 000	.526 840	.700 000	.700 000	.700 000	.700 000
.675 278	.524 619	.560 274	.526 843	.600 000	.600 000
.654 085	.522 134	.511 934	.501 948	.538 095	.538 005
.635 874	.519 632	.500 737	.500 014	.511 741	.517 228
.620 176	.517 245	.500 003	.500 000	.502 010	.504 121
.606 598	.515 041			.500 130	.504 121
.594 811	.513 048			.500 002	.502 073
.584 541	.511 271				.501 049
.575 562	.509 704				.500 533
.567 682	.508 333				.500 271
.560 746	.507 139				.500 138
.550 620	.506 104				.500 070
.549 194	.505 211				
.544 375	.504 442				
.540 084	.503 782				
.536 255	.503 217				
.532 829	.502 733				
.529 758	.502 320				
.527 001	.501 967				
.524 521	.501 667				
.522 286	.501 412				
.520 270	.501 195				
.518 449	.501 011				
.516 801	.500 855				
.515 310	.500 722				
.513 957					
.512 730					
.511 616					
.510 604					
.509 683					
.508 845					
.508 082					
.507 387					
.506 753					
.506 175					
.505 648					
.505 166					
.504 727					
.504 326					
.503 959					
.503 624					
.503 317					
.503 037					
.502 781					
.502 547					
.502 332					
.502 136					
.501 955					
.501 792					
.501 647					

Thus with the aid of the δ^2-process we have improved the rate of convergence without evaluating the derivative $F^{(1)}(x)$, but we have not achieved the second-order rate of convergence we seek. However, the Aitken δ^2-process itself suggests an additional modification which may (and, we shall subsequently prove, does) obtain the quadratic convergence rate we set out to achieve. In defining the Aitken δ^2-process we defined a new independent sequence by

$$x^1_{(i-1)} = x_{(i+1)} - \frac{(\Delta x_{(i)})^2}{\delta^2 x_{(i)}} \qquad i = 1, 2, \ldots \tag{21}$$

However, from just $x_{(0)}$, $x_{(1)}$, and $x_{(2)}$, we are able to calculate a new $x^1_{(0)}$ which is a better approximation to the required root X than $x_{(0)}$, $x_{(1)}$, or $x_{(2)}$. So let us use it as the argument of our iterating function f to calculate $x^1_{(1)}$ and thence $x^1_{(2)}$. We can then take the linear combination specified by Eq. (21) with $i = 1$. This then defines a new $x^2_{(0)}$ (say) which can be used as the starting value for our simple iteration. Algorithmically we can represent this modification to the δ^2-process as

$$x_{(0)} = x_{(0)} \qquad x^1_{(0)} = \frac{x_{(0)} x_{(2)} - x^2_{(1)}}{x_{(2)} - 2x_{(1)} + x_{(0)}} \qquad x^2_{(0)} = \frac{x_{(0)} x_{(2)} - (x^1_{(1)})^2}{x^1_{(2)} - 2x^1_{(1)} + x^1_{(0)}} \qquad \text{etc}$$

$$x_{(1)} = f(x_{(0)})$$

$$x^1_{(1)} = f(x^1_{(0)}) \qquad\qquad x^2_{(1)} = f(x^2_{(0)})$$

$$x_{(2)} = f(x_{(1)})$$

$$x^1_{(2)} = f(x^1_{(1)}) \qquad\qquad x^2_{(2)} = f(x^2_{(1)})$$

We note that we may write $x_{(2)} = f(x^1_{(1)}) = f(f(x_{(0)}))$, so that each starting value $x^i_{(0)}$ may be written as

$$x^i_{(0)} = \frac{x^{i-1}_{(0)} f(f(x^{i-1}_{(0)})) - f(x^{i-1}_{(0)}) f(x^{i-1}_{(0)})}{f(f(x^{i-1}_{(0)})) - 2f(x^{i-1}_{(0)}) + x^{i-1}_{(0)}} \qquad i = 1, 2, \ldots; x^0_{(0)} = x_{(0)} \tag{22}$$

Hence the modification to the Aitken δ^2-process can be considered to define a simple iteration procedure based upon the functional equation

$$x = g(x)$$

where

$$g(x) = \frac{x f(f(x)) - f(x) f(x)}{f(f(x)) - 2f(x) + x} \tag{23}$$

The iterative method $x_{(n+1)} = g(x_{(n)})$ with $g(x)$ defined by (23) is known as *Steffensen's method*. Thus in considering the convergence properties of the modified δ^2-process we can ignore the original simple iterative procedure and hence no longer require that it be a convergent process (as we did for the Aitken δ^2-process), as the convergence properties will be governed by (22).

In considering the convergence properties of (22) we have the following questions to consider:

   (i) Is the method (22) second-order convergent?
   (ii) What is the local condition for convergence for (22)?
   (iii) What are the global conditions for convergence for (22)?

To consider (i), assume the Aitken δ^2-process is applied to a first-order method $(f^{(1)}(x) \neq 0)$ for the equation $F(x) = 0$ which has only simple roots (i.e. $f^{(1)}(x) \neq 1$). In addition, assume the function f is three times continuously differentiable. Then to show the order of convergence is two, we must show that the root of Eq. (23) is indeed X and that in addition each root of Eq. (23) is also a root of Eq. (1).

Theorem 6   *Whenever Eq. (1) possesses a root X, then X is also a root of (23).*

Proof   At $x = X$ in Eq. (23), the function $g(x)$ is clearly indeterminate. Hence, apply L'Hospital's rule (see [27]). Assuming $f^{(1)}(X) \neq 1$, we find on substituting $x = X$ in Eq. (23),

$$g(X) = \frac{f(f(X)) + Xf^{(1)}(f(X))f^{(1)}(X) - 2f(X)f^{(1)}(X)}{f^{(1)}(f(X))f^{(1)}(X) - 2f^{(1)}(X) + 1}$$

Using $X = f(X)$ we see that

$$g(X) = X \text{ and hence } X \text{ is also a root of Eq. (23)}$$

Theorem 7   *Each root of Eq. (21) is also a root of Eq. (1).*

Proof   We see that two situations can occur,

either

$$\text{(a)} \quad f(f(x)) - 2f(x) + x \neq 0$$

or

$$\text{(b)} \quad f(f(x)) - 2f(x) + x = 0$$

For case (a) we have

$$x[f(f(x)) - 2f(x) + x] = xf(f(x)) - f(x)f(x)$$

from Eq. (23). Thus $x^2 - 2xf(x) + f^2(x) = 0$, i.e. $[x - f(x)]^2 = 0$ and so $x = f(x)$. Thus each root of Eq. (23) will also be a root of Eq. (1).

   In case (b), if x is to be finite (by assumption), the vanishing of the denominator must coincide with the vanishing of the numerator. This implies that both

$$f(f(x)) - 2f(x) + x = 0$$

and
$$xf(f(x)) - f(x)f(x) = 0$$

Substituting for $f(f(x))$, we have in this latter equation
$$2xf(x) - x^2 - f^2(x) = 0$$

Solve this quadratic in $f(x)$, namely
$$[f(x) - x]^2 = 0$$

so that $f(x) = x$, which indicates that x is a root of Eq. (1) and the theorem is proved.

We now require to prove that Steffensen's method has quadratic convergence.

Theorem 8 *If $f^{(1)}(x)$, $f^{(2)}(x)$ exist and are bounded and $|f^{(1)}(x)| \neq 1$ for $x \in [a, b]$ then Steffensen's method applied to the first-order simple iteration method based on $x = f(x)$ has quadratic convergence.*

Proof Consider the point $X + E \in [a, b]$ where E is a small error. Then by Taylor's series,
$$f(X + E) = f(X) + Ef^{(1)}(X) + \tfrac{1}{2}E^2 f^{(2)}(\zeta) \qquad \zeta \in (X, X + E)$$

Denoting $f^{(1)}(X)$ by A and $\tfrac{1}{2}f^{(2)}(\zeta)$ by B, then
$$f(X + E) = X + (A + BE)E$$

Let $(A + BE)E = \phi$.

By assumption, A and B are bounded and we may therefore choose an E sufficiently small so that $X \pm \phi \in [a, b]$.

Thus
$$f(X + \phi) = f(X) + \phi A + \tfrac{1}{2}\phi^2 f^{(2)}(\xi) \qquad \xi \in (X, X + \phi)$$

Therefore
$$f(X + \phi) = X + A\phi + B'\phi^2 \qquad B' = \tfrac{1}{2}f^{(2)}(\xi)$$

On substituting $x = X + E$ into Eq. (23), we obtain
$$g(X + E) = \frac{(X + E)f(X + \phi) - (X + \phi)^2}{f(X + \phi) - 2(X + \phi) + X + E} \tag{24}$$

Rearranging the right-hand side of Eq. (24), we obtain
$$g(X + E) = \frac{-(\phi^2 - A\phi E - B'E\phi^2)}{f(X + \phi) - 2(X + \phi) + X + E} + X \tag{25}$$

substituting $\phi = (A + BE)E$ and rearranging Eq. (25), we obtain
$$g(X + E) = \frac{-(A + BE)[(A + BE)E - AE - B'E(A + BE)E]}{E + (A - 2)(A + BE)E + B'(A + BE)^2 E^2} + X$$

which leads to

$$g(X+E) = \frac{-(A+BE)E^2[B-B'A-B'BE]}{[(1-A)^2(2B-BA-B'A^2E+2B'BAE^2+B'B^2E^3]} + X$$

As $E \to 0$, $B \to B' \to \frac{1}{2}f^{(2)}(X)$, and since $|f^{(1)}(X)| \neq 1$ we have

$$g(X+E) = X - \frac{f^{(1)}(X)f^{(2)}(X)}{2(1-f^{(1)}(X))}E^2 + \text{higher order terms in } E$$

Hence the Steffensen algorithm is second-order convergent.

We have left to prove that the Steffensen's method is convergent. As with the Newton–Raphson method we find that we can obtain convergence under two alternative sets of conditions, namely conditions which give the local convergence results and conditions which guarantee global convergence (in an interval $[a, b]$).

Theorem 9 *Consider $F(x) = x - g(x) = 0$ where $g(x)$ is defined by Eq. (23) and by Theorem 8, $g^{(1)}(X) = 0$ (the method is second order). Assume $f(x)$ and $f^{(1)}(x)$ are continuous for $x \in (a, b)$ and assume that for $x \in (a, b)$, $g(x) \in (a, b)$. Then there exists a number $z_0 > 0$ such that the iterative method (22) converges to the solution X for any $x_{(0)}$ satisfying*

$$|x_{(0)} - X| \leq z_0$$

Proof This theorem is essentially the same as Theorem 4 and hence the proof given there may be applied to g. The result then follows.

Theorem 10 *(global convergence)* *Let I denote the semi-infinite interval (α, ∞) and let the function f defined by writing $F(x) \equiv x - f(x)$ satisfy the following conditions:*
  (a) f is defined and twice continuously differentiable on I
  (b) $f(x) > 0$ for all $x \in I$
  (c) $f^{(1)}(x) < 0$ for all $x \in I$
  (d) $f^{(2)}(x) > 0$ for all $x \in I$
  Under these assumptions the given equation $x = f(x)$ has a unique solution in I and the Steffensen algorithm is convergent for any initial guess $x_{(0)} \in I$.

Proof The function satisfying conditions (a), (b), (c), and (d) may be interpreted graphically as shown in Figure 8. Under these assumptions it is clear that the equation $x = f(x)$ has a unique solution in I. Let this be X. We will prove convergence in two parts. First, let us assume the starting value $x_{(0)}$ is chosen so that $x_{(0)} > X$. By assumption (c), $f^{(1)}(x) < 0$ and so for $x > X$, $f(x) < X$ and $f(f(x)) > X$. Apply the mean value theorem to $f(x) - X$. Hence

$$f(x) - X = f(x) - f(X) = (x - X)f^{(1)}(\zeta) \qquad \zeta \in (X, x) \qquad (26)$$

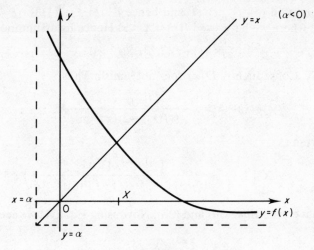

Figure 8

Similarly,

$$f(f(x)) - X = f^{(1)}(\xi)(f(x) - X) \qquad \xi \in (f(x), X)$$

so that

$$f(f(x)) - X = f^{(1)}(\xi) f^{(1)}(\zeta)(x - X)$$
$$\xi \in (f(x), X) \qquad \zeta \in (X, x) \tag{27}$$

Now

$$f(x) - x = f(x) - f(X) + X - x$$
$$= (x - X)[f^{(1)}(\zeta) - 1] \tag{28}$$

and

$$f(f(x)) - f(x) = f(f(x)) - X + X - f(x)$$
$$= (x - X)[f^{(1)}(\xi) - 1] f^{(1)}(\zeta) \tag{29}$$

Let

$$D(x) = f(f(x)) - 2f(x) + x$$

Then

$$D(x) = f(f(x)) - f(x) - (f(x) - x)$$

which may be written as

$$D(x) = (x - X)[f^{(1)}(\xi) - 1] f^{(1)}(\zeta) - (x - X)[f^{(1)}(\zeta) - 1]$$

where we have used Eqs. (28) and (29). Thus

$$D(x) = (x - X)([f^{(1)}(\zeta) - 1]^2 + [f^{(1)}(\xi) - f^{(1)}(\zeta)] f^{(1)}(\zeta)) \tag{30}$$

But by assumption (d), $f^{(2)}(x) > 0$ and hence $f^{(1)}(\xi) < f^{(1)}(\zeta)$ for $\xi < \zeta$ defined by Eq. (27), where we have used $f(x) < X < x$. Hence by assumption (c),

$$[f^{(1)}(\xi) - f^{(1)}(\zeta)]f^{(1)}(\zeta) > 0 \qquad (31)$$

and so $x > X$. Consequently $D(x)$ does not vanish. Thus

$$g(x) - X = x - \frac{[f(x) - x]^2}{f(f(x)) - 2f(x) + x} - X$$

may be written as

$$(x - X)\left(1 - \frac{[f^{(1)}(\zeta) - 1]^2}{[f^{(1)}(\zeta) - 1]^2 + [f^{(1)}(\xi) - f^{(1)}(\zeta)]f^{(1)}(\zeta)}\right)$$

where we have used Eqs. (30) and (28). Now using Eq. (31) we see that

$$0 < \frac{[f^{(1)}(\zeta) - 1]^2}{([f^{(1)}(\zeta) - 1]^2 + [f^{(1)}(\xi) - f^{(1)}(\zeta)]f^{(1)}(\zeta))} < 1$$

from which it follows that

$$0 < g(x) - X < x - X$$

and thus

$$X < g(x) < x.$$

Hence the sequence generated by Eq. (23) with $x_{(0)} > X$ decreases monotonically and is bounded by X. Hence (see Jones and Jordan [27]) the sequence converges to the unique solution X.

If $x = X$ then clearly the convergence is trivial since by definition $g(x) = x = X$ and we remain at the root X.

Assume now $x_{(0)} < X$. In this case for $x < X$ we find $f(x) > X$ and $f(f(x)) < X$. Hence Eq. (26) now has $\zeta \in (x, X)$ and Eq. (27) defines $\xi \in (X, f(x))$. Following the steps outlined by Eqs. (28)–(30) we then find

$$[f^{(1)}(\xi) - f^{(1)}(\zeta)]f^{(1)}(\zeta) < 0 \qquad (32)$$

and once again the denominator of Eq. (23) does not vanish. Hence we obtain, as before,

$$g(x) - X = (x - X)\left(1 - \frac{[f^{(1)}(\zeta) - 1]^2}{([f^{(1)}(\zeta) - 1]^2 + [f^{(1)}(\xi) - f^{(1)}(\zeta)]f^{(1)}(\zeta))}\right)$$

where using Eq. (32) we may write this as

$$g(x) - X = (x - X)(1 - a^2/(a^2 - b^2))$$

with

$$a^2 = [f^{(1)}(\zeta) - 1]^2 \quad \text{and} \quad b^2 = [f^{(1)}(\zeta) - f^{(1)}(\xi)]f^{(1)}(\zeta)$$

Consider

$$a^2 - b^2 = 1 - 2f^{(1)}(\zeta) + f^{(1)}(\xi)f^{(1)}(\zeta) \tag{33}$$

Now since $\xi < \zeta$, $f^{(1)}(\xi) < f^{(1)}(\zeta)$ by condition (d) of the theorem. Hence from Eq. (33) we find

$$a^2 - b^2 > 1 - 2f^{(1)}(\zeta) + f^{(1)^2}(\zeta) = [1 - f^{(1)}(\zeta)]^2 > 0$$

hence $a^2/(a^2 - b^2) > 1$ and hence $g(x) > X$.

Thus starting with an $x_{(0)} < X$ we produce $x_{(1)} = g(x_{(0)}) > X$ and the sequence converges by the first part of the theorem. Thus the proof is complete.

Example 8   Obtain the root of $x \exp(\sqrt{x}) - 1 = 0$ using Steffensen's method.

We write the equation as $x = \exp(-\sqrt{x})$ so that in the notation of this section $f(x) = \exp(-\sqrt{x})$ satisfies the conditions of Theorem 10. Hence Steffensen's method becomes

$$x_{(n+1)} = g(x_{(n)}) = \frac{x_{(n)} \exp(-\sqrt{(\exp(-\sqrt{x}_{(n)}))}) - \exp(-2\sqrt{x}_{(n)})}{\exp(-\sqrt{(\exp(-\sqrt{x}_{(n)}))}) - 2\exp(-\sqrt{x}_{(n)}) + x_{(n)}} \quad n = 0, 1, \ldots$$

where $x_{(0)}$ is chosen as 1.0. The results using simple iteration, the Aitken δ^2-process, and Steffensen's method are shown in Table 2.

Example 9   The Steffensen's method converges under conditions other than those implied by Theorem 10. To show this, the root required in Example 7 was solved using Steffensen's method. The result is shown in column 4 of Table 1. The quadratic convergence is apparent.

Table 2

First order	Aitken	Steffensen	Secant	Regula falsi
.367 879 44	.506 379 11	.506 379 11	.564 114 95	.564 114 95
.445 239 21	.496 418 82	.494 876 68	.505 924 68	.505 924 68
.477 876 40	.495 052 79	.494 866 44	.495 224 36	.495 224 36
.500 931 43	.494 889 72	.494 845 34	.494 868 33	.494 868 33
.492 744 20	.494 869 29	.494 866 42	.494 866 41	.494 866 41
.495 614 23	.494 866 77			
.494 603 55	.494 866 46			
.494 958 89	.494 866 42			
.494 833 89	.494 866 42			
.494 877 86	.494 866 41			
.494 862 39				
.494 867 83				
.494 865 92				
.494 866 59				
.494 866 35				
.494 866 44				
.494 866 41				

Exercises

17. Develop a procedure (subroutine) for the method of simple iteration for an arbitrary function $f(x)$. The procedure (subroutine) should have dummy parameters to represent the maximum number of iterations to be allowed, the tolerance required before convergence has occurred, and the starting value $x_{(0)}$. If convergence does not occur then a printed message is to be output from within the procedure (subroutine). If convergence does occur then the final iterate is to be placed in the variable $x_{(0)}$. The iterates calculated by the method should be stored in an array (subscripted variable) which is an additional parameter.

18. Use the procedure (subroutine) of Exercise 17 in a main program to calculate the root of the equation $x^2 - 2.1x + 1.0$ which is nearer to zero. The accuracy required is three decimal places.

19. Develop a procedure (subroutine) for the Aitken δ^2-process for a sequence of iterates stored in a parameter array (subscripted variable).

20. Incorporate the procedure (subroutine) of Exercise 19 into a main program of Exercise 18 to accelerate the convergence of the sequence produced by the method of simple iteration.

21. Use the procedures of Exercises 17 and 19 to calculate the zero of the function $-2\pi/3 + \sqrt{3} + 2\sin(x) + \sqrt{x}$ correct to five decimal places.

22. Develop a procedure (subroutine) for Steffensen's method for an arbitrary function $f(x)$. The procedure (subroutine) should have parameters for the initial value $x_{(0)}$, the tolerance ε, and the maximum number of iterations m. The final iterate should be stored in $x_{(0)}$ on exit from the procedure (subroutine) and a printed message output from within the procedure (subroutine) if convergence does not occur within the specified number of iterations.

23. Use the procedure of Exercise 22 in a main program to calculate the specified roots of the equations in Exercises 18 and 21.

24. Show that $\exp(-\sqrt{x})$ satisfies the conditions (a)–(d) of Theorem 10 in the interval $(0, \infty)$.

25. Using the procedures (subroutines) developed in Exercises 19 and 22 repeat the calculations carried out in Example 8 using $f(x) = (\ln(x))^2$. Any comments?

The method of accelerating the first-order method described above can in fact be generalized so as to produce higher-order convergent methods. We omit a discussion of such methods here, but refer the interested reader to Isaacson and Keller [24].

We now want to consider the alternative approach to gaining an increase in the rate of convergence suggested at the beginning of this section. Namely, we require to approximate the function $F(x)$ by a divided difference. If we assume the derivative $F^{(1)}(x)$ is linear in a given interval (a, b), then the divided difference

$$\frac{F(x_{(n)}) - F(x_{(n-1)})}{(x_{(n)} - x_{(n-1)})} \tag{34}$$

will exactly represent $F^{(1)}(x)$ for $x \in (a, b)$. Since in general $F^{(1)}(x)$ will not be linear, (34) will only be an approximation to $F^{(1)}(x)$. Thus instead of $F^{(1)}(x)$ in the Newton–Raphson method, insert the relation (34). Hence the iterative method becomes

$$x_{(n+1)} = x_{(n)} - \frac{(x_{(n)} - x_{(n-1)})F(x_{(n)})}{F(x_{(n)}) - F(x_{(n-1)})} \tag{35}$$

which may be rewritten in an obvious fashion as

$$x_{(n+1)} = \frac{x_{(n-1)}F(x_{(n)}) - x_{(n)}F(x_{(n-1)})}{F(x_{(n)}) - F(x_{(n-1)})} \qquad n = 1, 2, \ldots \qquad (36)$$

Equation (36) is called the *secant* method. The reader will note several new points about this method which have not arisen in the methods we have discussed to date. First, the method expresses the new iterate $x_{(n+1)}$ in terms of two previous levels $x_{(n)}$ and $x_{(n-1)}$. The secant method is called a one-pont method with memory, since in addition to the most recently calculated iterate $x_{(n)}$ the previous iterate $x_{(n-1)}$ must be stored (see Traub [43]). Previous methods described in this chapter, with the possible exception of Steffensen's method, have been one-point methods without memory. A seeming disadvantage of methods with memory is that in order to calculate $x_{(n+1)}$ two levels of data ($x_{(n)}$ and $x_{(n-1)}$) have to be stored. However, this increase in storage is offset by not requiring the evaluation of the derivative $F^{(1)}(x)$. Provided $F^{(1)}(x)$ is nonzero, the denominator of Eq. (36) will be nonzero and hence the secant method will be well defined. On heuristic grounds, since we are approximating $F^{(1)}(x)$ but do not exactly represent $F^{(1)}(x)$ by the divided difference, we would suspect that the rate of convergence of the secant method is somewhere between linear and quadratic convergent. This is supported by the numerical results quoted in column 5 of Table 1, where the number of iterates required to converge to the required root of $x^3 - 3x^2 + 2x - 0.375 = 0$ is more than that of the Newton–Raphson method but significantly less than that of the first-order method. In fact is can be shown that the rate of convergence of the secant method is 1.62 (see Householder [23]). A seeming disadvantage of the secant method is that as the root is approached, $F(x_{(n)}) - F(x_{(n-1)})$ becomes small and hence multiple-precision arithmetic would be required. However, by retaining the original form of Eq. (35),

$$\frac{(x_{(n)} - x_{(n-1)})F(x_{(n)})}{F(x_{(n)}) - F(x_{(n-1)})}$$

may be considered to be a small correction term near the root X and will therefore require few significant digits as the root X is approached in order to compute the solution X to almost full single-precision accuracy. Consequently, multiple-precision arithmetic will not in general be necessary.

Example 10   The root nearer to zero of $x^3 - 3x^2 + 2x - 0.375 = 0$ is required. The secant method in this case is defined as

$$x_{(n+1)} = \frac{\begin{aligned}x_{(n-1)}(x_{(n)}^3 - 3x_{(n)}^2 + 2x_{(n)} - 0.375)\\ - x_{(n)}(x_{(n-1)}^3 - 3x_{(n-1)}^2 + 2x_{(n-1)} - 0.375)\end{aligned}}{\begin{aligned}(x_{(n)}^3 - 3x_{(n)}^2 + 2x_{(n)} - 0.375)\\ - (x_{(n-1)}^3 - 3x_{(n-1)}^2 + 2x_{(n-1)} - 0.375)\end{aligned}} \qquad (37)$$

Starting with $x_{(0)} = 0.7$, $x_{(1)} = 0.6$, the results obtained using (37) are given in Table 1, column 5.

Example 11   For the root of $x \exp \sqrt{x} - 1 = 0$, the secant method becomes

$$x_{(n+1)} = \frac{x_{(n-1)}(x_{(n)} \exp (\sqrt{x_{(n)}}) - 1) - x_{(n)}(x_{(n-1)} \exp (\sqrt{x_{(n-1)}}) - 1)}{x_{(n)} \exp (\sqrt{x_{(n)}}) - x_{(n-1)} \exp (\sqrt{x_{(n-1)}})} \tag{38}$$

Using starting values $x_{(0)} = 1.0$, $x_{(1)} = 0.9$, the iterates obtained are given in column 4 of Table 2.

Although we have derived the secant method by substituting an approximation to the derivative $F^{(1)}(x)$, it is easy to see that Eq. (36) is just the linear inverse interpolant which we developed in Chapter 4.

Convergence of the secant method is guaranteed by the following theorem.

Theorem 11   *Let $F(x) = 0$ have a simple root X in $[a, b]$. Let $F^{(1)}(x)$ and $F^{(2)}(x)$ be defined and non-vanishing on $[a, b]$. Then for starting values $x_{(0)}$, $x_{(1)} \in [a, b]$ the secant method produces a sequence of iterates which converges to the solution X.*

When the conditions of this theorem are violated it is easily seen that the secant method does not converge—the function depicted in Figure 9 is an example in point. However, a simple modification to the secant method can prevent the situation depicted in Figure 9 from arising. We note that provided we ensure that the two iterates $x_{(n)}$ and $x_{(n-1)}$ enclose the required root X, the new approximation $x_{(n+1)}$ will always be a better approximation for continuous functions $F(x)$. To ensure the latest iterates used in the secant method do enclose the root, in updating the iterates to be used in the right-hand side of Eq. (36) we use the rule:

after calculating $x_{(n+1)}$, $n = 1, 2, \ldots,$

$$x_{(n-1)} = x_{(n-1)} \text{ if } F(x_{(n-1)}F(x_{(n+1)}) < 0 \tag{39}$$

else

$$x_{(n-1)} = x_{(n)}$$

and

$$x_{(n)} = x_{(n+1)}$$

and then Eq. (36) is used once again. The test in Eq. (39) ensures that we omit from the iterates to be used at the next stage that one which is on the same side of the root X as the latest iterate $x_{(n+1)}$.

This modification of the secant method is called the *regula falsi* method. For the function depicted in Figure 9 the regula falsi method produces the sequence of iterates depicted in Figure 10.

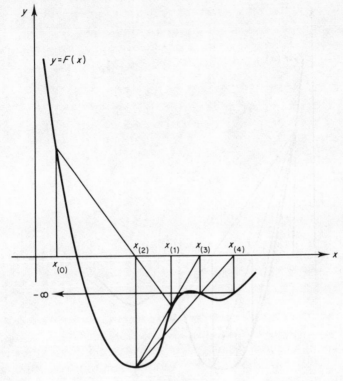

Figure 9

The iterates are discarded in the order $x_{(1)}, x_{(2)}, x_{(3)}, x_{(4)}, \ldots$, with the iterates ultimately converging to X.

The global convergence of the regula falsi method is, unfortunately, not gained without cost. Consideration of the error term for the linear inverse interpolation for this method indicates that the convergence rate is linear (see, for example Ralston [34]).

Thus the regula falsi method would appear to be most useful in the role of providing an accurate starting value for a method which possesses a higher rate of convergence, in the manner explained in section 1 of this chapter.

Example 12   The equations of Examples 7 and 11 were also solved using the regula falsi method. The results are given in column 6 of Table 1 and column 5 of Table 2 respectively.

Exercises

26. Develop a procedure (subroutine) for the secant method. The procedure should be provided with parameters to represent the function F, the starting values $x_{(0)}$ and $x_{(1)}$, the tolerance ε, and the maximum number of iterations to be allowed m. The final value of the iteration should be stored in an output parameter $x_{(2)}$

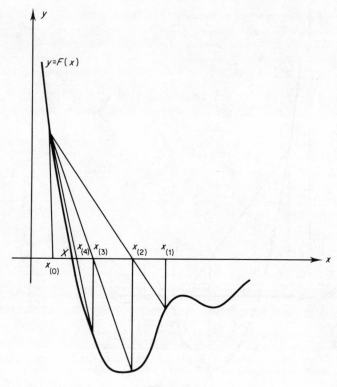

Figure 10

and a printed message output if convergence does not occur within the specified number of iterations.

27. Construct a similar procedure (subroutine) to Exercise 26 for the regula falsi method.

28. Use the procedures of Exercises 26 and 27 in a main program to calculate the root of the following equations correct to five decimal places:

   (i) $-2\pi/3 + \sqrt{3} + 2\sin(x) + \sqrt{x} = 0$

   (ii) $x - (\ln(x))^2 = 0$

   (iii) $x \exp(\sqrt{x}) - 1 = 0$

Compare your results with those of Exercises 21, 25, and Example 8 respectively.

29. In deriving the secant method we replaced the derivative $F^{(1)}(x)$ in the Newton–Raphson method by a first divided difference based upon the assumption that $F^{(1)}(x)$ was linear in the interval. If instead we assume the derivative $F^{(1)}(x)$ is a constant C (say), give an alternative method to the secant method. Derive conditions upon C so that the method is convergent. What would be a reasonable choice of C? Construct a program to use this method and compare the results obtained from it with those obtained in Exercise 28.

30. Replace the derivative $F^{(1)}(x)$ by a second divided difference on the assumption that $F^{(1)}(x)$ is a quadratic polynomial. Comment upon the properties of the method. Program the algorithm and test it upon the equations given in Exercise 28.

31. The Newton–Raphson method requires the derivative $F^{(1)}(x)$ to be evaluated anew each time a new iterate is calculated. Program the Newton–Raphson method

so that the derivative $F^{(1)}(x)$ is evaluated only every ith iterate, where i is to be given as a parameter, and remains constant (C_i say) in the succeeding iterates. Test the algorithm on the equations of Exercise 28. When would you expect this approach to be particularly effective?

32. In deriving the Newton–Raphson method for the nonlinear equation $F(x) = 0$ we have used the functional equation in its given form rather than in the form obtained by writing $F(x) \equiv x - f(x)$ as was done for the method of simple iteration. Clearly we can apply the Newton–Raphson method to $F(x)$ written in this latter form, so that

$$x_{(n+1)} = x_{(n)} - \frac{(x_{(n)} - f(x_{(n)}))}{(1 - f^{(1)}(x_{(n)}))} \qquad n = 0, 1, 2, \ldots$$

If f is chosen so that the method of simple iteration converges, then $x_{(n+1)} = f(x_{(n)})$ is a better approximation to X than is $x_{(n)}$. Hence we would surmise that using $f(x_{(n)})$ as the argument of $f^{(1)}$ would produce a more accurate iteration than the original Newton–Raphson method. Thus we propose

$$x_{(n+1)} = x_{(n)} - \frac{x_{(n)} - (x_{(n)} - f(x_{(n)}))}{1 - f^{(1)}(f(x_{(n)}))} \qquad n = 0, 1, 2, \ldots$$

as a modified Newton–Raphson method (this is *Stirling's method*). Program the algorithm of Stirling and compare the results, for a suitably defined f, for the examples of Exercise 28 with those obtained by the Newton–Raphson method.

7 The Newton–Raphson Method Where $F^{(1)}(x) = 0$ Close to the Root X

Before we progress to the situation where $F^{(1)}(x) = 0$ at the root, thereby indicating a root of multiplicity greater than 1, we wish to consider a modification of the Newton–Raphson method made necessary by the vanishing of $F^{(1)}(x)$ close to the root X. Such a situation is depicted in Figure 11. Let us assume $F^{(2)}(x) \neq 0$. With reference to Figure 11, let X be the required root. Let $x = \lambda$ be the point at which $F^{(1)}(x) = 0$. Let $x = \lambda + \varepsilon$ where ε is small ($\neq 0$). Since X is a root, $0 = F(X) = F(\lambda + \varepsilon)$, which by Taylor's series

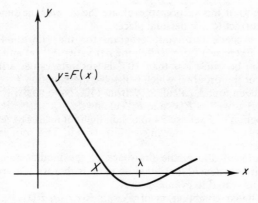

Figure 11

yields

$$0 = F(\lambda) + \varepsilon F^{(1)}(\lambda) + \frac{\varepsilon^2}{2} F^{(2)}(\lambda) + \frac{\varepsilon^3}{6} F^{(3)}(\xi) \qquad \text{with } \lambda < \xi < \varepsilon + \lambda$$

But by definition $F^{(1)}(\lambda) = 0$, so

$$F(\lambda) = -\tfrac{1}{2}\varepsilon^2 F^{(2)}(\lambda) + \text{higher-order terms in } \varepsilon \qquad (40)$$

From Eq. (40) we may solve for ε, namely

$$\varepsilon = \pm\sqrt{\frac{-2F(\lambda)}{F^{(2)}(\lambda)}} \qquad \text{approximately} \qquad (41)$$

The required root (for Figure 11) is then found from

$$\lambda - \sqrt{\frac{-2F(\lambda)}{F^{(2)}(\lambda)}} \qquad (42)$$

The other root close to X is clearly given by

$$\lambda + \sqrt{\frac{-2F(\lambda)}{F^{(2)}(\lambda)}}$$

Thus to determine the required root X of $F(x) = 0$ we first solve the equation $F^{(1)}(x) = 0$ using the Newton–Raphson method: this yields λ. We then substitute λ into Eq. (42) to obtain X. A drawback of this method is clearly the need to be able to evaluate $F^{(2)}(x)$, which for many examples may not be feasible. If $F^{(2)}(x) = 0$ near to X as well, then to obtain λ we would once again face the problem of trying to use the Newton–Raphson method with a (near) zero denominator. However, it is clear that we could repeat the procedure discussed above for $F^{(1)}(x) = 0$ for $F^{(2)}(x) = 0$.

Exercises

33. Use the method of this section to evaluate the zero of the function $2\pi/3 + \sqrt{3} + 2\sin(x) + x$ correct to five decimal places.
34. Incorporate a check into your program for the Newton–Raphson method, developed in Exercise 14, for testing the derivative $F^{(1)}(x)$ and comparing it with zero. If $F^{(1)}(x)$ becomes less than 10^{-4} in absolute value, arrange a jump into another part of the program which will solve the equation $F^{(1)}(x) = 0$. When this solution has been found, calculate X from (42). Note you will have to provide a form of $F^{(2)}(x)$ as well as $F(x)$ and $F^{(1)}(x)$ now.
35. Use your program of Exercise 34 to obtain the root nearer to zero of the equation $x^3 - 3x^2 + 2.99x - 0.99 = 0$ starting with $x_{(0)} = 0.0$. The root is required to five decimal places of accuracy.
36. As was mentioned above, the drawback of the modification to the Newton–Raphson method when $F^{(1)}(x) = 0$ close to X is that we require $F^{(2)}(x)$. Assume this is too complicated to evaluate.
    By taking a linear combination of $F(x_{(i)})$, $F(x_{(i+1)})$, $F(x_{(i-1)})$, $x_{(i)}$, $x_{(i+1)}$, $x_{(i-1)}$, determined by the Newton–Raphson method before $F^{(1)}(x) = 0$ was detected and

hence suitably close to λ, derive an approximation to $F^{(2)}(x)$ which can then be used in the Newton–Raphson method (instead of $F^{(2)}(x)$) and also in the formula for λ.

37. Program your algorithm of Exercise 36. Test it on the equation of Exercise 33. How does it compare with the modification to the Newton–Raphson method of this section?

38. We can use the same motivation in not evaluating $F^{(2)}(x)$ to propose using the secant method instead of the Newton–Raphson method to determine the root λ of $F^{(1)}(x) = 0$. Set up the secant method for $F^{(1)}(x) = 0$ and determine the root of Exercise 33. How do the results compare with those of Exercises 33 and 37?

39. If in addition to $F^{(1)}(x)$ being zero, $F^{(2)}$ is also zero close to X, we may then obtain a similar algorithm to that described in this section. If λ is again that x for which $F^{(1)}(x) = 0$, then let μ be the point where $F^{(2)}(x) = 0$ where we assume $\mu \neq \lambda$ (why?), then by repeating the steps outlined above, obtain the expression by Taylor series for μ which may then be used to provide λ, which in turn produces X.

40. Construct a program for the method you developed in Exercise 39 and apply it to the equation $20x^3 - x^2 - 0.1x + 0.02 = 0$, finding the root closest to zero to five decimal places accuracy.

41. If $\mu = \lambda$ in Exercise 39, how should you go about devising an algorithm to find X, assuming $F^{(3)} \neq 0$ in the neighbourhood of X? *Hint*: Repeat the analysis retaining terms $O(\varepsilon^3)$. Remember λ is also the required root of $F^{(2)}(x) = 0$.

42. Develop a program for your method of Exercise 41 and use it to determine the root of the equation

$$\frac{(610 - 10\sqrt{424})}{2} x^2 + x + 2\left(x - \frac{1}{610 - 10\sqrt{424}}\right) = 0$$

which is closer to -0.25 to an accuracy of five decimal places.

43. Replace the terms $F^{(2)}(x)$ and $F^{(3)}(x)$ needed in the method for Exercise 41 by suitable divided differences.

44. Program the method of Exercise 43 and apply it to the example of Exercise 42. How do the answers compare? Would you advocate the use of divided differences or the actual derivatives? Why?

8 Multiple Roots: The Case Where $F(X) = F^{(1)}(X) = \cdots = F^{(p-1)}(X) = 0$

In this section we want to consider briefly the problems of determining the root of the equation $F(x) = 0$ in the case that X is a multiple root of multiplicity p. For example, $F(x)$ may be of the form

$$F(x) = (x - X)^p g(x) \tag{43}$$

where $g(x)$ is a given function such that $g(X) \neq 0$. For such a function it is clear that at the multiple root X the Newton–Raphson method is not strictly defined. However, we may view the case of multiple roots for the Newton–Raphson method in one of two ways. *Either*: if in computing x we find $F(x) = 0$ then we need not evaluate the quotient; *or*: we assume since the iterative process cannot converge to X in a finite number of iterations and we assume there is no other point at which $F^{(1)}(x) = 0$ in the neighborhood of X, that we never actually need to evaluate $F(X)/F^{(1)}(X)$. However, as we would expect, the Newton–Raphson method is no longer quadratically convergent;

this point we shall now prove. Consider the Newton–Raphson method for the function (43). Then if at the nth iteration $x_{(n)} = X + \varepsilon_{(n)}$, we have

$$X + \varepsilon_{(n+1)} = X + \varepsilon_{(n)} - \frac{F(X + \varepsilon_{(n)})}{F^{(1)}(X + \varepsilon_{(n)})} \tag{44}$$

Expanding the quotient in Eq. (44) by Taylor series, we obtain

$$\varepsilon_{(n+1)} = \varepsilon_{(n)} - \frac{F(X) + \varepsilon_{(n)}F^{(1)}(X) + \cdots + \varepsilon_{(n)}^{(p)}F^{(p)}(X)/p! + \varepsilon_{(n)}^{(p+1)}F^{(p+1)}(\zeta_1)/(p+1)!}{F^{(1)}(X) + \varepsilon_{(n)}F^{(2)}(X) + \cdots + \varepsilon_{(n)}^{(p-1)}F^{(p)}(X)/(p-1)! + \varepsilon_{(n)}^{(p)}F^{(p+1)}(\zeta_2)/p!}$$

where $\zeta_1, \zeta_2 \in (X, \varepsilon_{(n)} + X)$. Using Eq. (43) we then have

$$\varepsilon_{(n+1)} = \varepsilon_{(n)} - [\varepsilon_{(n)}/p + \varepsilon_{(n)}^2 F_{(\zeta_1)}^{(p+1)}/p(p+1)/F^{(p)}(X)]$$
$$\times [1 + \varepsilon_{(n)} F_{(\zeta_2)}^{(p+1)}/pF^{(p)}(X)]^{-1} \tag{45}$$

Expanding the last square bracketed expression in Eq. (45) by the bionomial theorem we obtain

$$\varepsilon_{(n+1)} = \varepsilon_{(n)} - \frac{\varepsilon_{(n)}}{p} + \text{term of order } \varepsilon_{(n)}^2$$

i.e.

$$\varepsilon_{(n+1)} = \varepsilon_{(n)}\left(1 - \frac{1}{p}\right) + \text{higher order terms of } \varepsilon_{(n)}$$

Thus unless p were 1 (which by assumption it is not) the rate of convergence is linear. Note also that since $1 - 1/p$ is less than 1, the method is convergent for multiple roots for a starting value $x_{(0)}$ sufficiently close to the root X.

In deriving the rate of convergence we note that a factor $1/p\varepsilon_{(n)}$ arises from the function $F(x_{(n)})/F^{(1)}(x_{(n)})$. Clearly if we replace this latter term by $pF(x_{(n)})/F^{(1)}(x_{(n)})$ in the Newton–Raphson method, there would result a factor $p/p\varepsilon_{(n)}$, that is $\varepsilon_{(n)}$ which would then cancel with the remaining $\varepsilon_{(n)}$. In such a case the Newton–Raphson method thus modified becomes quadratically convergent. Thus for known multiplicity p the algorithm

$$x_{(n+1)} = x_{(n)} - pF(x_{(n)})/F^{(1)}(x_{(n)}) \tag{46}$$

is quadratically convergent for the root X of multiplicity p. Clearly, if p is not known, Eq. (46) is of little use. However, an alternative modification of the Newton–Raphson method which is second-order convergent can be obtained without a priori knowing the multiplicity p. For example, for the $F(x)$ defined by Eq. (43) we note that

$$F(x)/F^{(1)}(x) = (x - X)g(x)/(pg(x) + (x - X)g^{(1)}(x))$$

Clearly this vanishes at $x = X$, but more important $x - X$ is a simple root of $F(x)/F^{(1)}(x)$. Hence by applying our previous methods to the function

$F(x)/F^{(1)}(x)$ instead of $F(x)$, we again obtain a function with simple roots. Let

$$u(x) = F(x)/F^{(1)}(x)$$

Then the Newton–Raphson method applied to u is

$$x_{(n+1)} = x_{(n)} - u(x_{(n)})/u^{(1)}(x_{(n)})$$

namely

$$x_{(n+1)} = \frac{x_{(n)} - F(x_{(n)})F^{(1)}(x_{(n)})}{[F^{(1)}(x_{(n)})]^2 - F(x_{(n)})F^{(2)}(x_{(n)})} \tag{47}$$

This modification (47), however, requires us to evaluate the second derivative of $F(x)$ and is consequently bound to be less efficient than the original Newton–Raphson method applied to $F(x)$.

Example 13   Determine the root of the equation $(\exp(x) - 2)^3 = 0$ to four decimal places using the Newton–Raphson method and its variants (46) ($p = 3$) and (47). A starting value $x = 0.5$ is to be used.

The results of applying the three methods are reported in Table 3. It can be seen from the results in the first column that the Newton–Raphson method has not converged after 20 iterations whereas the modifications given by Eqs. (46) and (47) converge after just three iterations.

Table 3

Newton–Raphson	Eq. (46) ($p = 3$)	Eq. (47)
0.571 02	0.713 06	0.675 64
0.614 32	0.693 34	0.692 99
0.641 66	0.693 15	0.693 15
0.659 27	0.693 15	0.693 15
0.670 76		
0.678 30		
0.683 29		
0.686 59		
0.688 88		
0.690 24		
0.691 21		
0.691 86		
0.692 29		
0.692 57		
0.692 77		
0.692 89		
0.692 98		
0.693 03		
0.693 07		
0.693 10		

The discussion of this section is necessarily brief. The interested reader may find a comprehensive discussion of multiple roots in Traub [43].

Exercises

45. Program the modified Newton–Raphson method (46) for known multiplicity p. Use your program to evaluate the root of multiplicity 2 of the equation

$$\text{(a)} \quad (\cos (x) - x^{1/2})^2 = 0$$

and of multiplicity 3 of the equation

$$\text{(b)} \quad (\cos (x) - x^{1/2})^3 = 0$$

starting with $x_{(0)} = 0.5$. Five decimal places of accuracy is required. How does the Newton–Raphson method perform on the functions of (a) and (b)?
46. Program the Newton–Raphson method applied to $u = F(x)/F^{(1)}(x)$ given by Eq. (47). Use your program for the functions of Exercise 45 where the same starting value is used and the same accuracy is required.
47. In what cases of multiplicity p is
    (a) the regula falsi method
    (b) the secant method
    likely to work/not work? Why? Test your hypothesis by employing your programs of Exercises 26 and 27 for the equations of Exercise 45 using the starting value $x_{(0)} = 0.6$, $x_{(1)} = 0.8$.
48. The secant method can be applied like the Newton–Raphson method to the function $u(x)$. Program the secant method for this case and solve the equations of Exercise 47. Compare your results with those obtained in Exercises 45–47.

9 Systems of Nonlinear Equations

In this section we will do little more than introduce the reader to the problem of solving systems of nonlinear equations since this topic tends to be rather difficult and one which the reader will likely meet in a more advanced course or text.

In contrast to the single equation of a single variable x given by $F(x) = 0$ which we met in previous sections of this chapter, we now consider x and F to be column vectors with N components. To distinguish these from our single equation variables we write as before, the vectors in bold type. Hence

$$\mathbf{x} = (x_1, x_2, \ldots, x_N)^{\text{T}}$$

and $\mathbf{F}(\mathbf{x}) = (F_1(\mathbf{x}), F_2(\mathbf{x}), \ldots, F_N(\mathbf{x}))^{\text{T}}$

so that the system of nonlinear equations we intend to consider can be written as

$$\mathbf{F}(\mathbf{x}) = \mathbf{0} \tag{48}$$

where $\mathbf{0}$ is the null vector (i.e., the vector with every component equal to zero).

Example 14 For $N = 2$, a system of two nonlinear equations in the two variables x_1, x_2 is

$$x_1 + x_2 - x_2^{1/2} - 0.25 = 0$$
$$8x_1^2 + 16x_2 - 8x_1x_2 - 5 = 0 \tag{49}$$

so that

$$F_1(x_1, x_2) \equiv F_1(\mathbf{x}) = x_1 + x_2 - x_2^{1/2} - 0.25$$

and

$$F_2(x_1, x_2) \equiv F_2(\mathbf{x}) = 8x_1^2 + 16x_2 - 8x_1x_2 - 5$$

We will denote a solution of Eq. (48) as a vector $\mathbf{X}$ with components $(X_1, X_2, \ldots, X_N)$ such that $\mathbf{F}(\mathbf{X}) = \mathbf{0}$. Further, we will assume that $\mathbf{F}$ possesses as many partial derivatives with respect to the independent variables $x_1, \ldots, x_N$ as required in the neighborhood of the solution $\mathbf{X}$. We will denote by J the Jacobian matrix of $\mathbf{F}$ with respect to its components $x_1, \ldots x_N$, i.e.

$$J = \begin{bmatrix} \dfrac{\partial F_1}{\partial x_1} & \dfrac{\partial F_1}{\partial x_2} & \cdots & \dfrac{\partial F_1}{\partial x_N} \\[2mm] \dfrac{\partial F_2}{\partial x_1} & \dfrac{\partial F_2}{\partial x_2} & \cdots & \dfrac{\partial F_2}{\partial x_N} \\[1mm] \vdots & \vdots & & \\[1mm] \dfrac{\partial F_N}{\partial x_1} & \dfrac{\partial F_N}{\partial x_2} & \cdots & \dfrac{\partial F_N}{\partial x_N} \end{bmatrix} \tag{50}$$

and assume J is nonsingular in the neighborhood of $\mathbf{X}$.

It is often a simple matter then to propose methods for Eq. (48) based on our knowledge of algorithms for $F(x) = 0$. For example, the method of simple iteration of section 2 suggests the method

$$\mathbf{x}_{(n+1)} = \mathbf{f}(\mathbf{x}_{(n)}) \qquad n = 0, 1, 2, \ldots \tag{51}$$

where $\mathbf{x}_{(n)} = (x_1, x_2, \ldots, x_N)_{(n)}^{\mathrm{T}}$ is the vector of iterates obtained at the nth iteration and $\mathbf{x}_{(0)} = (x_1, x_2, \ldots, x_N)_{(0)}^{\mathrm{T}}$ is an initial starting vector. The function $\mathbf{f}$ has components $f_1, f_2, \ldots, f_N$ and is obtained by rewriting $\mathbf{F}(\mathbf{x})$ as $\mathbf{x} - \mathbf{f}(\mathbf{x})$.

Example 15 Consider the system of nonlinear equations in Example 14. These equations can clearly be written as

$$\text{(i)} \quad \begin{bmatrix} x_1 \\ x_2 \end{bmatrix} - \begin{bmatrix} 0.25 + x_2^{1/2} - x_2 \\ \frac{1}{16}(5 + 8x_1x_2 - 8x_1^2) \end{bmatrix} = \begin{bmatrix} 0 \\ 0 \end{bmatrix} \tag{52}$$

in which case $f_1 = 0.25 + x_2^{1/2} - x_2$; $f_2 = (5 + 8x_1x_2 - 8x_1^2)/16$; or

$$\text{(ii)} \quad \begin{bmatrix} x_1 \\ x_2 \end{bmatrix} - \begin{bmatrix} (5 - 16x_2 - 8x_1^2)/8x_2 \\ 0.25 + x_2^{1/2} - x_1 \end{bmatrix} = \begin{bmatrix} 0 \\ 0 \end{bmatrix} \tag{53}$$

when $f_1 = (5 - 16x_2 - 8x_1^2)/8x_2$; $f_2 = 0.25 + x_2^{1/2} - x_1$; or

$$\text{(iii)} \quad \begin{bmatrix} x_1 \\ x_2 \end{bmatrix} - \begin{bmatrix} (5 - 16x_2 - 8x_1^2)/8x_2 \\ (x_1 + x_2 - 0.25)^2 \end{bmatrix} = \begin{bmatrix} 0 \\ 0 \end{bmatrix} \tag{54}$$

when $f_1 = (5 - 16x_2 - 8x_1^2)/8x_2$; $f_2 = (x_1 + x_2 - 0.25)^2$.

Note that we do not insist that f_i be formed from F_i. As in Eq. (52), it may be more convenient to define f_i from another component, say F_j.

Example 16 The method of simple iteration applied to the system (49) with $f_1 \equiv 0.25 + x_2^{1/2} - x_2$; $f_2 \equiv (5 + 8x_1x_2 - 8x_1^2)/16$ has the form

$$x_{1_{(n+1)}} = 0.25 + (x_{2_{(n)}})^{1/2} - x_{2_{(n)}}$$

$$x_{2_{(n+1)}} = (5 + 8x_{1_{(n)}}x_{2_{(n)}} - 8(x_{1_{(n)}})^2)/16 \qquad n = 0, 1, 2, \ldots \tag{55}$$

where $x_{1_{(0)}}$ and $x_{2_{(0)}}$ are initial guesses of the solution. The results of applying Eq. (55) and $x_{1_{(0)}} = 0.4$, $x_{2_{(0)}} = 0.2$ are given in the first two columns of Table 4. The sequences clearly converge to the solution $x_1 = 0.5$, $x_2 = 0.25$.

Table 4

x_1	x_2	x_1^*	x_2^*
0.497 214	0.272 500	0.497 214	0.238 611
0.499 515	0.256 635	0.499 867	0.247 203
0.499 957	0.251 834	0.499 998	0.249 304
0.499 997	0.250 476	0.499 999	0.249 826
0.499 999(77)	0.250 120	0.500 000	0.249 957
0.499 999(99)	0.250 030		0.249 987
	0.250 008		0.250 000
	0.250 002		
	0.250 000		

Exercises

49. Derive other forms of f_1 and f_2 for the function $\mathbf{F}$ of Eq. (49).
50. Program the method of simple iteration based upon the functions of Eqs. (52), (53), (54) and the functions of Exercise 49. Compare the behavior of the resulting algorithms for the starting values $x_{1_{(0)}} = 0.4$, $x_{2_{(0)}} = 0.2$.
51. Suggest a method of simple iteration for the system of three nonlinear equations in the three variables x_1, x_2, x_3 given by

$$(x_1^2 + x_2^2/2)\pi/3 - x_3 = 0$$

$$x_1^2 + x_2/2 + 2\cos(x_3) = 0 \tag{56}$$

$$x_1x_2 - \sin(2x_3)\cos(x_2) - 2 = 0$$

52. Starting with $x_{1_{(0)}} = 0.8$, $x_{2_{(0)}} = 1.8$, $x_{3_{(0)}} = 3.0$, obtain the solution to Eq. (56) to four decimal places.

The reader will note that the amount of choice we have in defining the algorithm is considerably greater than for the corresponding method of simple iteration for a single equation. In addition to the greater choice of functions, however, we have also the choice of using, or not using, the latest obtained iterate in subsequent components of the algorithm. For example, for the results quoted in Table 4 the algorithm (55) was used. If instead of using $x_{1_{(n)}}$ in the second component of (55) we use $x_{1_{(n+1)}}$ which has just been calculated from the first equation, we obtain the modified algorithm

$$x_{1_{(n+1)}} = 0.25 + (x_{2_{(n)}})^{1/2} - x_{2_{(n)}}$$
$$x_{2_{(n+1)}} = (5 + 8x_{1_{(n+1)}}x_{2_{(n)}} - 8(x_{1_{(n+1)}})^2)/16 \tag{57}$$

This modified method (57) produced the results x_1^* and x_2^* indicated in Table 4 starting with $x_{1_{(0)}} = 0.4$, $x_{2_{(0)}} = 0.2$.

The nonlinear system counterpart of the Newton–Raphson method for the single equation $F(x) = 0$ may once again be derived from the method of simple iteration by requiring the vector function $\mathbf{f}$ obtained in writing $\mathbf{F}(\mathbf{x})$ as $\mathbf{x} - \mathbf{f}(\mathbf{x})$ to possess a "zero derivative." Of course, since $\mathbf{f}$ has N components, its derivative is the Jacobian matrix of $\mathbf{f}$ with respect to the components of $\mathbf{x}$. Hence following the derivation of section 4 we write

$$\mathbf{F}(\mathbf{x}) = H^{-1}(\mathbf{x})(\mathbf{x} - \mathbf{x}) + \mathbf{F}(\mathbf{x}) \tag{58}$$

where $H(\mathbf{x})$ is a matrix which is to be determined to produce zero J for $\mathbf{f}$, and $H^{-1}(\mathbf{x})$ represents the inverse of $H(\mathbf{x})$. This function therefore suggests the iterative method

$$\mathbf{x}_{(n+1)} = \mathbf{x}_{(n)} - H(\mathbf{x}_{(n)})\mathbf{F}(\mathbf{x}_{(n)})$$

where $\mathbf{f}(\mathbf{x})$ for the method of simple iteration is clearly $\mathbf{x} - H(\mathbf{x})\mathbf{F}(\mathbf{x})$. For this $\mathbf{f}$, differentiate the function with respect to each component of $\mathbf{x}$ to obtain the matrix J_f (say) corresponding to J in Eq. (50) for $\mathbf{F}$, namely

$$J_f = \begin{bmatrix} 1 - \sum_{j=1}^{N} \dfrac{\partial}{\partial x_1} H_{1j}F_j & \sum_{j=1}^{N} \dfrac{\partial}{\partial x_2} H_{1j}F_j & \cdots & \sum_{j=1}^{N} \dfrac{\partial}{\partial x_N} H_{1j}F_j \\ \sum_{j=1}^{N} \dfrac{\partial}{\partial x_1} H_{2j}F_j & 1 - \sum_{j=1}^{N} \dfrac{\partial}{\partial x_2} H_{2j}F_j & \cdots & \sum_{j=1}^{N} \dfrac{\partial}{\partial x_N} H_{2j}F_j \\ \vdots & & & \\ \sum_{j=1}^{N} \dfrac{\partial}{\partial x_1} H_{Nj}F_j & \cdots & & 1 - \sum_{j=1}^{N} \dfrac{\partial}{\partial x_N} H_{Nj}F_j \end{bmatrix} = \mathbf{0}$$

where we have assumed the matrix H has components H_{ij}, $i = 1, 2, \ldots, N$; $j = 1, 2, \ldots, N$, and we have suppressed the obvious dependence of the functions on the argument $\mathbf{X}$. On differentiating the products under the summation signs we obtain, for example,

$$\sum_{j=1}^{N} \frac{\partial}{\partial x_i} H_{kj}F_j = \sum_{j=1}^{N} \frac{(\partial H_{kj})}{\partial x_i} F_j + \sum_{j=1}^{N} H_{kj} \frac{\partial}{\partial x_i} F_j \qquad i, k = 1, 2, \ldots N$$

But the functions H_{kj} and F_j are evaluated at $\mathbf{x} = \mathbf{X}$. Hence by definition $\mathbf{F}(\mathbf{X}) = \mathbf{0}$ and so we may write

$$
J_f = \begin{bmatrix} 1 - \sum\limits_{j=1}^{N} H_{1j}\dfrac{\partial F_j}{\partial x_1} & \sum\limits_{j=1}^{N} H_{1j}\dfrac{\partial F_j}{\partial x_2} & \cdots & \sum\limits_{j=1}^{N} H_{1j}\dfrac{\partial F_j}{\partial x_N} \\ \sum\limits_{j=1}^{N} H_{2j}\dfrac{\partial F_j}{\partial x_1} & 1 - \sum\limits_{j=1}^{N} H_{2j}\dfrac{\partial F_j}{\partial x_2} & \cdots & \\ \vdots & & & \\ \sum\limits_{j=1}^{N} H_{Nj}\dfrac{\partial F_j}{\partial x_1} & \cdots & & 1 - \sum\limits_{j=1}^{N} H_{Nj}\dfrac{\partial F_j}{\partial x_N} \end{bmatrix} = \mathbf{0}
$$

where in turn J_f may be written, using $\cdot$ to represent the usual multiplication of two matrices, as

$$
I - \begin{bmatrix} H_{11} & H_{12} & & H_{1N} \\ H_{21} & H_{22} & & H_{2N} \\ \vdots & & \ddots & \\ H_{N1} & H_{N2} & & H_{NN} \end{bmatrix} \cdot \begin{bmatrix} \dfrac{\partial F_1}{\partial x_1} & \dfrac{\partial F_1}{\partial x_2} & & \dfrac{\partial F_1}{\partial x_N} \\ \dfrac{\partial F_2}{\partial x_1} & \dfrac{\partial F_2}{\partial x_2} & & \dfrac{\partial F_2}{\partial x_N} \\ \vdots & & \ddots & \\ \dfrac{\partial F_N}{\partial x_1} & \dfrac{\partial F_N}{\partial x_2} & & \dfrac{\partial F_N}{\partial x_N} \end{bmatrix} \tag{59}
$$

where I is the unit matrix of order N. Clearly (59) is just

$$
I - H(X)J(X)
$$

so that we merely require $H(\mathbf{X})$ to be the inverse of $J(\mathbf{X})$. As in section 4, we find that a suitable choice of H which satisfies this requirement is to choose for any $\mathbf{x}$, $H(\mathbf{x}) = J^{-1}(\mathbf{x})$. Hence the Newton–Raphson method for systems of nonlinear equations becomes

$$
\mathbf{x}_{(n+1)} = \mathbf{x}_{(n)} - J^{-1}(\mathbf{x}_{(n)})F(\mathbf{x}_{(n)}) \tag{60}
$$

with $\mathbf{x}_{(0)}$ a vector of initial guesses to $\mathbf{X}$.

Note that to calculate a new iterate we must solve for the correction $J^{-1}(\mathbf{x}_{(n)})F(\mathbf{x}_{(n)})$. This can be (and usually is) a formidable task and so if we were reluctant to calculate $F^{-1}(x)$ in the scalar problem, we will certainly be reluctant to solve for the correction term for this system. Therefore, as in the scalar equation we may prefer to approximate $J^{-1}(\mathbf{x})$ by some other matrix which is easier to calculate. For example, the system counterpart of the algorithm of Exercise 29 is

$$
\mathbf{x}_{(n+1)} = \mathbf{x}_{(n)} - J^{-1}(\mathbf{x}_{(I)})F(\mathbf{x}_{(n)}) \tag{61}
$$

or

$$
\mathbf{x}_{(n+1)} = \mathbf{x}_{(n)} - J^*(\mathbf{x}_{(I)})F(\mathbf{x}_{(n)}) \tag{62}
$$

where $\mathbf{x}_{(I)}$ is some iterate (prior to $\mathbf{x}_{(n+1)}$) at which for Eq. (61) the inverse of the Jacobian J was last evaluated, and for Eq. (62) the approximation J^* was last modified. For example, the matrices J^{-1} or J^* may be held constant for ten iterations at a time and then modified. I in Eqs. (60) and (61) would then be the multiple of ten closest to and less than $n+1$ (see Example 17).

To describe the *order of accuracy* of the methods for systems of equations we have to generalize the concept of order introduced in section 4 for the scalar problem. We define an iterative method

$$\mathbf{x}_{(n+1)} = \mathbf{f}(\mathbf{x}_{(n)})$$

for solving the nonlinear equations $\mathbf{F}(\mathbf{x}) = \mathbf{0}$ with solution $\mathbf{X}$ to be convergent with integral order p if $\mathbf{f}(\mathbf{X}) = \mathbf{X}$ and all partial derivatives of $\mathbf{f}$ with respect to the components of $\mathbf{x}$ of order less than p vanish at $\mathbf{X}$ and at least one partial derivative of order p of $\mathbf{f}$ with respect to the components of $\mathbf{x}$ does not vanish at $\mathbf{X}$. That is,

$$\frac{\partial^k f_i(\mathbf{X})}{\partial x_{j_1} \partial x_{j_2} \cdots \partial x_{j_k}} = 0 \qquad \text{for all } 1 \le k \le p-1 \text{ and } 1 \le i, j_1, j_2, \ldots, j_k \le n$$

and

$$\frac{\partial^p f_i(\mathbf{X})}{\partial x_{j_1} \partial x_{j_2} \cdots \partial x_{j_p}} \ne 0 \qquad \text{for some value of } i, j_1, j_2, \ldots, j_p$$

The corresponding expression for the *asymptotic error constant* of section 4 for the scalar expression is

$$\frac{(1/p!)\partial^p f_i(\mathbf{X})}{\partial x_{j_1} \partial x_{j_2} \cdots \partial x_{j_p}} \qquad \text{with } 1 \le i, j_1, j_2, \ldots, j_p \le n$$

for a pth-order algorithm for the system of nonlinear equations. Using this definition we may easily verify that the method of simple iteration for the system of equations is first-order convergent. Consider at the nth iteration the iterate $\mathbf{x}_{(n)}$ and let $\mathbf{X} = \mathbf{x}_{(n)} - \boldsymbol{\varepsilon}_{(n)}$ where $\boldsymbol{\varepsilon}_{(n)}$ is a vector $(\varepsilon_1, \varepsilon_2, \ldots, \varepsilon_N)^{\mathrm{T}}_{(n)}$ of errors. Then by Eq. (51) we obtain

$$\mathbf{X} + \boldsymbol{\varepsilon}_{(n)} = \mathbf{f}(\mathbf{X} + \boldsymbol{\varepsilon}_{(n-1)}) \tag{63}$$

Using the mean value theorem applied to the vector functions (see Blum [2]), we obtain

$$\boldsymbol{\varepsilon}_{(n)} = J_f(\boldsymbol{\zeta})\boldsymbol{\varepsilon}_{(n-1)} \tag{64}$$

where $J_f(\boldsymbol{\zeta})$ is the Jacobian matrix of $\mathbf{f}$ with respect to the components of $\mathbf{x}$ evaluated at some intermediate (vector) point $\boldsymbol{\zeta} \in (\mathbf{X}, \mathbf{X} + \boldsymbol{\varepsilon}_{(n-1)})$. Equation (64) is clearly the vector counterpart of Eq. (7). Thus by definition of order, the method of simple iteration is first-order convergent. We see that a choice of $\mathbf{f}$ which produces $J_f = 0$, produces a second-order convergent method. But the matrix H was constructed to this end and hence the Newton–Raphson method is second-order convergent.

On taking (vector) norms of both sides of (64) we obtain

$$\|\boldsymbol{\varepsilon}_{(n)}\| = \|J_f(\boldsymbol{\zeta})\boldsymbol{\varepsilon}_{(n-1)}\| \le \|J_f(\boldsymbol{\zeta})\| \, \|\boldsymbol{\varepsilon}_{(n-1)}\| \tag{65}$$

Clearly if the error $\boldsymbol{\varepsilon}_{(n)}$ is not to grow, we require $\|J_f(\boldsymbol{\zeta})\| < 1$. This is then the corresponding condition for convergence of the method of simple iteration given by Theorem 3. We summarize this in Theorem 12.

Theorem 12   *Let R denote the region $a_i \leq x_i \leq b_i$, $i = 1, 2, \ldots, N$, in Euclidean N-space.*
Let f satisfy the conditions:
   (i) *f is defined and continuous on R*
   (ii) $\|J_f(\mathbf{x})\| \leq L < 1$
   (iii) *For each $\mathbf{x} \in R$, $\mathbf{f}(\mathbf{x})$ also lies in R*

Then for any $\mathbf{x}_{(0)}$ in R the sequence of iterates $\{\mathbf{x}_{(n)}\}$ defined by (51) converges to the unique solution $\mathbf{X}$ of Eq. (48).

Proof   We first prove that the solution $\mathbf{X}$ is the unique solution. Let us assume there are two solutions in R represented by $\mathbf{X}_1$ and $\mathbf{X}_2$. Then we have, in a similar fashion to Theorem 1,

$$(\mathbf{X}_1 - \mathbf{X}_2) = \mathbf{f}(\mathbf{X}_1) - \mathbf{f}(\mathbf{X}_2) = J_f(\boldsymbol{\zeta})(\mathbf{X}_1 - \mathbf{X}_2) \tag{66}$$

where we have used the multidimensional form of the mean value theorem. Taking norms of both sides of (66), we obtain

$$\|\mathbf{X}_1 - \mathbf{X}_2\| \leq \|J_f(\boldsymbol{\zeta})\| \, \|\mathbf{X}_1 - \mathbf{X}_2\| < \|\mathbf{X}_1 - \mathbf{X}_2\|$$

where we have used condition (ii) of our theorem.

We thus have a contradiction and hence $\mathbf{X}_1 \equiv \mathbf{X}_2$, i.e. the solution $\mathbf{X}$ is unique in R.

We next require to show that if the method converges, it does so to $\mathbf{X}$. We note from the continuity of $\mathbf{f}$ that assuming convergence,

$$\lim_{n \to \infty} \mathbf{f}(\mathbf{x}_{(n)}) = \mathbf{f}(\mathbf{X})$$

Hence

$$\mathbf{X} = \lim_{n \to \infty} \mathbf{x}_{(n+1)} = \lim_{n \to \infty} \mathbf{f}(\mathbf{x}_{(n)}) = \mathbf{f}(\mathbf{X})$$

Finally we require to show that if condition (ii) is satisfied then the method converges to $\mathbf{X}$. This is easy since by Eq. (65) we have shown that

$$\|\boldsymbol{\varepsilon}_{(n)}\| \leq \|J_f(\boldsymbol{\zeta})\| \, \|\boldsymbol{\varepsilon}_{(n-1)}\|$$

Applying this inequality n times, we obtain

$$\|\boldsymbol{\varepsilon}_{(n)}\| \leq \prod_{i=1}^{n} \|J_f(\boldsymbol{\zeta}_i)\| \, \|\boldsymbol{\varepsilon}_{(0)}\| \qquad \text{where } \boldsymbol{\zeta}_i \in (\mathbf{X}, \mathbf{X} + \boldsymbol{\varepsilon}_{(i)})$$

If we let $\max_i \|J_f(\boldsymbol{\zeta}_i)\| = \|J_f(\boldsymbol{\zeta})\|$, $\boldsymbol{\zeta} \in R$ we then have that

$$\|\boldsymbol{\varepsilon}_{(n)}\| \leq \|J_f(\boldsymbol{\zeta})\|^n \|\boldsymbol{\varepsilon}_{(0)}\| < L^n \|\boldsymbol{\varepsilon}_{(0)}\|$$

by assumption (ii) of our theorem.
   Thus

$$\lim_{n \to \infty} \|\boldsymbol{\varepsilon}_{(n)}\| \leq \lim_{n \to \infty} L^n \|\boldsymbol{\varepsilon}_{(0)}\| = 0$$

since $L < 1$. The theorem is therefore proved.

Exercises

53. If instead of the assumption that the Jacobian $J_f(\mathbf{x})$ exists, we assume that for two points $\mathbf{x}_{(1)}$ and $\mathbf{x}_{(2)} \in R$ there exists a constant $L < 1$ such that

$$\|\mathbf{f}(\mathbf{x}_{(1)}) - \mathbf{f}(\mathbf{x}_{(2)})\| < L\|\mathbf{x}_{(1)} - \mathbf{x}_{(2)}\| \qquad (*)$$

prove Theorem 12 with condition (ii) replaced by the assumption $(*)$.

54. Considering convergence in the $p = 1$ and $p = \infty$ norms,[†] which of the algorithms based on the definition of $\mathbf{f}$ in Example 15 are theoretically convergent for $\mathbf{x}_{(0)} \in R = (0.4 \le x_1 \le 0.6; 0.2 \le x_2 \le 0.3)$?

55. What form does the Jacobian J_f take for the algorithm (57)?

56. For $p = 1, \infty$, and $\mathbf{x}_{(0)} \in R$ (defined in Exercise 54), is the algorithm (57) convergent?

Example 17 Consider the 2×2 system of nonlinear equations

$$\exp(x_1) + x_2 = 0$$
$$\cosh(x_2) - x_1 = 3.5 \qquad (67)$$

whose solution is $x_1 \in (-2, -1)$, $x_2 \in (-1, 0)$.

Defining the first-order algorithm as

$$\mathbf{x} = \mathbf{f}(\mathbf{x})$$

where $\mathbf{f} = (f_1, f_2)^{\mathrm{T}}$ and $f_1 = (\cosh(x_2) - 3.5)$; and $f_2 = -\exp(x_1)$, it is easy to see that

$$J_f = \begin{bmatrix} \dfrac{\partial f_1}{\partial x_1} & \dfrac{\partial f_1}{\partial x_2} \\[2mm] \dfrac{\partial f_2}{\partial x_1} & \dfrac{\partial f_2}{\partial x_2} \end{bmatrix} = \begin{bmatrix} 0 & \sinh(x_2) \\ -\exp(x_1) & 0 \end{bmatrix}$$

and so for $p = 1$ (and ∞)

$$\|J_f\| = \max\{|\sinh(x_2)|, |\exp(x_1)|\} < 1$$

for the given intervals of x_1 and x_2. Hence the first-order method would converge. Taking $x_{(0)} = (-2.0, -0.5)$, the results of applying the simple iteration method

$$x_{1_{(n+1)}} = \cosh(x_{2_{(n)}}) - 3.5$$
$$x_{2_{(n+1)}} = -\exp(x_{1_{(n)}}) \qquad (68)$$

are given in column 1 of Table 5. If instead we use the latest $x_{1_{(n+1)}}$ in place of $x_{1_{(n)}}$ in the second equation of (68), we obtain the results of column 2.

The Newton–Raphson method requires us to evaluate the Jacobian J of F for Eq. (67), and its inverse J^{-1}. It is easily seen that

$$J = \begin{bmatrix} \exp(x_1) & 1 \\ -1 & \sinh(x_2) \end{bmatrix}$$

---

† Norms were described in Chapter 3, section 7.

‡ Note that we have evaluated analytically the elements of J^{-1} in order to write the equations in the iteration scheme explicitly. In practice $(N > 2)$, the inversion would be performed by solving for the correction term.

Table 5

	Simple iteration	Simple iteration (updated)	Newton–Raphson	(61), $I=2$	(61), $I=6$	(62) with (70)	(62) with (71)
x_1	$-2.372\,37$	$-2.372\,37$	$-2.605\,07$	$-2.605\,07$	$-2.605\,07$	$-2.405\,00$	$-2.605\,07$
	$-2.490\,83$	$-2.495\,65$	$-2.499\,91$	$-2.527\,23$	$-2.527\,23$	$-2.493\,22$	$-2.527\,23$
	$-2.495\,65$	$-2.496\,60$	$-2.496\,78$	$-2.497\,13$	$-2.479\,84$	$-2.496\,51$	$-2.479\,84$
	$-2.496\,57$	$-2.496\,61$	$-2.496\,61$	$-2.496\,75$	$-2.495\,03$	$-2.496\,61$	$-2.495\,03$
	$-2.496\,60$		$-2.496\,61$	$-2.496\,61$	$-2.498\,68$	$-2.496\,61$	$-2.498\,68$
	$-2.496\,61$			$-2.496\,61$	$-2.496\,54$		$-2.496\,54$
					$-2.496\,61$		$-2.496\,33$
					$-2.496\,61$		$-2.496\,64$
							$-2.496\,63$
							$-2.496\,60$
							$-2.496\,60$
							$-2.496\,61$
							$-2.496\,61$
x_2	$-0.135\,34$	-9.3259×10^{-2}	$-0.161\,89$	$-0.161\,89$	$-0.161\,892$	$-0.163\,72$	$-0.161\,894$
	-9.3259×10^{-2}	-9.2443×10^{-2}	-6.3987×10^{-2}	-5.0011×10^{-2}	-5.0011×10^{-2}	-9.6493×10^{-2}	-5.0011×10^{-2}
	-8.2841×10^{-2}	-8.2365×10^{-2}	-8.2026×10^{-2}	-7.7716×10^{-2}	-7.8000×10^{-2}	-8.2724×10^{-2}	-7.8000×10^{-2}
	-8.2443×10^{-2}	-8.2364×10^{-2}	-8.2340×10^{-2}	-8.2327×10^{-2}	-8.6685×10^{-2}	-8.2376×10^{-2}	-8.6685×10^{-2}
	-8.2365×10^{-2}		-8.2364×10^{-2}	-8.2341×10^{-2}	-8.2353×10^{-2}	-8.2364×10^{-2}	-8.2353×10^{-2}
	-8.2364×10^{-2}		-8.2364×10^{-2}	-8.2364×10^{-2}	-8.1879×10^{-2}	-8.2364×10^{-2}	-8.1879×10^{-2}
					-8.2382×10^{-2}		-8.2422×10^{-2}
					-8.2364×10^{-2}		-8.2412×10^{-2}
					-8.2364×10^{-2}		-8.2352×10^{-2}
							-8.2360×10^{-2}
							-8.2366×10^{-2}
							-8.2364×10^{-2}
							-8.2364×10^{-2}

and hence

$$J^{-1} = \begin{bmatrix} \dfrac{\sinh(x_2)}{1+\exp(x_1)\sinh(x_2)} & \dfrac{-1}{1+\exp(x_1)\sinh(x_2)} \\[3mm] \dfrac{1}{1+\exp(x_1)\sinh(x_2)} & \dfrac{\exp(x_1)}{1+\exp(x_1)\sinh(x_2)} \end{bmatrix}^{\ddagger}$$

The results of applying the Newton–Raphson method (59) to Eq. (67) gives

$$x_{1_{(n+1)}} = x_{1_{(n)}} - \frac{\sinh(x_{2_{(n)}})}{1+\exp(x_{1_{(n)}})\sinh(x_{2_{(n)}})}(\exp(x_{1_{(n)}})+x_{2_{(n)}})$$

$$+ \frac{1}{(1+\exp(x_{1_{(n)}})\sinh(x_{2_{(n)}}))}(\cosh(x_{2_{(n)}})-x_{1_{(n)}}+3.5)$$

$$x_{2_{(n+1)}} = x_{2_{(n)}} - \frac{1}{(1+\exp(x_{1_{(n)}})\sinh(x_{2_{(n)}}))}(\exp(x_{1_{(n)}})+x_{2_{(n)}})$$

$$+ \frac{\exp(x_{1_{(n)}})}{(1+\exp(x_{1_{(n)}})\sinh(x_{2_{(n)}}))}(\cosh(x_{2_{(n)}})-x_{1_{(n)}}+3.5)$$

The results obtained using (60) with $\mathbf{x}_{(0)} = (-2, -0.5)^{\mathrm{T}}$ are given in column 3 of Table 5.

Applying the modification (61) of the Newton–Raphson method for $I = 2$ (it is usual to take $I = N$, the dimension of the problem), requires us to re-evaluate the elements of J^{-1} only after every two iterations. In contrast, taking $I = 6$ requires us to evaluate J^{-1} only after every six iterations. The results of these two strategies are reported in columns 4 and 5 respectively of Table 5.

Finally, we report the use of the modification (62) to the Newton–Raphson method. The constant approximation J^* to J^{-1} is taken first as

$$\begin{bmatrix} \dfrac{\sinh(X_2)}{a} & \dfrac{-1}{a} \\[3mm] \dfrac{1}{a} & \dfrac{\exp(X_1)}{a} \end{bmatrix} \tag{70}$$

where $a = 1/(1+\exp(X_1)\sinh(X_2))$ (a good approximation to J^{-1}!) and second as

$$\begin{bmatrix} \dfrac{\sinh(x_{2_{(0)}})}{a} & \dfrac{-1}{a} \\[3mm] \dfrac{1}{a} & \dfrac{\exp(x_{1_{(0)}})}{a} \end{bmatrix} \tag{71}$$

which is a poorer approximation to the inverse Jacobian but one which is frequently used in practice. The results using these two approximations are reported in columns 6 and 7 respectively of Table 5.

Exercises

57. What form does Stirling's method take for the system of nonlinear equations $F(\mathbf{x}) = \mathbf{0}$?
58. Prove that the sequence of iterates produced by Stirling's method converges to the solution $\mathbf{x}$ if $\|J\| \le \alpha < \frac{1}{3}$.

10 Zeros of Polynomial F

The methods we have described in previous sections have not taken into account any particular property of the function F (other than that necessary to ensure the algorithms work). Furthermore, the algorithms we have described have concentrated on finding a root in a given interval rather than finding all of the roots of the given equation. In this final section of this chapter we will introduce the reader to methods which are specially constructed for the polynomial F and we will set ourselves the goal of ultimately determining all the zeros. Furthermore, we now allow the zeros to be complex and hence use a different symbol from x, namely z, to denote the independent variable. The polynomial F, however, has *real* coefficients. Assume

$$F(z) = \sum_{i=0}^{N} a_i z^{N-i} \tag{72}$$

is a polynomial of degree N with real coefficients a_i. Let us assume we are able to determine a quadratic factor of $F(z)$ denoted by

$$r_1(z) = z^2 - p_1 z - q_1 \tag{73}$$

i.e. we are able to determine constants p_1 and q_1 such that

$$F(z) = r_1(z) s_1(z) \tag{74}$$

where $s_1(z)$ is a polynomial of degree $N - 2$. To determine the coefficients of $s_1(z)$ we clearly need to divide the given polynomial $F(z)$ by the known quadratic $r_1(z)$. Such a process is termed *synthetic division*. Let

$$s_1(z) = \sum_{i=0}^{N-2} b_i z^{N-2-i} \tag{75}$$

where b_i, $i = 0, 1, \ldots, N-2$, are to be determined. Substituting $s_1(z)$ and $r_1(z)$ into Eq. (74) and comparing the resulting coefficients with Eq. (72), we find

$$b_0 = a_0$$
$$b_1 = a_1 + b_0 p_1$$
$$b_2 = a_2 + b_1 p_1 + b_0 q_1$$

and in general

$$b_j = a_j + b_{j-1} p_1 + b_{j-2} q_1 \qquad j = 2, 3, \ldots, N-2 \tag{76}$$

If we denote $b_{-2} = b_{-1} = 0$, then Eq. (76) becomes

$$b_j = a_j + b_{j-1} p_1 + b_{j-2} q_1 \qquad j = 0, 1, \ldots, N-2 \tag{77}$$

and we therefore have a recurrence formula for defining the coefficients of the reduced polynomial $s_1(z)$. If we can now find another quadratic factor

$$r_2(z) = z^2 - p_2 z - q_2$$

of $s_1(z)$ such that

$$s_1(z) = r_2(z) s_2(z)$$

where $s_2(z)$ is of degree $N - 4$, we can repeat the process of synthetic division. Clearly the recurrence relation (77) with the appropriate range of j can be applied to define the coefficients of $s_2(z)$ in terms of those of $s_1(z)$. Therefore, provided we can determine the coefficients p and q in $r(z)$, we can successively reduce the degree of the polynomials $s(z)$ so that ultimately we are left with a simple linear factor (when N is odd) or a quadratic factor (when N is even). At each stage therefore, we merely have to solve a quadratic equation $r(z) = 0$ from which two zeros of the original $F(z)$ are found. It remains to derive a means of determining the coefficients p and q of the quadratic factor $r(z)$. To do this let us assume that we require to find p_1, q_1 such that $r_1(z) = z^2 - p_1 z - q_1$ is a quadratic factor of $F(z)$. For *arbitrary* p, q we will have

$$F(z) = (z^2 - pz - q) s_1(z) + b_{N-1}(z - p) + b_N \tag{78}$$

where the coefficients of $s_1(z)$ are functions of p and q defined through relation (77) and where, now $j = 0, 1, \ldots, N$. It is easy to prove that if $z^2 - pz - q$ is to be a quadratic factor of $F(z)$, then it is necessary and sufficient that $b_{N-1} = b_N = 0$ (see Exercise 61). But by the recurrence (77) with $j = 0, 1, \ldots, N$, b_{N-1} and b_N are functions of the parameters p and q. Hence for $z^2 - pz - q$ to be a quadratic factor of $F(z)$ we require those values of p and q (namely p_1 and q_1) such that

$$\begin{aligned} b_{N-1}(p, q) &= 0 \\ b_N(p, q) &= 0 \end{aligned} \tag{79}$$

Clearly (79) is a (2×2) case of our system of nonlinear equations discussed in the previous section. The use of the Newton–Raphson method to solve this system (79) using the particular properties of the nonlinear functions $b_{N-1}(p, q)$ and $b_N(p, q)$ is termed the *Bairstow* method. To use the Newton–Raphson method we need the partial derivatives of b_{N-1} and b_N with respect to p and q. To obtain these we will use the recurrence relation (77). Differentiating (77) with respect to p, we obtain

$$\frac{\partial b_0}{\partial p} = 0 \qquad \frac{\partial b_1}{\partial p} = b_0$$

and in general

$$\frac{\partial b_j}{\partial p} = b_{j-1} + p \frac{\partial b_{j-1}}{\partial p} + q \frac{\partial b_{j-2}}{\partial p} \qquad j = 2, \ldots, N$$

A simple recursion may be obtained for the derivatives if we introduce the notation

$$c_j = \frac{\partial b_{j+1}}{\partial p}$$

whence

$$c_{-1} = 0 \qquad c_0 = b_0 \qquad c_j = b_j + pc_{j-1} + qc_{j-2} \qquad j = 1, 2, \ldots, N-1 \quad (80)$$

Similarly, differentiating the recursion (77) with respect to q, we obtain, on introducing the notation

$$d_j = \frac{\partial b_{j+2}}{\partial q}$$

$$d_1 = 0 \qquad d_0 = b_0 \qquad d_j = b_j + pd_{j-1} + qd_{j-2} \qquad j = 1, 2, \ldots, N-2 \quad (81)$$

which is exactly (80). Therefore the partial derivatives of b with respect to p and q are identical. From (80), the particular partial derivatives we require are given by

$$\begin{align}
\frac{\partial b_{N-1}}{\partial p} = c_{N-2} \qquad & \frac{\partial b_N}{\partial p} = c_{N-1} \\
& \\
\frac{\partial b_{N-1}}{\partial q} = c_{N-3} \qquad & \frac{\partial b_N}{\partial q} = c_{N-2}
\end{align} \tag{82}$$

The Newton–Raphson method for the system (79) may be written as

$$\binom{p}{q}_{(n+1)} = \binom{p}{q}_{(n)} - J^{-1} \binom{b_{N-1}(p_{(n)}, q_{(n)})}{b_N(p_{(n)}, q_{(n)})} \qquad n = 0, 1, 2, \ldots \tag{83}$$

where J is the Jacobian

$$\begin{pmatrix} \dfrac{\partial b_{N-1}}{\partial p} & \dfrac{\partial b_{N-1}}{\partial q} \\ \dfrac{\partial b_N}{\partial p} & \dfrac{\partial b_N}{\partial q} \end{pmatrix}$$

evaluated at $p = p_{(n)}$, $q = q_{(n)}$. Rearranging (83), we obtain

$$J \binom{p_{(n+1)} - p_{(n)}}{q_{(n+1)} - q_{(n)}} = - \binom{b_{N-1}(p_{(n)}, q_{(n)})}{b_N(p_{(n)}, q_{(n)})}$$

If we denote the change in p from iteration n to $n+1$ as δp and the change in q as δq, then these changes must satisfy the two equations

$$\begin{align}
c_{N-2}\delta p + c_{N-3}\delta q = -b_{N-1}(p_{(n)}, q_{(n)}) \\
c_{N-1}\delta p + c_{N-2}\delta q = -b_N(p_{(n)}, q_{(n)})
\end{align} \tag{84}$$

from which p and q may be obtained. We now have all the information to calculate the quadratic factor $r(z)$. The complete algorithm progresses as follows.

For a given starting (guessed) quadratic factor $r(z) = z^2 - p_{(0)}z - q_{(0)}$, calculate the coefficients b, c from (77) and (80) respectively using $p_{(0)}$ and $q_{(0)}$. With b_N, b_{N-1}, and c_{N-1}, c_{N-2}, c_{N-3} calculated in terms of $p_{(0)}$ and $q_{(0)}$, calculate δp and δq from (84), from which new estimates of p and q are obtained, namely

$$p_{(1)} = p_{(0)} + \delta p \qquad q_{(1)} = q_{(0)} + \delta q$$

Substitute these new coefficients into $r(z)$ and recalculate the recursions (77), (80). New δp, δq can be obtained, again using (84). The process is repeated until the coefficients b_{N-1}, b_N become zero. The corresponding values of p and q are the required parameters for $r(z)$. This is $r_1(z)$ of (73). Synthetic division may be now employed to reduce $F(z)$. If estimates for the quadratic factor $r_2(z)$ can be obtained, Bairstow's method can be employed once again.

It can be shown that Bairstow's method will converge to the quadratic factors if the starting approximation $z^2 - p_{(0)}z - q_{(0)}$ is sufficiently close to the actual quadratic factor $r(z)$ and the zeros of $r(z)$ are distinct and simple. In this case the method converges with the characteristic second-order rate of the Newton–Raphson method.

An important computational detail should be borne in mind in using either of the Newton–Raphson or Bairstow methods. In order to obtain successive zeros, as one (pair of) zero(s) is found it must be divided out of the original polynomial and the reduced polynomial used in the next step. In order that this procedure does not give rise to a build-up of round-off errors, it is necessary to find the zeros in *increasing* orders of magnitude. The main drawback, however, with the Bairstow method is that in practice good initial approximations to the quadratic factor $r(z)$ are rarely forthcoming.

Example 18 Determine the quadratic factor of $z^2 - 3.1z^2 + 3.7z - 1.6$ which is close to $z^2 - 2z + 1.5$. Hence determine the three zeros.

To construct the coefficients in the Bairstow method we tabulate the coefficients in the following way:

j	a_j	b_j	c_j	$p_{(0)} = 2$	$q_{(0)} = -1.5$
0	1	1	1		
1	-3.1	-1.1	0.9		
2	3.7	0	0.3		
3	-1.6	0.05			

where we have used (77) and (80). Using (84), we have to solve

$$0.9\delta p + \delta q = 0$$

$$0.3\delta p + 0.9\delta q = -0.05$$

which has the solution $\delta p = 0.098\,04$, $\delta q = -0.088\,24$, and so

$$p_{(1)} = p_{(0)} + \delta p = 2.098\,04, \quad q_{(1)} = q_{(0)} + \delta q = -1.588\,24$$

Using these new p and q we employ (77) and (82) again to calculate the new coefficients tabulated as

j	a_j	b_j	c_j	$p_{(1)} = 2.098\,04,\ q_{(1)} = -1.588\,24$
0	1	1	1	
1	−3.1	−1.001 96	1.096 08	
2	3.7	0.009 61	0.721 00	
3	−1.6	0.011 52		

We have to solve

$$1.096\,08\delta p + \delta q = -0.009\,61$$

$$0.721\,00\delta p + 1.096\,05\delta_q = 0.011\,52$$

which has the solution

$$\delta p = 0.001\,97 \qquad \delta q = -0.011\,84$$

so that

$$p_{(2)} = 2.100\,01 \quad \text{and} \quad q_{(2)} = -1.600\,08$$

Calculation of the coefficients from (77) and (80) now produces

j	a_j	b_j	c_j	$p_{(2)} = 2.100\,01,\ q_{(2)} = -1.600\,08$
0	1	1	1	
1	−3.1	−0.999 92	1.100 16	
2	3.7	0.000 00(4)	0.710 34	
3	−1.6	−0.000 03		

To four decimal places the coefficients b_3 and b_2 are zero and the iteration is deemed to have converged. The resulting quadratic factor to four decimal places is $z^2 - 2.1z + 1.6$, from which we obtain the complex conjugate roots

$$\frac{2.1 \pm i\sqrt{1.99}}{2}$$

The coefficients of the remaining linear factor to four decimal places are given by $b_1 = -1$, $b_0 = 1$, so that the third zero is $z = 1$. We have therefore obtained the three zeros of $z^3 - 3.1z^2 + 3.7z - 1.6$ as $1.05 \pm i0.7054, 1$.

Exercises

59. Program Bairstow's method for a given polynomial with coefficients $a_0, a_1, \ldots, a_N$ and given initial guesses for the quadratic factor's coefficients $p_{(0)}$ and $q_{(0)}$. Consider convergence to have occurred when b_N and b_{N-1} are less than 10^{-8} in absolute value. Convergence may be assumed not to occur if more than 10 iterations are needed without the convergence criterion being satisfied.

60. Using your program of Exercise 59 determine a quadratic factor of the quartic $z^4 - 4.9z^3 + 11.3z^2 - 14.6z + 7.2$ which is close to $z^2 - 2z + 3.5$. Determine correct to six decimal places the four zeros of the quartic.
61. Prove that the necessary and sufficient condition for $z^2 - pz - q$ to be a quadratic factor of $F(z)$ is that $b_{N-1} = b_N = 0$ in (78).
62. For a polynomial $F(z)$ given by (72), devise an algorithm for the synthetic division of $F(z)$ by the linear factor $z - z_1$.

We surmise that the Bairstow method would appear to be an excellent method if we can obtain sufficiently good initial approximations to the quadratic factors. From our discussions of previous sections, this would therefore suggest that we endeavor to devise a method which, although not rapidly convergent, does converge without requiring accurate initial approximations, and then use the results of this iteration as starting values for the Bairstow method. We will conclude this chapter with the description of two such methods which, when incorporated with the Bairstow or Newton–Raphson methods, can provide a general-purpose algorithm for finding all the zeros of polynomial F. The first of these algorithms is a classical method due to Bernoulli. The aim of this method is to produce the zero of maximum modulus, the so-called *dominant* zero, but it can also by a suitable reorganization be used to produce the zero of smallest modulus—see exercise 67. On using this approximation to the required zero as the initial value for the Newton–Raphson method and obtaining the zero to the required degree of accuracy, F can be reduced by synthetic division and the process repeated.

Let us initially assume that the zero of maximum modulus is real, unique and denote it by z_1. Further we assume that the remaining $N-1$ zeros, $z_2, \ldots, z_N$, are distinct. We therefore write as an alternative to (72)

$$F(z) = \prod_{i=1}^{N} (z - z_i) \qquad (85)$$

Differentiating (85) with respect to (a sufficiently large) z, we find

$$F^{(1)}(z) = \sum_{\substack{j=1 \\ }}^{N} \prod_{\substack{i=1 \\ i \neq j}}^{N} (z - z_i)$$

$$= \sum_{j=1}^{N} \prod_{i=1}^{N} \frac{(z - z_i)}{(z - z_j)}$$

$$= \prod_{i=1}^{N} (z - z_i) \sum_{j=1}^{N} \frac{1}{(z - z_j)}$$

Hence

$$F^{(1)}(z) = F(z) \sum_{j=1}^{N} (z - z_j)^{-1} \qquad (86)$$

Expanding the sum in (86), we find

$$F^{(1)}(z) = F(z) \left[Nz^{-1} + \sum_{j=1}^{N} z_j z^{-2} + \sum_{j=1}^{N} z_j^2 z^{-3} + \cdots \right]$$

which we write as

$$F^{(1)}(z) = F(z)[Nz^{-1} + s_1 z^{-2} + s_2 z^{-3} + \cdots] \tag{87}$$

where

$$s_k = \sum_{j=1}^{N} z_j^k \qquad k = 1, 2, \ldots \tag{88}$$

But we have an alternative expression to (87) for $F^{(1)}(z)$, using (72), given by

$$F^{(1)}(z) = \sum_{i=1}^{N} (N - i + 1)a_{i-1} z^{N-i} \tag{89}$$

Therefore, if we equate coefficients between (89) and (87) we obtain (see Exercise 63) the system of equations given by

$$a_0 s_{j+1} + a_1 s_j + \cdots + a_j s_1 + (j+1)a_{j+1} = 0 \tag{90}$$

(coefficients of z^{N-2-j}), $j = 0, 1, \ldots, N-1$

$$a_0 s_{N+j} + a_1 s_{N+j-1} + \cdots + a_N s_j = 0$$

(coefficients of z^{-j-1}), $j = 1, 2, \ldots$. $\tag{91}$

Equations (90), (91) are known as *Newton's identities*. These equations may be rewritten as

$$s_{N+j} = \frac{-1}{a_0} \sum_{k=1}^{N} a_k s_{N+j-k} \qquad j = 1, 2, \ldots \tag{92}$$

which is clearly an iterative formula for s_{N+j} in terms of *known* s_{N+j-k}. These latter Ns are known if we define the required initial values by the N equations (90). Consider the s_{N+j} generated by (92) for two successive values of j, say J and $J+1$. Then using (88),

$$\frac{s_{N+J+1}}{s_{N+J}} = \frac{\sum_{j=1}^{N} z_j^{N+J+1}}{\sum_{j=1}^{N} z_j^{N+J}} = z_i \left\{ \frac{1 + \sum_{j=1}^{N} (z_j/z_1)^{N+J+1}}{1 + \sum_{j=1}^{N} (z_j/z_1)^{N+J}} \right\} \tag{93}$$

If we now iterate with Eq. (92) for $J \to \infty$ then

$$\lim_{J \to \infty} \frac{s_{N+J+1}}{s_{N+J}} = z_1$$

since

$$\lim_{J \to \infty} \left(\frac{z_j}{z_1}\right)^{N+J} = 0 \qquad j = 2, 3, \ldots, N$$

by assumption that z_1 is the dominant zero.

Therefore, computationally, Bernoulli's method proceeds by defining N initial values by (the simple) equations (90) and then progressing calculating successive s using (92). The ratio of successive s is formed and once the difference between successive ratios is deemed small enough, the dominant root z_1 is obtained. The rate at which the method converges to z_1 is clearly dependent upon the rates at which the ratios (z_j/z_1) tend to zero with increasing J; the larger the difference between z_1 and the zero of second largest modulus, the faster will be the convergence.

The algorithm can also be shown to work if we relax the uniqueness condition upon z_1, that is, if we allow *real* z_1 with multiplicity > 1. (See Henrici [20, p. 152]).

For other cases of dominant zeros (e.g. simple complex dominant zeros) the Bernoulli method must be adapted in order that it works. For example, when the polynomial F has a single pair of complex conjugate dominant zeros we may achieve the adaptation in the following manner. Let the dominant zeros be denoted by z_1 and z_2. Let us employ the polar coordinate form of the complex numbers and represent $z_1 = r \exp(i\theta)$, $z_2 = r \exp(-i\theta)$, where $i = \sqrt{-1}$. In this case the solution of (92) subject to (90) given by (88) can be rewritten as

$$s_k = r^k (\exp(ik\theta) + \exp(-ik\theta)) \left[1 + \sum_{j=3}^{N} \frac{z_j^k}{(z_1^k + z_2^k)} \right]$$

$$= 2r^k \cos k\theta \left[1 + \sum_{j=3}^{n} \frac{z_j^k}{(z_1^k + z_2^k)} \right]$$

where r and θ are to be determined in order to obtain z_1 and z_2. By forming the quotients

$$\frac{s_k^2 - s_{k+1}s_{k-1}}{s_{k-1}^2 - s_k s_{k-2}} \quad \text{and} \quad \frac{s_k s_{k-1} - s_{k+1}s_{k-2}}{s_{k-1}^2 - s_k s_{k-2}} \tag{94}$$

it is not difficult to show (see Exercise 64) that the limits of these two quotients produce r^2 and $2r \cos \theta$ respectively as $k \to \infty$. Hence by iterating with (90) and (92) and by forming the ratios in (94), the dominant complex conjugate zeros z_1 and z_2 can be obtained.

Example 19   Let us determine, by Bernoulli's method, the dominant real zero of the polynomial $F(z) \equiv z^3 - (6+\alpha)z^2 + (5+6\alpha)z - 5\alpha$, for $\alpha = 0, 4, 5$. These values of α represent respectively the cases of:

   (i) a simple dominant zero substantially different in magnitude from the other zeros;
   (ii) a second root "close" to the dominant zero; and
   (iii) a dominant root of multiplicity 2.

The iterates obtained using (92) with starting values given by (90) depicted in bold type, are given in columns 2, 4, and 6 of Table 6 for the three values of α indicated. The results in columns 3, 5, and 7 are the corresponding ratios formed from successive iterates. For $\alpha = 0$, the ratios clearly converge quickly to the dominant root $z_1 = 5$. For $\alpha = 4$ the effect of two nearly equal dominant roots is already being felt where convergence to six decimal places occurs only after 46 iterations.

Table 6

α	0		4		5	
j	s_j	s_j/s_{j-1}	s_j	s_j/s_{j-1}	s_j	s_j/s_{j-1}
1	**0.600 000 00E 01**		**0.100 000 00E 02**		**0.110 000 00E 02**	
2	**0.260 000 00E 02**	0.433 333 33E 01	**0.420 000 00E 02**	0.420 000 00E 01	**0.510 000 00E 02**	0.463 636 36E 01
3	**0.126 000 00E 03**	0.484 615 39E 01	**0.190 000 00E 03**	0.452 380 96E 01	**0.251 000 00E 03**	0.492 156 87E 01
4	0.626 000 00E 03	0.496 825 40E 01	0.882 000 00E 03	0.464 210 53 E 01	0.125 100 00E 04	0.498 406 38E 01
5	0.312 600 00E 04	0.499 361 02E 01	0.415 000 00E 04	0.470 521 54E 01	0.625 100 00E 04	0.499 680 26E 01
6	0.156 260 00E 05	0.499 872 04E 01	0.197 220 00E 05	0.475 228 92E 01	0.312 510 00E 05	0.499 936 01E 01
7	0.781 260 00E 05	0.499 974 40E 01	0.945 100 00E 05	0.479 211 04E 01	0.156 251 00E 06	0.499 987 20E 01
8	0.390 626 00E 06	0.499 994 88E 01	0.456 162 00E 06	0.482 660 04E 01	0.781 251 00E 06	0.499 997 44E 01
9	0.195 312 60E 07	0.499 998 98E 01	0.221 527 00E 07	0.465 632 30E 01	0.390 625 10E 07	0.499 999 49E 01
10	0.976 562 61E 07	0.499 999 80E 01	0.108 142 02E 08	0.488 166 32E 01	0.195 312 51E 08	0.499 999 90E 01
11	0.488 281 26E 08	0.499 999 96E 01	0.530 224 30E 08	0.490 303 68E 01	0.976 562 52E 08	0.499 999 98E 01
12	0.244 140 63E 09	0.500 000 00E 01	0.260 917 84E 09	0.492 089 56E 01		
13			0.128 781 20E 10	0.493 569 93E 01		
14			0.637 195 12E 10	0.494 788 94E 01		
15			0.315 913 21E 11	0.495 787 25E 01		
16			0.156 882 87E 12	0.496 601 15E 01		
17			0.780 119 37E 12	0.497 262 32E 01		
18			0.388 341 70E 13	0.497 797 79E 01		
19			0.193 483 66E 14	0.498 230 45E 01		
20			0.964 669 54E 14	0.498 579 33E 01		
21			0.481 235 26E 15	0.498 860 22E 01		
22			0.245 177 83E 16	0.499 086 10E 01		
23			0.119 913 00E 17	0.499 267 54E 01		

24	0.598 861 29 E 17	0.499 413 17 E 01
25	0.299 149 18 E 18	0.499 529 99 E 01
26	0.149 462 00 E 19	0.499 623 64 E 01
27	0.746 859 67 E 19	0.499 698 89 E 01
28	0.373 249 70 E 20	0.499 758 81 E 01
29	0.186 552 80 E 21	0.499 806 96 E 01
30	0.932 475 77 E 21	0.499 845 51 E 01
31	0.466 122 61 E 22	0.499 876 37 E 01
32	0.233 515 19 E 23	0.499 901 67 E 01
33	0.116 489 15 E 24	0.499 920 83 E 01
34	0.582 371 95 E 24	0.499 936 66 E 01
35	0.291 156 46 E 25	0.499 949 33 E 01
36	0.145 566 43 E 26	0.499 959 46 E 01
37	0.727 784 92 E 26	0.499 967 57 E 01
38	0.363 873 58 E 27	0.499 974 65 E 01
39	0.131 929 23 E 28	0.499 979 24 E 01
40	0.909 615 96 E 28	0.499 988 39 E 01
41	0.454 795 91 E 29	0.499 986 73 E 01
42	0.227 393 12 E 30	0.499 989 38 E 01
43	0.113 694 63 E 31	0.499 991 50 E 01
44	0.568 465 42 E 31	0.499 993 20 E 01
45	0.284 229 62 E 32	0.499 994 57 E 01
46	0.142 113 57 E 33	0.499 995 66 E 01
47	0.710 562 93 E 33	0.499 996 53 E 01
48	0.355 279 49 E 34	0.499 997 22 E 01
49	0.177 638 95 E 35	0.499 997 77 E 01
50	0.888 191 61 E 35	0.499 998 22 E 01

For $\alpha = 5$ it can be seen that the dominant root, with multiplicity 2, is obtained by the algorithm with no loss in the convergence rate. Furthermore, as the number of iterations increases, the iterates s_j become increasingly large and can exceed the largest representable floating point number on the computer before the required convergence has occurred. This production of large numbers for the s_j is an unfortunate characteristic of Bernoulli's method for zeros larger than 1 in modulus. A normalization process would therefore seem in order. Such a process is described by Ralston [34]. In addition, it can be shown that Aitken's process is applicable to the sequence generated by Bernoulli's method so that the convergence can be accelerated. Details may be found in Henrici [20].

Other variants of Bernoulli's method can be devised for other cases of dominant zeros. However, we omit further discussion here in favor of a description of the second method referred to on page 313, which is a more recent adaptation of Bernoulli's method due to Rutishauser [37] known as the quotient difference (QD) algorithm which furnishes all the zeros of the given polynomial F, *simultaneously*.†

Exercises

63. Using equations (87) and (89) verify that s_k satisfy the systems of equations (90) and (91).
64. Show that the limits as $k \to \infty$ of the quotients in (94) are r^2 and $2r \cos \theta$ respectively.
65. Program Bernoulli's method for the case of real dominant zeros. Hence calculate correct to four decimal places the dominant zero of F in Example 19 when $\alpha = 50$, -50. Comment on the convergence.
66. Program the adaptation of Bernoulli's method using (94). Hence determine the dominant complex conjugate zeros of $F(z) = z^3 - 2z - 4$.
67. Devise Bernoulli's method for calculating the zero of minimum modulus assuming this is real and unique.
68. In the light of the slow convergence of the polynomial in Example 19 with $\alpha = 4$, use your method of Exercise 67 and then use linear synthetic division to derive the resulting quadratic polynomial from which the two larger zeros can be easily obtained.
69. Apply Aitken's δ^2-process (section 6) to the sequence in columns 3, 5, and 7 of Example 19, Table 6.

The analysis of the QD algorithm is beyond the scope of this book and hence we will content ourselves with a description of the method.‡ Consider the polynomial $F(z)$ given by (72), where for the moment we assume all a_i,

---

† An alternative powerful algorithm which we might have discussed is the Lehmer–Schur method [31]. Descriptive details of this algorithm may be found in Ralston [34].

‡ There are two variants of the QD algorithm—a stable version and an unstable version. We describe the stable method, which is more commonly called the progressive QD algorithm.

$i = 0, 1, \ldots, N$, are nonzero. Define the sequences

$$q_0^{\{1\}} = \frac{-a_1}{a_0}; \quad q_{1-k}^{\{k\}} = 0 \qquad k = 2, 3, \ldots, N \tag{95}$$

$$d_{1-k}^{\{k\}} = \frac{a_{k+1}}{a_k} \qquad k = 1, 2, \ldots, N \tag{96}$$

and

$$q_{j+1}^{\{k\}} = d_j^{\{k\}} - d_{j+1}^{\{k-1\}} + q_j^{\{k\}} \qquad \begin{array}{l} k = 1, 2, \ldots, N \\ j = -k+2, -k+3, \ldots \end{array} \tag{97}$$

$$d_{j+1}^{\{k\}} = \frac{q_j^{\{k+1\}}}{q_j^{\{k\}}} d_j^{\{k\}} \qquad \begin{array}{l} k = 1, 2, \ldots, N-1 \\ j = -k+2, -k+3, \ldots \end{array} \tag{98}$$

where also

$$d_j^{\{0\}} = d_j^{\{N\}} = 0 \qquad j = 1, 2, \ldots \tag{99}$$

Conveniently these entities may be tabulated as in Table 7.

Table 7

	$-a_1/a_0$		0		0		0
0		a_2/a_1		a_3/a_2		a_4/a_3	
	$q_1^{\{1\}}$		$q_0^{\{2\}}$		$q_{-1}^{\{3\}}$		$q_{-2}^{\{4\}}$
0		$d_1^{\{1\}}$		$d_0^{\{2\}}$		$d_{-1}^{\{3\}}$	
	$q_2^{\{1\}}$		$q_1^{\{2\}}$		$q_0^{\{3\}}$		$q_{-1}^{\{4\}}$
0		$d_2^{\{1\}}$		$d_1^{\{2\}}$		$d_0^{\{3\}}$	
	$\vdots$		$\vdots$				

In the table the superscript denotes the column number (of q or d columns) and the subscript denotes the row number (of q or d rows).

If the algorithm (95)–(99) exists, i.e. if all the elements are well defined (no divisions by zero) then the limits of the q-columns in Table 7 bear a very important relation to the zeros of F. The particular relationship will depend upon the properties of the zeros. If the zeros of F satisfy

$$|z_1| > |z_2| > \cdots > |z_N| > 0 \tag{100}$$

then it can be shown that

$$\lim_{j \to \infty} q_j^{\{k\}} = z_k \qquad k = 1, 2, \ldots, N$$

i.e. the limit of the kth column of q is the kth (in modulus) zero of F. In addition, in this case

$$\lim_{j \to \infty} d_j^{\{k\}} = 0 \qquad k = 1, 2, \ldots, N$$

i.e. distinct simple zeros are characterized by alternate columns tending to zero.

Alternatively, if several zeros have the same modulus, the limits of the columns are somewhat different. If we assume

$$|z_1| \geq |z_2| \geq |z_3| \geq \cdots \geq |z_N| > 0$$

it can be shown that for every k such that $|z_{k-1}| > |z_k| > |z_{k+1}|$, $k = 1, 2, \ldots, N$, where for convenience we have put $|z_0| = \infty$ and $|z_{N+1}| = 0$, then

$$\lim_{j \to \infty} q_j^{\{k\}} = z_k$$

and for every k such that $|z_k| > |z_{k+1}|$

$$\lim_{j \to \infty} d_j^{\{k\}} = 0$$

This latter feature divides the table into subtables, the *number* of q-columns contained in each subtable corresponding to the number of zeros of equal modulus. Consequently, if a subtable contains just one q-column then there is a unique zero of the modulus to which the sequence converges. Within the subtables of more than q-column, the zeros can be extracted as follows. Assume within a particular subtable that there are l equal-moduli roots such that $|z_k| > |z_{k+1}| = |z_{k+2}| = \cdots = |z_{k+l}| > |z_{k+l+1}|$. Define the sequence of polynomials

$$\{p_j^{\{i\}}(z)\} \qquad i = k, k+1, \ldots, k+l$$

by the recurrence relations

$$p_j^{\{k\}}(z) = 1 \qquad j = 0, 1, 2, \ldots \tag{101}$$

$$p_j^{\{i\}}(z) = z p_{j+1}^{\{i-1\}}(z) - q_j^{\{i\}} p_j^{\{i-1\}}(z) \qquad i = 1, 2, \ldots, l; j = 0, 1, 2, \ldots$$

where $q_j^{\{i\}}$ are defined through (95)–(99). In terms of these polynomials the zeros of equal moduli of F are related by

$$\lim_{j \to \infty} p_j^{\{k+l\}}(z) = \prod_{r=k+1}^{k+l} (z - z_r)$$

Clearly this is most useful in the (most frequent) practical case when $l = 2$ whence we obtain from (101) the coefficients of the quadratic which has z_{k+1} and z_{k+2} as its zeros.

We have left to *state* conditions under which the QD algorithm exists. It can be shown that the QD method exists if the *Hankel* determinants

$$H_j^{\{k\}} = \begin{vmatrix} s_j & s_{j+1} & \cdots & s_{j+k-1} \\ s_{j+1} & s_{j+2} & \cdots & s_{j+k} \\ \vdots & & & \\ s_{j+k-1} & s_{j+k} & \cdots & s_{j+2k-2} \end{vmatrix} \qquad j = -1, -2, \ldots, -N+1 \quad j > -k$$

are nonzero, where the elements s_j defined from (92) have starting values $s_j = 0$, $j = -1, -2, \ldots, -N+1$ (instead of (90)).

Above we assumed that the coefficients a_i of F were all nonzero. If some of these coefficients are zero then it is a simple matter to apply a transformation by introducing the change of variable

$$z^* = z - a \qquad (102)$$

and choosing a accordingly whereby a new polynomial F^* is defined in which all the coefficients are non zero (see Exercise 70). F^* is then used in the QD algorithm and the original zeros obtained by adding a to each limit of the q's.

Example 20 Let us determine the zeros of $F(z) = z^3 - 6z^2 + 11z - 6$ correct to three decimals by the QD algorithm. Using equations (95)–(99) we tabulate the calculated results in Table 8. Each column of d's clearly converges to zero, indicating zeros of distinct modulus. These are given to three decimal places after 18 applications of the recurrence relations. These results will confirm what the thoughtful reader will have already suspected, namely that since the Bernoulli method is a special case of the QD algorithm, the rate of convergence can be no better than that of the former method.

Table 8

D	Q	D	Q	D	Q
+0.000 00	+6.000 00	−1.833 33	+0.000 00	−0.545 45	+0.000 00
+0.000 00	+4.166 67	−0.566 67	+1.287 88	−0.231 02	+0.545 45
+0.000 00	+3.600 00	−0.255 56	+1.623 53	−0.110 49	+0.776 47
+0.000 00	+3.344 44	−0.135 14	+1.768 60	−0.055 41	+0.886 96
+0.000 00	+3.209 30	−0.077 83	+1.848 33	−0.028 25	+0.942 37
+0.000 00	+3.131 47	−0.047 17	+1.897 91	−0.014 45	+0.970 62
+0.000 00	+3.084 30	−0.029 53	+1.930 64	−0.007 37	+0.985 06
+0.000 00	+3.054 77	−0.018 88	+1.952 80	−0.003 75	+0.992 43
+0.000 00	+3.035 89	−0.012 24	+1.967 93	−0.001 90	+0.996 18
+0.000 00	+3.023 66	−0.008 01	+1.978 27	−0.000 96	+0.998 08
+0.000 00	+3.015 65	−0.005 27	+1.985 31	−0.000 48	+0.999 03
+0.000 00	+3.010 38	−0.003 48	+1.990 10	−0.000 24	+0.999 52
+0.000 00	+3.006 90	−0.002 31	+1.993 35	−0.000 12	+0.999 76
+0.000 00	+3.004 59	−0.001 53	+1.995 53	−0.000 06	+0.999 88
+0.000 00	+3.003 05	−0.001 02	+1.997 01	−0.000 03	+0.999 94
+0.000 00	+3.002 03	−0.000 68	+1.998 00	−0.000 02	+0.999 97
+0.000 00	+3.001 36	−0.000 45	+1.998 66	−0.000 01	+0.999 98
+0.000 00	+3.000 90	−0.000 30	+1.999 10	−0.000 00	+0.999 99

Consequently, it is not recommended to use the QD algorithm to determine the zeros of polynomial F to any degree of accuracy. Rather, the method should be used in conjunction with the Newton–Raphson or Bairstow methods to give *good* initial approximations as starting values for these methods. In this way convergence will then occur two or three iterations later, as exemplified in Example 21.

Example 21 When successive iterates agree to within 0.2, the most recent values in the q columns of Table 8 were used as starting values in the Newton–Raphson method. Within at most 4 iterations, 8 decimal places of accuracy has been achieved, as is indicated in the columns of Table 9. The figures in bold type were obtained from Table 8.

Table 9

Root 1	Root 2	Root 3
3.209 300 00	**1.848 330 00**	**0.942 370 00**
3.044 212 33	2.007 495 23	0.995 607 99
3.002 658 15	1.999 999 16	0.999 971 36
3.000 010 53	2.000 000 00	1.000 000 00
3.000 000 00		

Exercises

70. Transform the cubic polynomial $F(z) = z^3 - 7z + 6$ into a corresponding cubic polynomial in which all of the coefficients are nonzero.
71. Program the QD method assuming all of the zeros are of distinct modulus.
72. Use your program of Exercise 71 to calculate all of the roots of Exercise 70 correct to: two decimal places; six decimal places.
73. In Exercise 71 arrange for a subroutine defining the Newton–Raphson method to be called if the difference between successive elements of the q-column of the QD algorithm is less than a prescribed parameter EPS, *or* the number of applications of the QD recurrence relations exceeds a second prescribed parameter ITMAX. Cater also for the return to the QD algorithm from the Newton–Raphson method if convergence to a specified tolerance EPS1 does not occur within NITMAX iterations. Note this return may require the suitable redefinition of EPS or ITMAX.
74. Use your program of Exercise 73 for F given in Exercise 70 with:
    (i) EPS $= 0.5 * 10^{-1}$ ITMAX $= 5$ EPS1 $= 0.5 * 10^{-3}$ NITMAX $= 3$
    (ii) EPS $= 0.5 * 10^{-1}$ ITMAX $= 5$ EPS1 $= 0.5 * 10^{-6}$ NITMAX $= 6$
    (iii) EPS $= 0.5 * 10^{-2}$ ITMAX $= 10$ EPS1 $= 0.5 * 10^{-3}$ NITMAX $= 2$
    (iv) EPS $= 0.5 * 10^{-2}$ ITMAX $= 10$ EPS1 $= 0.5 * 10^{-6}$ NITMAX $= 6$
75. Modify your programs of Exercises 71 and 73 to handle the case of zeros of equal modulus occurring, assuming that no more than two zeros ever have equal moduli.
76. Use your program of Exercise 73 to calculate correct to six decimal places the zeros of the polynomial $F(z) = z^3 - 3z + 2$. Comment, as a result of your computer results, on the significance of the constant a in (102) for the case where the original F has roots of equal modulus. In what circumstances might such a translation be useful even in the case where F has nonzero coefficients?

In our discussions of methods for determining the zeros of a polynomial we have indicated possible methods and their shortcomings. In trying to develop a general-purpose routine for a general polynomial, it has become clear that difficulties can arise from two points. First, the convergence conditions of the algorithm must be ensured. Second, special cases of zeros may have to be dealt with. If these were not enough, there is in addition an extremely important consideration upon which we have not touched, namely the so-called *conditioning* of the given polynomials. This property of a polynomial is associated with the way in which zeros of polynomials change when small changes occur in the coefficients of the polynomial (see Chapter 2). Recall that a polynomial is said to be *well-conditioned* if a small change in any (or all of its coefficients results in only a small change in the zeros. In contrast, if a small change in the coefficients creates a large change in the zeros, the polynomial is said to be *badly conditioned*. This property of badly or *ill-conditioned* polynomials is frequently encountered when trying to determine the eigenvalues of a matrix by solving for the zeros of its characteristic polynomial (see Chapter 10). Likewise, if F is of high degree (say degree 20) then the polynomial is very likely to be badly conditioned. For such badly conditioned polynomials, the small changes in the coefficients which change the zeros occur as a result of rounding processes in the computer calculation of the coefficients. The effect of these rounding errors is thoroughly analyzed in Wilkinson [48] to which the reader is referred. To point out the seriousness of the conditioning of a polynomial of degree 20, we quote a (now) classic example due to Wilkinson [48].

Example 22 Consider determining the zeros of the polynomial

$$F(z) = \prod_{j=1}^{20} (z+j)$$

which clearly has the distinct zeros $-j$, $j = 1, 2, \ldots, 20$. If the coefficient of x^{19} is perturbed by the "trivial" amount 2^{-23}, the zeros of the resulting perturbed polynomial are given as shown in Table 10. It is clear that not only have some of the zeros changed in size, but some have even become complex. To minimize such an effect of round-off in a badly conditioned polynomial, *multiple* precision arithmetic should be used in conjunction with the algorithms we have described. The reader is referred to [48] for fuller details.

Table 10

$-1.000\,000\,000$	$-10.095\,266\,145 \pm 0.643\,500\,904i$
$-2.000\,000\,000$	$-11.793\,633\,881 \pm 1.652\,329\,728i$
$-3.000\,000\,000$	$-13.992\,358\,137 \pm 2.518\,830\,070i$
$-4.000\,000\,000$	$-16.730\,737\,466 \pm 2.812\,624\,894i$
$-4.999\,999\,928$	$-19.502\,439\,400 \pm 1.940\,330\,347i$
$-6.000\,006\,944$	
$-6.999\,697\,234$	
$-8.007\,267\,603$	
$-8.917\,250\,249$	
$-20.846\,098\,101$	

CHAPTER 10
The Algebraic Eigenvalue Problem

1 Introduction

In this section we will provide the reader with an overview of some of the basic theory which will be required to comprehend the material of this chapter. Although for the algebraic eigenvalue problem matrices with complex elements do exist, the overwhelming majority of matrices have real elements. For this reason we restrict our attention to matrices with real elements. For the more general situation the reader is referred to Wilkinson [49] and Stewart [39], as well as Gourlay and Watson [18].

In simple terms the problem is one of determining one or many (all) values of a scalar λ for which the system of N homogeneous linear equations in N unknowns

$$A\mathbf{x} = \lambda\mathbf{x} \tag{1}$$

has a nontrivial solution $\mathbf{x}$. For our purposes A is a real $N \times N$ matrix.

For this we require

$$\det(A - \lambda I) = 0 \tag{2}$$

A direct cofactor expansion of Eq. (1) indicates that

$$\det(A - \lambda I) = (-1)^N \lambda^N + C_1 \lambda^{N-1} + \cdots + C_{N-1}\lambda + C_N \tag{3}$$

where $\{C_i\}$ are coefficients determined from the elements of A. The polynomial in Eq. (3) is the so-called *characteristic polynomial* of A. The N roots of Eq. (2) are called the *characteristic roots*, *latent roots*, or more conventionally, *eigenvalues* of A. As with any real-coefficient polynomial, the zeros may be distinct or not. Where an eigenvalue occurs exactly m times, it is said to have *multiplicity m*. In our considerations the eigenvalues may be real or complex. Where they are complex we need to distinguish between an eigenvalue of multiplicity 2 and a pair of complex conjugate eigenvalues (which are clearly distinct) but which have the same modulus.

Corresponding to each eigenvalue λ satisfying Eq. (2) there is a nontrivial **x** (which may have complex elements) called the *eigenvector*. Such eigenvectors are arbitrary up to a constant multiplier, which is often chosen to normalize the eigenvector in some appropriate manner; for example, so that $\mathbf{x}^*\mathbf{x} = 1$ where $\mathbf{x}^*$ is the complex conjugate transpose of **x**.

If A possesses N linearly independent eigenvectors, they are said to form a *complete* set.

Theorem 1 *If A has distinct eigenvalues, then there exists a complete set of linearly independent eigenvectors, unique up to a multiplicative constant.*

Since the determinant of A equals the determinant of A^T, it is easy to see that λ is also an eigenvalue of A^T.

If **y** is an eigenvector of A^T corresponding to λ, then $\mathbf{y}^T A = \lambda \mathbf{y}^T$

For this reason $\mathbf{y}^T$ is sometimes referred to as the *left* eigenvector of A, whereas **x** in Eq. (1) is referred to as the *right* eigenvector of A.

Theorem 2 (*Cayley–Hamilton theorem*) *The matrix A satisfies its own characteristic equation, i.e. if $p(x)$ is a polynomial in x, then*

$$p(A) = p(\lambda)$$

where λ is an eigenvalue of A.

Similarity transformation

Transformations of the matrix A of the form $R^{-1}AR$, where R is nonsingular, are of fundamental importance from both a theoretical and practical point of view. These are known as *similarity transformations*.

Definition A *real* matrix Q is *orthogonal* if $Q^T Q = I$.

Definition If Q is orthogonal, then A and $Q^{-1}AQ$ are said to be *orthogonally similar*.

The usefulness of similarity transformations is a direct result of the following simple observation: if

$$A\mathbf{x} = \lambda\mathbf{x}$$

then assuming the existence of a nonsingular Q,

$$A x = A Q Q^{-1}\mathbf{x} = \lambda\mathbf{x}$$

so that

$$Q^{-1}AQQ^{-1}\mathbf{x} = Q^{-1}\lambda\mathbf{x}$$

Thus the eigenvalues are *preserved* under the similarity transformation.

Some of the most important methods for determining eigenvalues are based on the reduction of a matrix by a similarity transformation (or by a sequence of such transformations) to a matrix of special *canonical* form whose eigensolution may be obtained more easily.

Theorem 3 *Any square matrix is orthogonally similar to a triangular matrix with the eigenvalues on the diagonal.*

Note: This is the best we can say, i.e. we *cannot* say that a general matrix can be reduced to a diagonal matrix. The most "compact" form of triangular matrix comes about from considering the *Jordan canonical form*.

Define

$$J_1(\lambda) = [\lambda]$$
$$\vdots$$

$$J_r(\lambda) = \begin{bmatrix} \lambda & 1 & & & \\ & \lambda & 1 & & \\ & & \ddots & \ddots & \\ & & & & 1 \\ & & & & \lambda \end{bmatrix} \qquad \text{(an } (r \times r) \text{ matrix)}$$

where λ is an eigenvalue of the matrix A with multiplicity r with only one eigenvector $\mathbf{x} = \mathbf{e}_1$.

The matrix $J_r(\lambda)$ is called a simple *Jordan* submatrix of order r. Suppose A has s distinct eigenvalues $\lambda_1, \lambda_2, \ldots, \lambda_s$ of multiplicities $m_1, m_2, \ldots, m_s$, so that

$$\sum_{i=1}^{s} m_i = N$$

Then we have the following result.

Theorem 4 *There exists a nonsingular matrix R such that $R^{-1}AR$ has simple Jordan submatrices $J_r(\lambda_i)$ isolated along the diagonal with all other elements equal to zero.*

If there are p submatrices of orders r_j, $j = 1, 2, \ldots, p$, associated with any λ_i, then

$$\sum_{j=1}^{p} r_j = m_i$$

The matrix $R^{-1}AR$ is called the *Jordan canonical form* of A and is unique apart from the possible ordering of the submatrices along the diagonal. The number of independent eigenvectors equals the number of submatrices.

Example 1 Consider the matrix A of order 7 which has a Jordan canonical form given by

$$\begin{bmatrix} \lambda_1 & 1 & & & & & \\ & \lambda_1 & 1 & & & & \\ & & \lambda_1 & & & & \\ & & & \lambda_2 & 1 & & \\ & & & & \lambda_2 & & \\ & & & & & \lambda_3 & \\ & & & & & & \lambda_4 \end{bmatrix}$$

Then A has four independent eigenvectors, one each for the eigenvalues λ_1, λ_2, λ_3, and λ_4.

For a given Jordan canonical form J we can form the matrix $J - \lambda I$ whose determinant is easily seen to be $\Pi_r \det (J_r(\lambda_i - \lambda))$. The individual determinants $\det J_r(\lambda_i - \lambda)$ are called *elementary divisors*.

Example 2 The elementary divisors for Example 1 are

$$(\lambda_1 - \lambda)^3, (\lambda_2 - \lambda)^2, (\lambda_3 - \lambda), (\lambda_4 - \lambda)$$

When the exponent in the elementary divisor is 1, the divisor is said to be *linear*. Otherwise it is *nonlinear*.

A matrix with distinct eigenvalues has linear divisors. If the eigenvalues are not distinct, the divisors may or may not be linear.

A matrix A is said to be *defective* if it has an eigenvalue with multiplicity k having fewer than k linearly independent eigenvectors.

Orthogonal transformations of a matrix

To determine the eigenvalues of a given matrix A it is frequently desirable to convert the matrix A through similarity transformations (so the eigenvalues are preserved) into another matrix whose eigensystem is more easily obtained. To do this we will employ elementary orthogonal matrices or reflectors, commonly called *Householder* transformations.

Definition An *elementary reflector* is a matrix of the form

$$Q = I - 2\mathbf{u}\mathbf{u}^T \tag{4}$$

where

$$\mathbf{u}^T\mathbf{u} = 1 \tag{5}$$

It is a straightforward matter to verify that such matrices are:
  (a) symmetric
  (b) orthogonal.

In practice the vector **u** is chosen to annihilate elements from a given vector. To facilitate the computation, however, it is frequently more convenient to recast Eq. (4) as

$$Q = I - \frac{2\mathbf{u}\mathbf{u}^T}{\|\mathbf{u}\|_2^2} \tag{6}$$

where **u** is not normalized according to (5).

If it is required to annihilate all elements but x_1 in $\mathbf{x} = (x_1, x_2, \ldots, x_N)^T$, then choosing

$$\mathbf{u} = \mathbf{x} + \sigma\mathbf{e}_1 \tag{7}$$

where $\mathbf{e}_1 = (1, 0, \ldots, 0)^T$, will achieve this when we form $Q\mathbf{x}$. The parameter σ is chosen in order that Q so defined is an elementary reflector. Through direct computation it is simple to show that the choice $\sigma = \pm\|\mathbf{x}\|_2$ ensures that Q is an elementary reflector. Therefore, σ is defined up to the sign. In practice the sign $(\pm)$ is chosen equal to the sign of x_1.

Example 3 Determine the elementary reflector which annihilates x_2 and x_3 in $\mathbf{x} = (1, 2, 2)^T$. Write down the transformed **x**. By simple calculation $\|\mathbf{x}\|_2 = 3$ and $\sigma = +3$

$$\mathbf{u} = \mathbf{x} + \sigma\mathbf{e}_1 = (4, 2, 2)^T$$

Then

$$Q = \begin{bmatrix} 1 & 0 & 0 \\ 0 & 1 & 0 \\ 0 & 0 & 1 \end{bmatrix} - \frac{1}{12} \begin{bmatrix} 4 \\ 2 \\ 2 \end{bmatrix} [4, 2, 2]$$

$$= \begin{bmatrix} -1/3 & -2/3 & -2/3 \\ -2/3 & 2/3 & -1/3 \\ -2/3 & -1/3 & 2/3 \end{bmatrix}$$

and

$$Q\mathbf{x} = \begin{bmatrix} -3 \\ 0 \\ 0 \end{bmatrix}$$

Example 4 By generating a sequence of elementary reflectors, triangulate the matrix

$$A = \begin{bmatrix} 1 & 0 & 1 \\ 2 & 1 & 0 \\ 2 & -1 & 1 \end{bmatrix}$$

From Example 3 we already know the elementary reflector Q_1 (say) which reduces the first column of A. Premultiplying A by Q_1, we see that

$$Q_1A = \begin{bmatrix} -3 & 0 & -1 \\ 0 & 1 & -1 \\ 0 & -1 & 0 \end{bmatrix}$$

We now require to generate a second reflector Q_2 which has the form

$$Q_2 = \begin{bmatrix} 1 & 0 & 0 \\ 0 & & \\ 0 & & Q'_2 \end{bmatrix}$$

where $Q'_2 = I - 2\mathbf{u}\mathbf{u}^T/\|\mathbf{u}\|_2^2$ is 2×2. We propose

$$\mathbf{u} = \begin{pmatrix} 1 \\ -1 \end{pmatrix} + \sigma \begin{pmatrix} 1 \\ 0 \end{pmatrix} \quad \text{with } \sigma = \sqrt{2}$$

so that

$$\|\mathbf{u}\|_2^2 = 4 + 2\sqrt{2}$$

Hence

$$Q'_2 = \begin{bmatrix} -\dfrac{(1+\sqrt{2})}{2+\sqrt{2}} & \dfrac{1+\sqrt{2}}{2+\sqrt{2}} \\ \dfrac{1+\sqrt{2}}{2+\sqrt{2}} & \dfrac{1+\sqrt{2}}{2+\sqrt{2}} \end{bmatrix}$$

Substituting Q'_2 into Q_2 and forming Q_2Q_1A, we see that

$$Q_2Q_1A = \begin{bmatrix} -3 & 0 & -1 \\ 0 & \dfrac{-2(1+\sqrt{2})}{2+\sqrt{2}} & \dfrac{1+\sqrt{2}}{2+\sqrt{2}} \\ 0 & 0 & \dfrac{-(1+\sqrt{2})}{2+\sqrt{2}} \end{bmatrix}$$

which is the required triangulated matrix.

With such elementary reflectors it is clear that we could use such a process for solving systems of linear equations as an alternative to the Gaussian elimination and its variants described in Chapter 3. The elements of $\mathbf{u}$ at each stage and the scalar σ can be stored, from which the matrices Q_j could be constructed, in the lower triangle of the matrix A and one extra row for the σ's. However, the matrix so defined (the lower triangle) does not contain the product matrix

$$Q = Q_{N-1}Q_{N-2} \cdots Q_2Q_1$$

or make its calculation readily available, as was the case in the LU factorization of Chapter 3.

Whereas we could solve systems of linear equations using this triangular decomposition, it is our purpose here to introduce the concept of annihilation of the elements in columns of a given matrix as a step in the determination of the eigenvalues and eigenvectors of the given matrix A. So, at each stage at which we premultiply the elementary reflector, we require also to postmultiply by the same elementary reflector to effect a similarity transformation which preserves the eigenvalues of A. As we will show in Example 5, this will introduce, in general, nonzero elements on the subdiagonal. For this reason we will be able only to generate an upper *Hessenberg* matrix for general A or a *tridiagonal* matrix for symmetric A.

Example 5 Investigate the structure of the matrix resulting from premultiplying and postmultiplying the matrix

$$A = \begin{bmatrix} 0 & 1 & 0 & 0 \\ 1 & 0 & 0 & 0 \\ 0 & 0 & 1 & 0 \\ 0 & 0 & 0 & 1 \end{bmatrix}$$

by the elementary reflector

$$Q = \begin{bmatrix} 0 & -1 & 0 & 0 \\ -1 & 0 & 0 & 0 \\ 0 & 0 & 1 & 0 \\ 0 & 0 & 0 & 1 \end{bmatrix}$$

(The reader may verify that Q is the elementary reflector obtained from annihilating the nonzero element in the first column of A.)

So

$$QA = \begin{bmatrix} -1 & 0 & 0 & 0 \\ 0 & -1 & 0 & 0 \\ 0 & 0 & 1 & 0 \\ 0 & 0 & 0 & 1 \end{bmatrix}$$

and then

$$QAQ^{\mathrm{T}} = \begin{bmatrix} 0 & 1 & 0 & 0 \\ 1 & 0 & 0 & 0 \\ 0 & 0 & 1 & 0 \\ 0 & 0 & 0 & 1 \end{bmatrix}$$

which is tridiagonal. The example is of course contrived (we do not as a rule obtain the tridiagonal matrix after constructing just one elementary reflector). In general, the postmultiplication by Q^{T} reintroduces a nonzero in the subdiagonal element of the column being annihilated but leaves the remaining annihilated elements zero. When A is unsymmetric, in general all the elements on the upper triangle will remain nonzero so that the final matrix is upper Hessenberg.

Example 6   Apply Householder's transformations to reduce the symmetric matrix

$$A = \begin{bmatrix} 1 & 0 & 1 & 0 \\ 0 & 1 & 0 & 1 \\ 1 & 0 & 0 & 0 \\ 0 & 1 & 0 & 1 \end{bmatrix}$$

to tridiagonal form.

In this instance, since we know the subdiagonal element will not be eliminated, we annihilate only the $(3, 1)$ and $(4, 1)$ elements. Hence we construct the 3×3 matrix

$$Q_1' = I - 2\mathbf{u}\mathbf{u}^T / \|\mathbf{u}\|_2^2$$

where

$$\mathbf{u} = \begin{bmatrix} 0 \\ 1 \\ 0 \end{bmatrix} + \sigma \begin{bmatrix} 1 \\ 0 \\ 0 \end{bmatrix}$$

and $\sigma = 1$. Hence

$$\mathbf{u} = \begin{bmatrix} 1 \\ 1 \\ 0 \end{bmatrix} \quad \text{and} \quad \|\mathbf{u}\|_2^2 = 2$$

So

$$Q_1' = \begin{bmatrix} 0 & -1 & 0 \\ -1 & 0 & 0 \\ 0 & 0 & 1 \end{bmatrix}$$

So, forming

$$Q_1 = \begin{bmatrix} 1 & 0 & 0 & 0 \\ 0 & 0 & -1 & 0 \\ 0 & -1 & 0 & 0 \\ 0 & 0 & 0 & 1 \end{bmatrix}$$

and hence

$$Q_1 A = \begin{bmatrix} 1 & 0 & 1 & 0 \\ -1 & 0 & 0 & 0 \\ 0 & -1 & 0 & -1 \\ 0 & 1 & 0 & 1 \end{bmatrix}$$

we obtain

$$Q_1AQ_1^T = \begin{bmatrix} 1 & -1 & 0 & 0 \\ -1 & 0 & 0 & 0 \\ 0 & 0 & 1 & -1 \\ 0 & 0 & -1 & 1 \end{bmatrix}$$

This is also a somewhat contrived example, as with the construction of just one elementary reflector we have produced the required tridiagonal matrix. In general (for the 4×4 matrix), we would have to construct one more (2×2) elementary reflector and reduce the matrix by pre- and postmultiplication one more time.

Example 7 The following unsymmetric matrix exemplifies the reduction to upper Hessenberg form using the elementary reflectors.

$$A = \begin{bmatrix} 1 & 0 & 1 & 1 \\ 0 & 1 & 0 & 1 \\ 1 & 1 & 0 & 0 \\ 0 & 0 & 1 & 1 \end{bmatrix}$$

As in Example 6,

$$Q_1' = \begin{bmatrix} 0 & -1 & 0 \\ -1 & 0 & 0 \\ 0 & 0 & 1 \end{bmatrix}$$

Then

$$Q_1 = \begin{bmatrix} 1 & 0 & 0 & 0 \\ 0 & & & \\ 0 & & R_1' & \\ 0 & & & \end{bmatrix}$$

so that

$$A_1 = Q_1AQ_1^T = \begin{bmatrix} 1 & -1 & 0 & 1 \\ -1 & 0 & 1 & 0 \\ 0 & 0 & 1 & -1 \\ 0 & -1 & 0 & 1 \end{bmatrix}$$

Now construct

$$Q_2' = I - 2\mathbf{u}\mathbf{u}^T/\|\mathbf{u}\|_2^2$$

with

$$\mathbf{u} = \begin{bmatrix} 0 \\ -1 \end{bmatrix} + \sigma \begin{bmatrix} 1 \\ 0 \end{bmatrix}$$

where

$$\sigma = 1 \quad \text{so} \quad \mathbf{u} = \begin{bmatrix} 1 \\ -1 \end{bmatrix} \quad \text{and} \quad \|\mathbf{u}\|_2^2 = 2$$

So

$$Q_2' = \begin{bmatrix} 0 & 1 \\ 1 & 0 \end{bmatrix} \quad \text{and} \quad R_2 = \begin{bmatrix} 1 & 0 & 0 & 0 \\ 0 & 1 & 0 & 0 \\ 0 & 0 & 0 & 1 \\ 0 & 0 & 1 & 0 \end{bmatrix}$$

from which

$$Q_2 A_1 Q_2^{\mathrm{T}} = \begin{bmatrix} 1 & -1 & 1 & 0 \\ -1 & 0 & 0 & 1 \\ 0 & -1 & 1 & 0 \\ 0 & 0 & -1 & 1 \end{bmatrix}$$

which is upper Hessenberg.

As we have seen it is possible with the use of Householder transformations to produce for a given matrix A a similar matrix having at most a single subdiagonal. If we were given such a matrix (tridiagonal or Hessenberg) then applying a sequence of Householder transformations to annihilate all the elements below the main diagonal will require a considerable amount of wasted work, reducing to zero elements which are already zero. For this reason we will have cause to use alternative orthogonal transformations which will reduce a specified element to zero. These transformations are known as *rotation matrices*.

Define the rotation matrix $Q(j, k)$ as the unit matrix subject to the following alterations:

$$q_{jj} = \cos \theta \qquad q_{jk} = \sin \theta$$

$$q_{kj} = -\sin \theta \qquad q_{kk} = \cos \theta$$

where θ is a real parameter to be chosen.

It is a simple matter to verify that Q so defined is orthogonal. Consequently $Q^{-1} = Q^{\mathrm{T}}$.

Premultiplication (postmultiplication) of a matrix A by the matrix Q affects only rows (columns) j and k of A. Hence the only elements modified in the similarity transformation QAQ^{T} of A lie in the jth and kth rows and columns.

Suppose $\mathbf{x}$ is a column of a matrix of dimension 3 ($x_2 \neq 0$). Premultiply $\mathbf{x}$ by the rotation matrix $Q(1, 2)$, to produce the following:

$$\mathbf{y} = \begin{bmatrix} \cos \theta & \sin \theta & 0 \\ -\sin \theta & \cos \theta & 0 \\ 0 & 0 & 1 \end{bmatrix} \begin{bmatrix} x_1 \\ x_2 \\ x_3 \end{bmatrix} = \begin{bmatrix} \cos \theta \, x_1 + \sin \theta \, x_2 \\ -\sin \theta \, x_1 + \cos \theta \, x_2 \\ x_3 \end{bmatrix}$$

We here have the parameter θ and can choose it to annihilate one of the first two elements of **y**. Suppose we choose θ to make $y_1 = 0$. Then

$$
\mathbf{y} = \begin{bmatrix} \cos \theta\, x_1 + \sin \theta\, x_2 \\ -\sin \theta\, x_1 + \cos \theta\, x_2 \\ x_3 \end{bmatrix} = \begin{bmatrix} 0 \\ x_1 \sin \arctan(-x_1/x_2) + x_2 \cos \arctan(-x_1/x_2) \\ x_3 \end{bmatrix}
$$

Conversely, we could have chosen θ to annihilate y_2. With a little thought it should be clear that for a vector of length N a sequence of such rotations can be applied so that for each application an additional element of the vector is annihilated. This, by propitious choice of rotations, can be achieved without introducing nonzeros into elements previously annihilated.

So now we are implying we reduce elements in the column of a matrix using a sequence of premultiplications by rotation matrices instead of the elementary matrices used in Gaussian elimination.

Example 8   Decompose

$$
A = \begin{bmatrix} 1 & 1 & 1 \\ 1 & 2 & 1 \\ 1 & 1 & 2 \end{bmatrix}
$$

using rotation matrices.

The elements are annihilated column by column commencing in the leftmost column. Here there are two nonzero elements to annihilate, the $(2, 1)$ and $(3, 1)$ elements. Hence, in turn, we construct the rotation matrices $Q(1, 2)$ and $Q(1, 3)$ and premultiply by them. For $Q(1, 2)$ we see that

$$
A_1 = \begin{bmatrix} \cos \theta & \sin \theta & 0 \\ -\sin \theta & \cos \theta & 0 \\ 0 & 0 & 1 \end{bmatrix} \begin{bmatrix} 1 & 1 & 1 \\ 1 & 2 & 1 \\ 1 & 1 & 2 \end{bmatrix}
$$

requires us to choose $\sin \theta = \cos \theta = 1/\sqrt{2}$, so that

$$
Q(1, 2) = \begin{bmatrix} 1/\sqrt{2} & 1/\sqrt{2} & 0 \\ -1/\sqrt{2} & 1/\sqrt{2} & 0 \\ 0 & 0 & 1 \end{bmatrix} \quad \text{and} \quad A_1 = \begin{bmatrix} 2/\sqrt{2} & 3/\sqrt{2} & 2/\sqrt{2} \\ 0 & 1/\sqrt{2} & 0 \\ 1 & 1 & 2 \end{bmatrix}
$$

Similarly, in constructing $Q(1, 3)$,

$$
A_2 = \begin{bmatrix} \cos \theta & 0 & \sin \theta \\ 0 & 1 & 0 \\ -\sin \theta & 0 & \cos \theta \end{bmatrix} \begin{bmatrix} 2/\sqrt{2} & 3/\sqrt{2} & 2/\sqrt{2} \\ 0 & 1/\sqrt{2} & 0 \\ 1 & 1 & 2 \end{bmatrix}
$$

we require to choose $\sin \theta = \sqrt{1/3}$ and $\cos \theta = \sqrt{2/3}$. Then

$$
Q(1, 3) = \begin{bmatrix} \sqrt{2/3} & 0 & \sqrt{1/3} \\ 0 & 1 & 0 \\ \sqrt{1/3} & 0 & \sqrt{2/3} \end{bmatrix} \quad A_2 = \begin{bmatrix} 3/\sqrt{3} & 4/\sqrt{3} & 4/\sqrt{3} \\ 0 & 1/\sqrt{2} & 0 \\ 0 & -1/\sqrt{6} & 2/\sqrt{6} \end{bmatrix}
$$

As was intended we have successfully annihilated the (2, 1) and (3, 1) elements.

We now continue and annihilate the (3, 2) element by choosing θ appropriately in

$$Q(2,3) = \begin{bmatrix} 1 & 0 & 0 \\ 0 & \cos\theta & \sin\theta \\ 0 & -\sin\theta & \cos\theta \end{bmatrix}$$

so that $A_3 = Q(2,3)A_2$ is the required upper triangular matrix. Simple arithemtic produces $\sin\theta = 1/2$, $\cos\theta = -\sqrt{3/2}$, so that

$$Q(2,3) = \begin{bmatrix} 1 & 0 & 0 \\ 0 & -\sqrt{3}/2 & 1/2 \\ 0 & -1/2 & -\sqrt{3}/2 \end{bmatrix} \quad \text{and} \quad A_3 = \begin{bmatrix} 3/\sqrt{3} & 4/\sqrt{3} & 4/\sqrt{3} \\ 0 & -2/\sqrt{6} & 1/\sqrt{6} \\ 0 & 0 & 1/\sqrt{2} \end{bmatrix}$$

In general, the rotation matrix $Q(j, k)$ on premultiplying A has the form

$$\begin{bmatrix} 1 & \cdots & 0 & \cdots & 0 & \cdots & 0 & \cdots & 0 \\ 0 & 1 & & & & & & & \\ \vdots & & \ddots & & \vdots & & \vdots & & \vdots \\ 0 & \cdots & & \cos\theta & \cdots & \sin\theta & \cdots & & 0 \\ & & & & 1 & & & & \\ \vdots & & & & & \ddots & \vdots & & \vdots \\ & & & & & & 1 & & \\ 0 & \cdots & & -\sin\theta & \cdots & \cos\theta & \cdots & & 0 \\ & & & & & & & 1 & \\ \vdots & & & \vdots & & \vdots & & \ddots & \vdots \\ & & & & & & & & 1 \\ 0 & \cdots & & 0 & \cdots & 0 & \cdots & & 1 \end{bmatrix} \begin{bmatrix} a_{11} & a_{12} & \cdots & a_{1N} \\ \vdots & & & \\ a_{j1} & a_{j2} & \cdots & a_{jN} \\ \vdots & & & \\ a_{k1} & a_{k2} & \cdots & a_{kN} \\ \vdots & & & \\ a_{N1} & a_{N2} & \cdots & a_{NN} \end{bmatrix}$$

so that the elements of A are unaltered except for rows j and k, which in turn become:

row $j[a_{j1}\cos\theta + a_{k1}\sin\theta, a_{j2}\cos\theta + a_{k2}\sin\theta, \ldots, a_{jN}\cos\theta + a_{kN}\sin\theta]$

$$(8)$$

row $k[-a_{j1}\sin\theta + a_{k1}\cos\theta, -a_{j2}\sin\theta + a_{k2}\cos\theta, \ldots, -a_{jN}\sin\theta + a_{kN}\cos\theta]$

Hence, choosing θ to annihilate the required element in row k, say, the (k, i) element, we require

$$\tan\theta = \frac{a_{ki}}{a_{ji}}$$

so that

$$\sin \theta = \frac{a_{ki}}{\sqrt{a_{ki}^2 + a_{ji}^2}} \quad \text{and} \quad \cos \theta = \frac{a_{ji}}{\sqrt{a_{ki}^2 + a_{ji}^2}} \tag{9}$$

If $a_{ki} = a_{ji} = 0$, set $\cos \theta = 1$ and $\sin \theta = 0$.

Bounds on eigenvalues

In some instances the information required of the eigensystem of a given matrix takes the form of the following questions:

(1) Are all the eigenvalues positive (negative)?
(2) How many eigenvalues are contained in an interval $[a, b]$?
(3) Obtain an approximate value of the largest (in absolute value) eigenvalue.
(4) Provide an approximation to the eigenvalue closest to a given value p.

Such questions arise in the context of providing starting values for an iterative process for determining the required eigenvalue accurately or answering the question whether the matrix is positive definite, etc. In all of these cases we are required to provide quantitative information. For this purpose we have available two remarkable results which are extremely easy to use and in a simple fashion provide the answers to the type of questions asked above. These two results are presented in Theorems 5 and 6 below.

Theorem 5 (*Gerschgorin's theorem*)   Let A be a given $N \times N$ matrix† and let $C_i, i = 1, 2, \ldots, N$, be the discs with centres a_{ii} and radii

$$R_i = \sum_{\substack{k=1 \\ k \neq i}}^{N} |a_{ik}|$$

Let D denote the union of the discs C_i. Then all the eigenvalues of A lie within D.

Example 9   Locate the eigenvalues of

$$A = \begin{bmatrix} 1+i & i & 0 \\ i & 1+i & i \\ 0 & i & 1+i \end{bmatrix}$$

and confirm they lie within the region specified by Gerschgorin's theorem.

The eigenvalues of A are easily obtained by solving the cubic equation $\det(A - \lambda I) = 0$. They are $1+i$, $1+\sqrt{2}+i$, $1-\sqrt{2}+i$. Their locations are shown in Figure 1. The associated discs having centres $1+i$ and radii 1 and 2 (and 2) are also shown. The eigenvalues clearly lie in the union of the discs (in this case simply the larger of the two).

† Although we have restricted our attention to real matrices in this Chapter for completeness we give the theorem for the matrix whose elements may be complex.

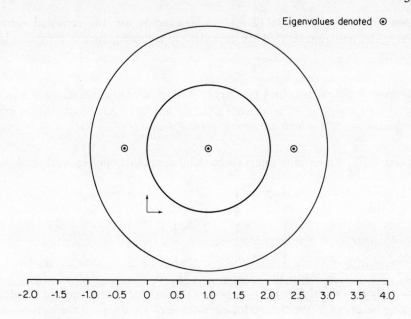

Figure 1

Sturm sequences

When the matrix A is symmetric tridiagonal (obtained as a result of House-holder transformations applied to a symmetric matrix), the location of the eigenvalues can be obtained in a particularly simple manner.

Suppose the tridiagonal matrix T (say) is given by

$$
T = \begin{bmatrix}
b_1 & c_1 & & & \\
c_1 & b_2 & c_2 & & \\
& c_2 & b_3 & c_3 & \\
& & & & c_{N-1} \\
& & & c_{N-1} & b_{N-1}
\end{bmatrix}
$$

where we assume $c_j \neq 0, j = 1, 2, \ldots, N-1$

Construct the principal minors of $T - \lambda I$ by the recurrence relation

$$p_0(\lambda) = 1$$

$$p_1(\lambda) = b_1 - \lambda$$

$$p_k(\lambda) = (b_k - \lambda)p_{k-1}(\lambda) - c_{k-1}^2 p_{k-2}(\lambda) \qquad k = 2, 3, \ldots, N \tag{10}$$

(see Exercise 8).

When $k = N$ we have $\det(T - \lambda I)$ whose zeros are the required eigenvalues. The polynomials (10) are important as a consequence of the following theorem.

Theorem 6 (*Sturm sequence property*) *The number of agreements in sign of successive numbers of the sequence $\{p_r(\hat{\lambda})\}$ for any given $\hat{\lambda}$ is equal to the number of eigenvalues of T which are strictly greater than $\hat{\lambda}$.*

Example 10 Determine intervals containing at most one eigenvalue of

$$
T = \begin{bmatrix}
2 & -1 & & & \\
-1 & 2 & -1 & & \\
& -1 & 2 & -1 & \\
& & -1 & 2 & -1 \\
& & & -1 & 2
\end{bmatrix}
$$

The polynomials of Eq. (10) can be evaluated for a sequence of carefully chosen values of $\hat{\lambda}$. We record for each value of $\hat{\lambda}$ the sign of the polynomial. If $p_i(\hat{\lambda}) = 0$, then we record the sign opposite to the sign of $p_{i-1}(\hat{\lambda})$. The sequence of signs for T for the given values of $\hat{\lambda}$ are recorded in Table 1.

Table 1

				sign of $p_k(\hat{\lambda})$					
K $\quad$ $\hat{\lambda}$	0	0.5	1.0	1.5	2.0	2.5	3.0	3.5	4.0
0	+	+	+	+	+	+	+	+	+
1	+	+	+	+	−	−	−	−	−
2	+	+	−	−	−	−	−	+	+
3	+	+	−	−	+	+	+	−	−
4	+	−	−	+	+	+	−	−	+
5	+	−	+	+	−	−	+	+	−
No. eigenvalues $> \hat{\lambda} =$	5	4	3	3	2	2	1	1	0

By considering the final row of Table 1 it is clear that there is a single eigenvalue in each of the intervals $[0, 0.5]$, $[0.5, 1.0]$, $[1.5, 2.0]$, $[2.5, 3.0]$, $[3.5, 4.0]$. An interesting conclusion can be drawn immediately from the first column of signs in Table 1—the matrix has 5 positive eigenvalues, i.e. it is positive definite. We could also have discerned this property easily using Gershgorin's Theorem 5.

Exercises

1. Develop a subroutine or procedure REFLECT (M, X) which on input has a vector X with M elements and on output contains the elements of **u** defined by Eq. (7).
2. Develop a subroutine or procedure HSHLDR (M, U, Q) which on input has the M-vector **U** and on output the $M \times M$ elementary reflector matrix Q defined by Eq. (6).
3. Using the subroutines REFLECT and HSHLDR, write a program to reduce the following matrices to either tridiagonal or upper Hessenberg form as appropriate:

(a) $A = \begin{bmatrix} 1 & 0 & 1 & 0 \\ 0 & 1 & 0 & 1 \\ 1 & 0 & 0 & 0 \\ 0 & 1 & 0 & 1 \end{bmatrix}$
(b) $A = \begin{bmatrix} 1 & 2 & 3 & 4 & 5 \\ 2 & 3 & 4 & 5 & 6 \\ 3 & 4 & 5 & 6 & 7 \\ 4 & 5 & 6 & 7 & 8 \\ 5 & 6 & 7 & 8 & 9 \end{bmatrix}$

(c) $A = \begin{bmatrix} 1 & 0 & 1 & 1 \\ 0 & 1 & 0 & 1 \\ 1 & 1 & 0 & 0 \\ 0 & 0 & 1 & 1 \end{bmatrix}$
(d) $A = \begin{bmatrix} 1 & 2 & 3 & 5 & 4 \\ 2 & 3 & 4 & 1 & 5 \\ 3 & 4 & 5 & 2 & 1 \\ 1 & 5 & 4 & 3 & 2 \\ 5 & 3 & 1 & 2 & 4 \end{bmatrix}$

4. Write a subroutine or procedure ROTATE $(M, B, C, L, \text{THETA})$ which calculates the angle THETA which annihilates the element $C(L)$ in the M-vector C, where B and C represent the jth and kth rows respectively of matrix A. The transformed vectors are to be returned in the vectors B and C and the associated angle θ in the variable THETA as defined in Eqs. (8) and (9).
5. Use your subroutine or procedure of Exercise 4 in a main program: (a) to triangulate the matrices of Exercise 3; (b) to triangulate the tridiagonal, or Hessenberg matrices which are the solutions to Exercise 3.
6. Develop a subroutine or procedure STURM (M, B, C, LA) which will produce as printed output the sequence of signs obtained when evaluating the polynomials (10) at the real number LA. The elements of the tridiagonal matrices are stored in the M-vectors B and C. Arrange also to print out a statement giving the number of eigenvalues strictly greater than LA.
7. Use your subroutine or procedure of Exercise 6 to provide intervals containing exactly one eigenvalue for the following matrices:

(a) $\begin{bmatrix} 3 & 1 & & & \\ 1 & 3 & 1 & & \\ & 1 & 3 & 1 & \\ & & 1 & 3 & 1 \\ & & & 1 & 3 \end{bmatrix}$
(b) $\begin{bmatrix} 1 & 0.5 & & \\ 0.5 & 2 & 0.5 & \\ & 0.5 & 3 & 0.5 \\ & & 0.5 & 4 \end{bmatrix}$

(c) $\begin{bmatrix} 1 & 0.5 & & & \\ 0.5 & 2 & 1.0 & & \\ & 1.0 & 3 & 2.0 & \\ & & 2.0 & 4 & 3.0 \\ & & & 3.0 & 5 \end{bmatrix}$
(d) $\begin{bmatrix} -1 & 0.1 & & \\ 0.1 & -2 & 0.2 & \\ & 0.2 & -3 & 0.3 \\ & & 0.3 & -4 \end{bmatrix}$

8. Verify that Eqs. (10) are true by expanding the determinants of the principal minors of T.

2 The Power Method

It is frequently the case that the dominant eigenvalue and corresponding eigenvector are required for a given real matrix A. In this case, where the dominant eigenvalue is unique, the *power* method is appropriate. As we will see, the actual behavior of the algorithm will very much depend on the relative closeness of the subdominant eigenvalues to the dominant eigenvalue.

Let A be a given $N \times N$ matrix with real elements. Let the eigenvalues of A be denoted by λ_j, $j = 1, 2, \ldots, N$, and defined so that

$$|\lambda_1| > |\lambda_2| \geq |\lambda_3| \geq \cdots \geq |\lambda_N| \tag{11}$$

Let the corresponding eigenvectors of A be designated $\mathbf{u}_j$ with $\mathbf{u}_j$ the right eigenvector corresponding to λ_j; that is,

$$A\mathbf{u}_j = \lambda_j \mathbf{u}_j \qquad j = 1, 2, \ldots, N \tag{12}$$

If we assume A has N linearly independent eigenvectors and that they therefore form a basis, we may write an arbitrary $\mathbf{x}_{(0)}$, say, as

$$\mathbf{x}_{(0)} = \sum_{j=1}^{N} \alpha_j \mathbf{u}_j \tag{13}$$

where the coefficients α_j are unique.

For such an $\mathbf{x}_{(0)}$ the iteration sequence

$$\mathbf{x}_{(k+1)} = A\mathbf{x}_{(k)} \qquad k = 0, 1, 2, \ldots \tag{14}$$

produces a sequence of vectors which, under appropriate assumptions, converges to a vector parallel to the dominant eigenvector. To see this, consider applying (14) using (12) and (13). Then

$$\mathbf{x}_{(k+1)} = A^{k+1}\mathbf{x}_{(0)} = A^{k+1} \sum_{j=1}^{N} \alpha_j \mathbf{u}_j$$

so that

$$\mathbf{x}_{(k+1)} = \sum_{j=1}^{N} \alpha_j \lambda_j^{k+1} \mathbf{u}_j \tag{15}$$

Recalling Eq. (11) and writing (15) as

$$\mathbf{x}_{(k+1)} = \lambda_1^{k+1} \left[\sum_{j=1}^{N} \alpha_j \left(\frac{\lambda_j}{\lambda_1}\right)^{k+1} \mathbf{u}_j \right] \tag{16}$$

we see that the vector $\mathbf{x}_{(k+1)}$ ultimately becomes parallel to $\mathbf{u}_1$. Also, if we consider (16) with $k+1$ replaced by k, we have

$$\mathbf{x}_{(k)} = \lambda_1^{k} \left[\sum_{j=1}^{N} \alpha_j \left(\frac{\lambda_j}{\lambda_1}\right)^{k} \mathbf{u}_j \right]$$

so that forming the ratio of any component of $\mathbf{x}_{(k+1)}$ and $\mathbf{x}_{(k)}$, say the sth component,

$$\frac{x_{s(k+1)}}{x_{s(k)}} = \frac{\lambda_1 \left[\alpha_1 u_{s1} + \sum_{j=2}^{N} \alpha_j \left(\frac{\lambda_j}{\lambda_1}\right)^{k+1} u_{sj} \right]}{\left[\alpha_1 u_{s1} + \sum_{j=2}^{N} \alpha_j \left(\frac{\lambda_j}{\lambda_1}\right)^{k} u_{sj} \right]} \tag{17}$$

where u_{sj} denotes the sth element of $\mathbf{u}_j$ and $x_{s(k)}$ denotes the sth element of the kth iterate $\mathbf{x}_{(k)}$.
Hence

$$\lim_{k \to \infty} \frac{x_{s(k+1)}}{x_{s(k)}} = \lambda_1$$

Instead of choosing an arbitrary element, it is conventional to search for the maximum element in $\mathbf{x}_{(k)}$ and form the ratio of it and the corresponding element in $\mathbf{x}_{(k+1)}$. This is done to minimize the growth of rounding errors. Even better, however, is to scale the iterates $\mathbf{x}_{(k)}$ at each stage. Thus we advocate the following, for arbitrary $\mathbf{x}_{(0)}$: $\quad x_{(k+1)} = A x_{(k)}$
Define

and

$$\left. \begin{aligned} \mathbf{y}_{(k+1)} &= A\mathbf{x}_{(k)} \\ \mathbf{x}_{(k+1)} &= \frac{1}{S_{k+1}} \mathbf{y}_{(k+1)} \end{aligned} \right\} \quad k = 0, 1, 2, \ldots \tag{18}$$

where S_{k+1} is a scaling factor. Commonly two alternate choices of S_{k+1} are used:

$$S_{k+1} = \max_{1 \le j \le N} |x_{j(k)}| = \|\mathbf{x}_{(k)}\|_\infty \tag{19}$$

or

$$S_{k+1} = \|\mathbf{x}_{(k)}\|_2 \tag{20}$$

In either case the ratio advocated by (17) will provide an estimate of the dominant eigenvalue. The sequence $\mathbf{x}_{(k+1)}$ will then converge to $\mathbf{u}_1$ normalized according to (19) or (20).
Using (17), it is not difficult to show that

$$\frac{x_{s(k+1)}}{x_{s(k)}} = \lambda_1 + O\left(\frac{\lambda_2}{\lambda_1}\right)^k$$

so that the ratio produces λ_1 with first-order convergence with a convergence ratio of at most $|\lambda_2/\lambda_1|$. If $|\lambda_2| \ll |\lambda_1|$ then the method will produce a sequence of iterates which converges rapidly to the solution. In contrast, if $|\lambda_2| \approx |\lambda_1|$ then the convergence can be quite slow.

If the dominant eigenvalue has _multiplicity_ r, say, we see by simple manipulation that (16) becomes

$$\mathbf{x}_{(k+1)} = \lambda_1^{k+1} \left[\sum_{j=1}^{r} \alpha_j \mathbf{u}_j + \sum_{j=r+1}^{N} \alpha_j \left(\frac{\lambda_j}{\lambda_1} \right)^{k+1} \mathbf{u}_j \right]$$

and

$$\frac{x_{s(k+1)}}{x_{s(k)}} = \lambda_1 + 0 \left(\frac{\lambda_{r+1}}{\lambda_1} \right)^{k}$$

In this case the dominant eigenvalue is produced, but $\mathbf{x}_{(k+1)}$ converges to a linear combination of the first r eigenvectors.

In the event the dominant eigenvalue is nonunique, i.e. $|\lambda_1| = |\lambda_2|$ ($\lambda_1 \neq \lambda_2$) which occurs either when λ_1 and λ_2 are of opposite sign or represent complex conjugates, then the power method, in its current form, breaks down. However, in this case we note that

$$\mathbf{x}_{(k+i)} = \lambda_1^{k+i} \alpha_1 \mathbf{u}_1 + \bar{\lambda}_1^{k+i} \alpha_2 \mathbf{u}_2 + \sum_{j=3}^{N} \alpha_j \lambda_j^{k+i} \mathbf{u}_j \qquad i = 0, 1, 2$$

where $\bar{\lambda}_1$ is the negative or complex conjugate of λ_1.

If the summation term is sufficiently small to be neglected, then we may take a linear combination of $\mathbf{x}_{(k+i)}$, $i = 0, 1, 2$, to produce

$$\mathbf{x}_{(k+2)} + b\mathbf{x}_{(k+1)} + c\mathbf{x}_{(k)} = (\lambda_1^{k+2} + b\lambda_1^{k+1} + c\lambda_1^{k}) \alpha_1 \mathbf{u}_1$$
$$+ (\bar{\lambda}_1^{k+2} + b\bar{\lambda}_1^{k+1} + c\bar{\lambda}_1^{k}) \alpha_2 \mathbf{u}_2$$

We can make the right-hand side zero if λ_1 and $\bar{\lambda}_1$ are the zeros of the quadratic $\lambda^2 + b\lambda + c$. Assuming this to be the case, knowing b and c produces λ_1 and $\bar{\lambda}_1$. To determine b and c we merely need to solve

$$\mathbf{x}_{(k+2)} + b\mathbf{x}_{(k+1)} + c\mathbf{x}_{(k)} = 0 \tag{21}$$

This is an overdetermined system of n equations in 2 unknowns which can be solved by the least squares technique described in Chapter 5. Alternatively, any two of the equations (21) could be chosen and the resulting system solved.

Example 11 Determine the dominant eigenvalue of the real symmetric matrix

$$A = \begin{bmatrix} 4 & 0.5 & 0 \\ 0.5 & 4 & 0.5 \\ 0 & 0.5 & 4 \end{bmatrix}$$

using the power method with an initial vector $\mathbf{x}_{(0)} = (1, 1, 1)^T$.

$$X_{(0)} = \sum_{j=1}^{N} d_j u_j$$

The actual eigenvalues of such a matrix are known theoretically (see for example Mitchell and Griffiths [32]) and are given by

$$\lambda_1 = 4 + \frac{1}{\sqrt{2}}; \; 4; \; 4 - \frac{1}{\sqrt{2}}$$

Hence the convergence ratio is $|\lambda_2/\lambda_1| = 4/4.7071$. We therefore expect the convergence to be slow. In Table 2 the ratio of the (unscaled) elements of $\mathbf{x}_{(k+1)}$ and $\mathbf{x}_{(k)}$ is shown for the first 3 iterations and when the ratio is correct to 5 significant figures. The corresponding (eigen) vector $\mathbf{x}_{(k+1)}$ is also shown.

It is observed that as the iterations increase, the magnitude of the elements of the vectors $\mathbf{x}_{(k)}$ increases substantially. The elements of $\mathbf{x}_{(26)}$ normalized by $\|\mathbf{x}_{(26)}\|_\infty$ are found to be (0.707 128 87, 1, 0.707 128 87), which we prefer to the vector $\mathbf{x}_{(26)}$ for obvious reasons!

Table 2

k		Ratios of elements			Eigenvector	
1	4.5	5	4.5	[4.5	5	$4.5]^{\mathrm{T}}$
2	4.555 556	4.9	4.555 556	[2.05E1	2.45E1	$2.05\text{E}1]^{\mathrm{T}}$
3	4.598 561	4.836 735	4.497 561	[9.425E1	1.1185E1	$9.425\text{E}1]^{\mathrm{T}}$
$\vdots$						
26	4.707 075	4.707 139	4.707 075	[2.647 54E18	3.744 07E18	$2.647\,54\text{E}18]^{\mathrm{T}}$

Example 12 Determine the eigenvalue of example 11 using the scaled power method when $S_{k+1} = \|\mathbf{x}_{(k)}\|_\infty$.

The convergence with the scaled method should appear similar to Example 11 except that the elements of the iterates $\mathbf{x}_{(k)}$ will now remain bounded by unity. The results are displayed in Table 3.

Table 3

k	S_{k+1}	Eigenvector		
1	5	[0.9	1	$0.9]^{\mathrm{T}}$
2	4.9	[0.836 735	1	$0.836\,735]^{\mathrm{T}}$
3	4.836 735	[0.795 359	1	$0.795\,359]^{\mathrm{T}}$
$\vdots$				
26	4.707 129	[0.707 129	1	$0.707\,129]^{\mathrm{T}}$

Example 13 Examine the effect on the number of iterations of Example 12 when the origin is shifted by $(\lambda_2 + \lambda_3)/2$ (the optimal shift—see Isaacson and Keller [24]), namely by 3.646 446 6.

The power method is applied to the matrix

$$B = \begin{bmatrix} 0.353\ 553\ 4 & 0.5 & 0 \\ 0.5 & 0.353\ 553\ 4 & 0.5 \\ 0 & 0.5 & 0.353\ 553\ 4 \end{bmatrix}$$

to determine its dominant eigenvalue. The constant 3.646 446 6 is then added to obtain the dominant eigenvalue of A. For the shifted matrix B we have a convergence ratio $|\lambda_2/\lambda_1|$ given by 0.33 (approximately) in comparison with 0.85 (approximately) given in Examples 11 and 12.

The results are given in Table 4.

Table 4

k	ratio	Eigenvector		
1	5	[0.630 602	1	0.630 602]^T
2	4.630 602	[0.734 591	1	0.734 591]^T
3	4.734 591	[0.698 177	1	0.698 177]^T
4	4.698 177	[0.710 109	1	0.710 109]^T
5	4.710 109	[0.706 109	1	0.706 109]^T
6	4.706 109	[0.707 440	1	0.707 440]^T
7	4.707 440	[0.706 996	1	0.706 996]^T
8	4.706 996	[0.707 144	1	0.707 144]^T
9	4.707 144	[0.707 094	1	0.707 094]^T
10	4.707 094	[0.707 111	1	0.707 111]^T

The faster convergence is evident from the table. Note also that the eigenvector has converged nicely in the 10 iterations.

The following interesting observation is noted in the event the matrix A is real and symmetric. In this case, the eigenvalues are real and the eigenvectors $\mathbf{u}_j$ orthonormal.

Forming the sequence (14) and considering the ratios

$$\beta_k = \mathbf{x}_{(k)}^T \mathbf{x}_{(k+1)} / \mathbf{x}_{(k)}^T \mathbf{x}_{(k)} \tag{22}$$

it is not difficult to show using (16) that

$$\beta_k = \lambda_1 \left[1 + 0 \left(\frac{\lambda_2}{\lambda_1} \right)^{2k} \right]$$

so that the $\{\beta_k\}$ converge quadratically to the dominant eignevalue λ_1. The ratio

$$\frac{\mathbf{x}_{(k)}^T \mathbf{x}_{(k+1)}}{\mathbf{x}_{(k)}^T \mathbf{x}_{(k)}} \left(\equiv \frac{\mathbf{x}_{(k)}^T A \mathbf{x}_{(k)}}{\mathbf{x}_{(k)}^T \mathbf{x}_{(k)}} \right)$$

is commonly called the *Rayleigh quotient*. For real symmetric matrices (and only these), the Rayleigh quotient provides a means of accelerating the convergence rate over that given by (17).

Example 14   For the matrix of Example 11 determine the dominant eigenvalue using the Rayleigh quotient. Does the eigenvector converge to the dominant eigenvector in a comparable manner to the dominant eigenvalue approximation?

Using $\mathbf{x}_0 = (1, 1, 1)^T$, the β_k defined by (22) are given in Table 5.

The convergence to the dominant eigenvalue is seen to occur after 11 iterations. However, it takes a further 14 iterations to produce the required eigenvector; normalized it is given as the final row of Table 5.

Table 5

k	β_k	Eigenvector		
1	4.666 67	[4.5	5.0	4.5]^T
2	4.687 023	[2.05E1	2.45E1	2.05E1]^T
3	4.697 206	[9.425E1	1.185E2	9.425E1]^T
4	4.702 244	[4.362 5E2	5.682 5E2	4.362 5E2]^T
11	4.707 074	[2.152 71E7	3.023 95E7	2.152 71E7]^T
25	4.707 107	[5.625 60E17	7.954 03E17	5.624 60E17]^T
		(0.707 112 86	1	0.707 112 86)^T

As is seen from Table 5, the elements of the eigenvector grow as the number of iterations increases. To avoid this, once again the elements of $\mathbf{x}_{(k)}$ should be scaled at each iteration so that $\|\mathbf{x}_{(k)}\|_2 = 1$. With this scaling the iteration is defined by

$$\mathbf{y}_{(k+1)} = A\mathbf{x}_{(k)}$$
$$\beta_k = \mathbf{y}_{(k+1)}^T\mathbf{x}_{(k)} \tag{23}$$
$$\mathbf{x}_{k+1} = \mathbf{y}_{(k+1)}^T/\|\mathbf{y}_{(k+1)}\|_2$$

Having noted the possibility of increasing the rate at which the algorithm converges in the case of symmetric A, the question naturally arises: can we do likewise for general A? The simplest device to improve the convergence of the sequence of approximations to λ_1 is known as the *shift of origin*. This is based on the simple observation that the convergence ratio $|\lambda_2/\lambda_1|$ essentially dictates the behavior of the sequence generated by (14) and if $|\lambda_2/\lambda_1| \simeq 1$ the number of iterations to obtain convergence can be large. If we could make $|\lambda_2/\lambda_1| \simeq 1$ the number of iterations to obtain convergence can be large. If we could make $|\lambda_2/\lambda_1|$ smaller, without changing the ordering of the eigenvalues (11), then this would improve the situation. Clearly the actual value of λ_1,

λ_2, etc., are determined by the particular coordinate system being used. If we change the origin in this manner to facilitate the convergence and then add back the shift of the determined eigenvalue, we should be able to choose the shift to introduce an optimum first term $|\lambda_2/\lambda_1|^{k+1}$ in (17).

Suppose we propose shifting the origin by an amount p. This is equivalent to determining the eigenvalues of the shifted matrix $B = A - pI$ (why?) Then the eigenvalues μ_j of B are related to those (λ_j) of A by

$$\mu_j = \lambda_j - p$$

To determine the dominant eigenvalue λ_1 we clearly must ensure that $\lambda_1 - p$ is still the dominant eigenvalue of B. In determining the eigenvalues of B we see the convergence is now governed by $(\lambda_2 - p)/(\lambda_1 - p)$, where, as we have said, p is to be chosen to make $|(\lambda_2 - p)/(\lambda_1 - p)|$ much smaller than $|\lambda_2/\lambda_1|$.

Unfortunately it is a difficult, if not impossible task to design an automatic procedure for determining the optimum p. In practice an ad hoc approach is taken on the basis that any p which improves the situation is better than no p at all!

The effect of choosing a suitable p is exemplified by the following. Suppose the eigenvalues of a matrix A are

$$10, 11, 12, 13, 14, \ldots, 19$$

The convergence rate for the eigenvalue 19 is essentially $(18/19)^k$, which would produce many iterations for convergence. By shifting the origin by 10 we make the eigenvalues of B into $0, 1, \ldots, 9$. The convergence is governed by $(8/9)^k$. This is an improvement, but we can do even better!

If we shift the origin a further 4, the eigenvalues of B are

$$-4, -3, -2, -1, 0, 1, 2, 3, 4, 5$$

and the convergence is now dictated by $(4/5)^k$.

Clearly the optimal shift for the original set would be $p = 14$. Any other value of p produces a ratio closer to 1 if the eigenvalue corresponding to the original 19 is to remain dominant.

Example 15 Compute the dominant eigenvalue of A in Example 11 using the Rayleigh quotient with a shift $p = 3.646\,446\,6$ and scaled eigenvalues, using (20).

The results are tabulated in Table 6.

The elements in the column β_k are obtained by determining the Rayleigh quotients of

$$B = \begin{bmatrix} 0.353\,553\,34 & 0.5 & 0 \\ 0.5 & 0.353\,553\,34 & 0.5 \\ 0 & 0.5 & 0.353\,553\,34 \end{bmatrix}$$

and then adding $3.646\,446\,6$. The eigenvectors are normalized with respect to the l_2-norm; this also applying to $\mathbf{x}_{(0)}$.

Table 6

k	β_k	Eigenvector		
1	4.666 667	[0.492 799	0.781 474	0.492 799]^{T}
2	4.702 496	[0.539 558	0.734 502	0.539 558]^{T}
3	4.206 593	[0.526 864	0.754 629	0.526 864]^{T}
4	4.707 050	[0.531 443	0.748 396	0.531 443]^{T}
5	4.707 100	[0.529 954	0.750 528	0.529 954]^{T}
6	4.707 106	[0.530 455	0.749 823	0.530 455]^{T}
7	4.707 107	[0.530 288	0.75	0.530 288]^{T}
8	4.707 107	[0.530 344	0.749 98	0.530 344]^{T}
		(0.707 144	1	0.707 144)

The fast convergence of the dominant eigenvalue approximation is apparent as is, in this case, the eigenvector approximation.

In Chapter 9 we described the Aitken δ^2-process to accelerate the convergence of a linearly convergent sequence—see Eq. (21) of Chapter 9. By identifying the ratios $x_{s(k+1)}/x_{s(k)}$ in Eq. (17) of this chapter with $x_{(i)}$ in Eq. (21) of Chapter 9, a faster convergent sequence of approximations to the dominant eigenvalue λ_1 can be generated. The details are left to the reader as an exercise.

Exercises

9. Develop a subroutine or procedure POWER (N, A, X, EPS, MAX, SYMM, EIG) where A is the real $N \times N$ array representing the matrix whose dominant eigenvalue is required; X on input is the initial approximation to the eigenvector and on output is the latest approximation to the dominant eigenvector. EPS is a real parameter which gives the tolerance required for accepting the approximation to the eigenvalue; MAX an integer giving the maximum number of iterations allowed before terminating the iteration. EIG contains on output the latest approximation to the dominant eigenvalue. The parameter SYM is used to indicate whether the matrix A is symmetric or not. If A is symmetric, SYM is to be set to TRUE, otherwise it is to be FALSE.

POWER is to use the first-order method (18) and (19) if SYMM is FALSE and the Rayleigh quotient (23) if SYMM is TRUE.

A test should be included to make sure $\mathbf{x}_{(0)}$ given as input in X is normalized appropriately, and if not X should be normalized.

The iteration is to continue until

$$\left| \frac{\lambda_{(k+1)} - \lambda_{(k)}}{\lambda_{(k)}} \right| \leq \text{EPS} \qquad \lambda_{(k)} \neq 0.$$

or

$$k > \text{MAX}$$

whichever is the sooner. If $\lambda_{(k)} = 0$ then the absolute error tolerance $|\lambda_{(k+1)} - \lambda_{(k)}| \leq$ EPS is to be used.

If convergence fails, NOT SYMM should be the output value of SYMM. If convergence occurs, then SYMM is to retain its input value.

10. Use your subroutine of Exercise 9 to determine the dominant eigenvalue of the following matrix:

$$A = \begin{bmatrix} 5 & 1 & 0 & 0 & 0 \\ 1 & 5 & 1 & 0 & 0 \\ 0 & 1 & 5 & 1 & 0 \\ 0 & 0 & 1 & 5 & 1 \\ 0 & 0 & 0 & 1 & 5 \end{bmatrix}$$

Given that the eigenvalues of this matrix are defined by

$$\lambda_j = 5 + 2\cos{(j\pi/6)} \qquad j = 1, 2, \ldots, 5$$

determine the optimal shift and determine again the dominant eigenvalue. Compare the number of iterations in each case to obtain 4 decimal places of accuracy.

11. Given the eigenvalues in Exercise 10, apply two separate shifts to obtain, successively, two eigenvalues of A.

12. Use your subroutine of Exercise 9 to determine the dominant eigenvalue of:

$$\text{(a)} \quad A = \begin{bmatrix} 7 & 2 & 1 \\ 2 & 4 & 1 \\ 1 & 1 & 2 \end{bmatrix}$$

$$\text{(b)} \quad A = \begin{bmatrix} 0 & 1 & 1 \\ -1 & 0 & 1 \\ -1 & -1 & 0 \end{bmatrix}$$

using in both cases $\mathbf{x}_{(0)} = [1, 1, 1]^T$

13. For the matrix of Exercise 12(a) record the approximations obtained for the dominant eigenvalue and apply Aitken's δ^2-process. How many iterations are required to provide a dominant eigenvalue correct to 4 decimal places?

3 Inverse Iteration

Despite the efforts to speed up the convergence of the power method, the algorithm is frequently slow in practice. The method of this section, in general, converges more rapidly. It turns out that the method of the current section is one of the most powerful methods known for finding a specified eigenvalue.

Suppose

$$A\mathbf{x} = \lambda\mathbf{x} \text{ where } A \text{ is nonsingular}$$

Then

$$A^{-1}A\mathbf{x} = \lambda A^{-1}\mathbf{x}$$

or

$$\frac{1}{\lambda}\mathbf{x} = A^{-1}\mathbf{x}$$

says that $1/\lambda$ is an eigenvalue of A^{-1}.

Consequently, if $\lambda_1, \lambda_2, \ldots, \lambda_N$ are the eigenvalues of A, $1/\lambda_1$, $1/\lambda_2, \ldots, 1/\lambda_N$ are the corresponding eigenvalues of A^{-1}. Then for any arbitrary vector

$$\mathbf{x}_{(0)} = \sum_{j=1}^{N} \alpha_j \mathbf{u}_j \tag{24}$$

where $\{\mathbf{u}_j\}$ are the eigenvectors of A,

$$\mathbf{x}_{(1)} = A^{-1}\mathbf{x}_{(0)} = \sum_{j=1}^{N} \frac{\alpha_j}{\lambda_j} \mathbf{u}_j \tag{25}$$

Consequently,

$$\mathbf{x}_{(k)} = A^{-k}\mathbf{x}_{(0)} = \sum_{j=1}^{N} \frac{\alpha_j}{\lambda_j^k} \mathbf{u}_j \tag{26}$$

Then assuming $\alpha_N \neq 0$, if λ_N is the *smallest* (in modulus) eigenvalue of A, then

$$\frac{\alpha_N}{\lambda_N^k} \mathbf{u}_N$$

will dominate the expression (26).

Thus considering (25) to be simply the power method for A^{-1}, we see that, repeating the discussion of the previous section,

$$\mathbf{x}_{(k)} \rightarrow \alpha \mathbf{u}_N \qquad (\alpha \text{ a constant of proportionality})$$

and

$$\lim_{k \to \infty} \frac{x_{j(k+1)}}{x_{j(k)}} = \frac{1}{\lambda_N}$$

In proposing (26), it goes without saying that we do not form the matrix A^{-k} but, of course, *solve*

$$A\mathbf{x}_{(k+1)} = \mathbf{x}_{(k)}$$

with $\mathbf{x}_{(k+1)}$ suitably normalized in the manner described in section 2.

We may also employ the shift of origin described in section 2 to isolate a particular eigenvalue (e.g. we may wish to determine the subdominant eigenvalue λ_2). Suppose the eigenvalue λ_r is required and p, the shift, is chosen in order that p is much closer to λ_r than to any other eigenvalue. Then

$$(A - pI)\mathbf{x}_{(k+1)} = \mathbf{x}_{(k)}$$

will produce

$$\mathbf{x}_{(k+1)} = \sum_{j=1}^{N} \alpha_j (\lambda_j - p)^{-k} \mathbf{u}_j$$

which will be parallel to $\mathbf{u}_r$, and

$$\lim_{k \to \infty} \frac{x_{j(k+1)}}{x_{j(k)}} = \frac{1}{(\lambda_r - p)} \qquad j = 1, 2, \ldots, N. \tag{27}$$

As an alternative to the ratio given in (27), it is possible to use the ratio

$$\beta_k = \mathbf{x}_{(k+1)}^{\mathrm{T}}\mathbf{x}_{(k)}/\mathbf{x}_{(k)}^{\mathrm{T}}\mathbf{x}_{(k)} \qquad (28)$$

as was used in section 2, whence

$$\beta_k \text{ converges to } \frac{1}{(\lambda_r - p)}$$

or more specifically

$$\lim_{k \to \infty} \frac{1}{\beta_k} + p = \lambda_r$$

Example 16 Determine the maximum eigenvalue of A of Example 11 using the inverse power method using the ratios given by (27) and the β_k given by (28).

The results of Examples 11, 12, 13 suggest that $p = 4.7$ would be a reasonable shift of origin to translate λ_1 into the smallest eigenvalue of $A - pI$ and hence so that $1/\lambda_1$ is the maximum eigenvalue of $(A - pI)^{-1}$.

Forming $A = pI$ we have, say,

$$B = \begin{bmatrix} -0.7 & 0.5 & 0 \\ 0.5 & -0.7 & 0.5 \\ 0 & 0.5 & -0.7 \end{bmatrix}$$

We compute the LU factorization of B initially (and store the factors). With $\mathbf{x}_{(0)} = (1, 1, 1)^{\mathrm{T}}$ the sequence of iterates of values of the maximum value of (27) and β_k given by (28) are reported in Table 7. We also give the corresponding values of the approximate eigenvalue.

Table 7

k	$(27)_{\max}$	λ_1 (approx.)	β_k	λ_1 (approx.)
1	170.000 084	4.705 882	136.666 735	4.707 317
2	140.833 406	4.707 101	140.710 644	4.707 107
3	140.711 369	4.707 107	140.710 751	4.707 107

The (normalized) eigenvector determined at iteration 3 was found to be

$$(0.707\,106\,75, 1, 0.707\,106\,75)^{\mathrm{T}}$$

It is clear that the result has been obtained very quickly starting from $(1, 1, 1)^{\mathrm{T}}$ having chosen p appropriately. The reader should appreciate that the convergence will depend on the difference $\lambda_1 - p$, so that the better is the approximation initially, the better will be the convergence.

Exercises

14. Develop a subroutine or procedure INVIT (N, ALU, X, EPS, MAX, EIG) which is to implement the method of inverse iteration described in this section. On input ALU contains the LU-factored matrix with the lower triangle containing the multipliers (you may assume for this example that an actual physical interchange has taken place, if necessary). The integer N indicates the dimension of the problem. The N-vector X contains the initial vector $x_{(0)}$ which is to be normalized on entry to the subroutine. INVIT is to iterate using inverse iteration until

$$\left|\frac{\lambda_{(k+1)} - \lambda_{(k)}}{\lambda_{(k)}}\right| \le \text{EPS} \qquad (\lambda_{(k)} \ne 0)$$

or

$$k > \text{MAX}$$

whichever is the sooner. If $\lambda_{(k)} = 0$ then the absolute error tolerance $|\lambda_{(k+1)} - \lambda_{(k)}| <$ EPS is to be used.

If convergence fails a message should be output before leaving the subroutine.

At each iteration calculate β_k given by (28) and normalize the approximation to the eigenvector using the l_2-norm.

On exit the most recently obtained approximation to the eigenvalue should be stored in EIG and the corresponding eigenvector in X.

15. The eigenvalues of the matrix

$$A = \begin{bmatrix} 5 & 1 & 0 & 0 & 0 \\ 1 & 5 & 1 & 0 & 0 \\ 0 & 1 & 5 & 1 & 0 \\ 0 & 0 & 1 & 5 & 1 \\ 0 & 0 & 0 & 1 & 5 \end{bmatrix}$$

are to be determined by inverse iteration using in turn shifts of origin

$$p = 3.3, 6.7, 5.3, 4.05, 6.05$$

Obtain the LU factorization using a routine of chapter 3 (see page 58) for each matrix and using the routine INVIT obtain the required eigenvalues correct to 4 decimal places.

16. A modification to the algorithm employing a single shift to determine a particular eigenvalue is the following. Since the initial shift may be considered to have come from some (other) initial approximation, the new approximation, if it is a better one than the initial approximation, can be assumed to provide a more appropriate shift.

Using the value of EIG at each iteration, modify your main program of Exercise 12 to allow a redefinition and factorization of the matrix in each iteration. Obtain, if possible, the eigenvalues of the matrix in Exercise 15. How does the number of iterations compare, for each shift given in Exercise 15, using this approach as opposed to that of Exercise 15?

Note: this iteration (for symmetric matrices) is known as the *Rayleigh quotient iteration*.

17. For the matrices in Exercise 7, employ your program written there to produce convenient intervals each containing a single eigenvalue which allow the definition of a suitable shift of origin. Using each of the shifts determine the eigenvalues of the matrices in Exercise 7 correct to three significant figures.

We have discussed the inverse iteration method for full matrices. As a result of applying Householder transformations, the matrix will be frequently tridiagonal or upper Hessenberg. In this case the method can be implemented more efficiently taking into account the structure of A.

4 The Shifted QR Algorithm for Symmetric Tridiagonal Matrices

The Householder transformations of section 1 when applied as advocated reduce a symmetric matrix to tridiagonal form. Suppose the real $(N \times N)$ matrix A is transformed to tridiagonal T. The QR algorithm then uses this matrix T as the starting point for determining the eigenvalues of A by constructing a sequence of matrices $\{T_{(i)}\}$, $i = 0, 1, 2, \ldots$ (say), which ultimately converges to a bidiagonal matrix whose diagonal elements constitute the required eigenvalues of A. We will describe the algorithm superficially first and then return to fill in the computational details.

Let T be denoted as $T_{(0)}$. For a given suitable shift of origin $p_{(0)}$, form the orthogonal decomposition of

$$T_{(0)} - p_{(0)}I$$

using a sequence of rotation matrices. Thus if in annihilating the subdiagonal element in position $(k + 1, k)$ the rotation matrix

$$Q_{k,k+1} \ (\equiv Q(k, k+1)\text{—see Example 8})$$

is constructed, then

$$R_{(1)} = \prod_{k=1}^{N-1} Q_{k,k+1}(T_{(0)} - p_{(0)}I) = Q_{(1)}^{\mathrm{T}}(T_{(0)} - p_{(0)}I)$$

is upper bidiagonal.

So

$$T_{(0)} - p_{(0)}I = Q_{(1)}R_{(1)}$$

is the orthogonal decomposition of $T_{(0)} - p_{(0)}I$.

With this decomposition, we define

$$T_{(1)} = R_{(1)}Q_{(1)} + p_{(0)}I$$

It is simple to see that this commuted product of $Q_{(1)}$ and $R_{(1)}$ produces (another) tridiagonal matrix. This in turn is decomposed using rotational transformations so that at step s we have for suitable shift $p_{(s-1)}$: decompose

$$T_{(s-1)} - p_{(s-1)}I = Q_{(s)}R_{(s)}$$

construct

$$T_{(s)} = R_{(s)}Q_{(s)} + p_{(s-1)}I$$

with elements $t_{ij}^{(s)}$.

It is simple to see that $T_{(s)}$ is similar to $T_{(s-1)}$ (and hence inductively to $T_{(0)}$), since

$$R_{(s)} = (T_{(s)} - p_{(s-1)}I)Q_{(s)}^{-1}$$

$$T_{(s-1)} - p_{(s-1)}I = Q_{(s)}(T_{(s)} - p_{(s-1)}I)Q_{(s)}^{-1}$$

and so $T_{(s-1)} = Q_{(s)}T_{(s)}Q_{(s)}^{-1}$.

Thus this iteration defines a sequence of similar matrices. Some comments are in order at this stage:

(i) The choice of similarity transformations based on rotational matrices is sensible since at each annihilation only one subdiagonal element is to be reduced to zero.

(ii) With a suitable choice of shift $p_{(s)}$ the subdiagonal element in the final row of $T_{(s)}$ rapidly approaches zero. In this case the (N, N) element is an approximation to the (largest in absolute value) eigenvalue of $T_{(0)}$ (and hence A).

(iii) If at any stage other subdiagonal elements are reduced to zero, the resulting matrix can conveniently be partitioned into submatrices each of which can be treated separately in a manner analogous to that currently being described.

(iv) We will not go into the details of the proof of the convergence here—the interested reader is referred to Gourlay and Watson [18].

(v) After applying the similarity transformations described above we will obtain a tridiagonal matrix of the following form.

$$\begin{bmatrix} \times & \times & & & & & \vert & \times \\ \times & \times & \times & & & & \vert & \times \\ & \times & \times & \times & & & \vert & \times \\ & & & \ddots & & & \vert & \vdots \\ & & & & \times & \times & \vert & \times \\ \hline 0 & \cdots & & & & 0 & \vert & \lambda \end{bmatrix}$$

(the $(N, N-1)$ element of course will not be identically zero but to a close approximation can be replaced by zero—see (viii) below).

Consequently a new sequence of iterations can be initiated for the leading principal $(N-1) \times (N-1)$ submatrix.

The number λ is an approximate eigenvalue of A.

(vi) Until the subdiagonal element becomes (acceptably) zero, at each iteration the shift $p_{(s)}$ is chosen to be the value of the eigenvalue of

$$\begin{bmatrix} t_{N-1N-1} & t_{N-1N} \\ t_{NN-1} & t_{NN} \end{bmatrix} \quad \text{closer to } t_{NN}.$$

(vii) At the stage in the algorithm where the matrix being considered has been *deflated* to an $(N-k) \times (N-k)$ matrix (say), the amount of work necessary to form and manipulate the rotation matrices is clearly considerably less as k increases. Further, the appropriate shift $p_{(s)}$ is always chosen as the value of the eigenvalue of the corresponding 2×2 matrix given in (vi).

(viii) Because of the possible size of the eigenvalues it is important not to assume off-diagonal elements are zero (and are therefore replaced by zero) too soon. There are several criteria for this; perhaps the safest is to accept at the s-iteration that

$$t_{k-1,k}^{(s)} \text{ is zero if } t_{k-1,k}^{(s)} \leq \varepsilon \ \min \{|t_{k-1,k-1}^{(s)}|, |t_{k,k}^{(s)}|\}$$

where ε is typically $10^{-\beta}$ where there are β significant digits in the calculation.

Let us now pay some attention to the implementation of the QR algorithm. At each iteration it would appear at first sight that we would need to form and store the unitary matrices $Q_{(s)}$. This would be expensive in time and storage. However, considering the effect of premultiplication and postmultiplication by the rotation matrices, we see that the rotation matrix $Q_{k,k+1}$ affects only rows k and $k+1$ on premultiplication, and columns k and $k+1$ on postmultiplication. Consequently, when T has been premultiplied by both $Q_{1,2}$ and $Q_{2,3}$, further premultiplications by successive rotational matrices will not affect columns 1 and 2. So after premultiplication by $Q_{2,3}$ we can arrange to postmultiply by $Q_{1,2}^T$. Following this postmultiplication, we premultiply by $Q_{3,4}$ and then postmultiply by $Q_{2,3}$, etc.

We therefore form the following sequence:

$$\tilde{T}_{(0)} = T_{(0)} - P_{(0)}I$$

$$Q_{2,3}Q_{1,2}\tilde{T}_{(0)} \to Q_{2,3}Q_{1,2}\tilde{T}_{(0)}Q_{1,2}^T \to Q_{3,4}Q_{2,3}Q_{1,2}\tilde{T}_{(0)}Q_{1,2}^T$$

$$\to \quad \cdots \quad \to Q_{N-1,N} \cdots Q_{1,2}\tilde{T}_{(0)}Q_{1,2}^T \cdots Q_{N-3,N-2}^T$$

$$\to Q_{(1)}\tilde{T}_{(0)}Q_{1,2}^T \cdots Q_{N-2,N-1}^T$$

$$\to Q_{(1)}\tilde{T}_{(0)}Q_{1,2}^T \cdots Q_{N-2,N-1}^T Q_{N-1,N}^T = Q_{(1)}\tilde{T}_{(0)}Q_{(1)}^T$$

Then

$$T_{(1)} = Q_{(1)}\tilde{T}_{(0)}Q_{(1)}^T + p_{(0)}I$$

Example 17 Consider the first "iteration" of the QR method to obtain $T_{(1)}$ for the matrix

$$T_{(0)} = \begin{bmatrix} 5.000 & 1.000 & 0 & 0 & 0 \\ 1.000 & 5.000 & 1.000 & 0 & 0 \\ 0 & 1.000 & 5.000 & 1.000 & 0 \\ 0 & 0 & 1.000 & 5.000 & 1.000 \\ 0 & 0 & 0 & 1.000 & 5.000 \end{bmatrix}$$

The initial shift $p_{(0)}$ is chosen to be the eigenvalue of

$$\begin{bmatrix} 5 & 1 \\ 1 & 5 \end{bmatrix}$$

nearer to the (5, 5) element. Here the eigenvalues are 4, 6, so we choose arbitrarily $p_{(0)} = 6$. Subtracting this from the diagonal elements of $T_{(0)}$ to define $\tilde{T}_{(0)}$, we then determine the rotation matrix $Q_{1,2}$, namely having $\sin \theta = \sqrt{2}$ and $\cos \theta = -\sqrt{2}$, and premultiply $\tilde{T}_{(0)}$ by it, storing the values of $\sin \theta$ and $\cos \theta$. Then

$$
Q_{1,2}\tilde{T}_{(0)} =
\begin{bmatrix}
1.414 & -1.414 & 0.707 & 0 & 0 \\
0 & 0 & -0.707 & 0 & 0 \\
0 & 1.000 & -1.000 & 1.000 & 0 \\
0 & 0 & 1.000 & -1.000 & 1.000 \\
0 & 0 & 0 & 1.000 & -1.000
\end{bmatrix}
$$

We report just three decimal places here to conserve space. The actual elements were calculated on a Honeywell 66/60, which in single precision has about 8 decimal places of accuracy. Continuing as outlined above we construct:

$$Q_{2,3}Q_{1,2}\tilde{T}_{(0)} \quad (\sin \theta = 1; \cos \theta = 0)$$

$$
\begin{bmatrix}
1.414 & -1.414 & 0.707 & 0 & 0 \\
0 & 1.000 & -1.000 & 1.000 & 0 \\
0 & 0 & 0.707 & 0 & 0 \\
0 & 0 & 1.000 & -1.000 & 1.000 \\
0 & 0 & 0 & 1.000 & -1.000
\end{bmatrix}
$$

$$Q_{2,3}Q_{1,2}\tilde{T}_{(0)}Q_{1,2}^{T}$$

$$
\begin{bmatrix}
-2.000 & 0.000 & 0.707 & 0 & 0 \\
0.707 & -0.707 & -1.000 & 1.000 & 0 \\
0 & 0 & 0.707 & 0 & 0 \\
0 & 0 & 1.000 & -1.000 & 1.000 \\
0 & 0 & 0 & 1.000 & -1.000
\end{bmatrix}
$$

$$Q_{3,4}Q_{2,3}Q_{1,2}\tilde{T}_{(0)}Q_{1,2}^{T}$$

$$(\sin \theta = 0.816\,496\,58$$
$$\cos \theta = 0.577\,350\,26)$$

$$
\begin{bmatrix}
-2.000 & 0.000 & 0.707 & 0 & 0 \\
0.707 & -0.707 & -1.000 & 1.000 & 0 \\
0 & 0 & 1.225 & -0.816 & 0.816 \\
0 & 0 & 0 & -0.577 & 0.577 \\
0 & 0 & 0 & 1.000 & -1.000
\end{bmatrix}
$$

$$Q_{3,4}Q_{2,3}Q_{1,2}\tilde{T}_{(0)}Q_{1,2}^{T}Q_{2,3}^{T}$$

$$\begin{bmatrix} -2.000 & 0.707 & -0.000 & 0 & 0 \\ 0.707 & -1.000 & 0.707 & 1.000 & 0 \\ 0 & 1.225 & 0 & -0.816 & 0.816 \\ 0 & 0 & 0 & -0.577 & 0.577 \\ 0 & 0 & 0 & 1.000 & -1.000 \end{bmatrix}$$

$$Q_{4,5}Q_{3,4}Q_{2,3}Q_{1,2}\tilde{T}_{(0)}Q_{1,2}^{T}Q_{2,3}^{T}$$

$(\sin \theta = 0.866\,025\,40$
$\cos \theta = 0.5)$

$$\begin{bmatrix} -2.000 & 0.707 & -0.000 & 0 & 0 \\ 0.707 & -1.000 & 0.707 & 1.000 & 0 \\ 0 & 1.225 & 0 & -0.816 & 0.816 \\ 0 & 0 & 0 & 1.155 & -1.155 \\ 0 & 0 & 0 & -0.000 & 0.000 \end{bmatrix}$$

$$Q_{4,5}Q_{3,4}Q_{2,3}Q_{1,2}\tilde{T}_{(0)}Q_{1,2}^{T}Q_{2,3}^{T}Q_{3,4}^{T}$$

$$\begin{bmatrix} -2.000 & 0.707 & -0.000 & 0.000 & 0 \\ 0.707 & -1.000 & 1.225 & -0.000 & 0 \\ 0 & 1.225 & -0.667 & -0.471 & 0.816 \\ 0 & 0 & 0.943 & 0.667 & -1.155 \\ 0 & 0 & 0.000 & 0.000 & 0.000 \end{bmatrix}$$

$$Q_{(1)}\tilde{T}_{(0)}Q_{(1)}^{T}$$

$$\begin{bmatrix} -2.000 & 0.707 & -0.000 & -0.000 & -0.000 \\ 0.707 & -1.000 & 1.225 & 0.000 & 0.000 \\ 0 & 1.225 & -0.667 & 0.943 & -0.000 \\ 0 & 0 & 0.943 & -1.333 & 0.000 \\ 0 & 0 & -0.000 & 0.000 & -0.000 \end{bmatrix}$$

$$T_{(1)}$$

$$\begin{bmatrix} 4.000 & 0.707 & -0.000 & -0.000 & -0.000 \\ 0.707 & 5.000 & 1.225 & 0.000 & 0.000 \\ 0 & 1.225 & 5.333 & 0.943 & -0.000 \\ 0 & 0 & 0.943 & 4.667 & 0.000 \\ 0 & 0 & -0.000 & 0.000 & 6.000 \end{bmatrix}$$

The sequence of pre- and postmultiplications indicate that the tridiagonal structure of the matrix is preserved. In this example we see that in just one iteration we have, in $T_{(1)}$, isolated one eigenvalue $\lambda = 6$, the last column and row being zero. The second iteration progresses with the principal 4×4 matrix of $T_{(1)}$ using a shift $p_{(1)} = 4.000$, being the eigenvalue of

$$\begin{bmatrix} 5.333 & 0.943 \\ 0.943 & 4.667 \end{bmatrix}$$

closer to 4.667.

Example 18 For the matrix of Example 17 continue the QR iteration to determine the remaining eigenvalues.

We continue the iteration from the matrix last reported in Example 17 using at the second stage the 4×4 principal submatrix given there. We report below the successive matrices obtained and the shifts $p_{(2)}$ used at each stage.

$$T_{(2)}; p_{(1)} = 4$$

$$\begin{bmatrix} 5.000 & 1.414 & -0.000 & -0.000 & -0.000 \\ 1.414 & 5.000 & 1.000 & -0.000 & 0.000 \\ -0.000 & 1.000 & 5.000 & -0.000 & -0.000 \\ -0.000 & -0.000 & -0.000 & 4.000 & 0.000 \\ 0 & 0 & -0.000 & 0.000 & 6.000 \end{bmatrix}$$

$$T_{(3)}; p_{(2)} = 6$$

$$\begin{bmatrix} 3.667 & 0.943 & 0.000 & -0.000 & -0.000 \\ 0.943 & 5.833 & 0.866 & -0.000 & 0.000 \\ -0.000 & 0.866 & 5.500 & -0.000 & -0.000 \\ -0.000 & -0.000 & -0.000 & 4.000 & 0.000 \\ 0 & 0 & -0.000 & 0.000 & 6.000 \end{bmatrix}$$

$$T_{(3)}; p_{(2)} = 4.784\ 749\ 51$$

$$\begin{bmatrix} 3.638 & 1.066 & 0.000 & -0.000 & -0.000 \\ 1.066 & 6.354 & -0.138 & -0.000 & 0.000 \\ -0.000 & -0.138 & 5.009 & -0.000 & -0.000 \\ -0.000 & -0.000 & -0.000 & 4.000 & 0.000 \\ 0 & 0 & -0.000 & 0.000 & 6.000 \end{bmatrix}$$

$$T_{(3)}; p_{(2)} = 4.994\ 719\ 74$$

$$\begin{bmatrix} 3.639 & 1.071 & 0.000 & -0.000 & -0.000 \\ 1.071 & 6.361 & 0.000 & -0.000 & 0.000 \\ -0.000 & 0.000 & 5.000 & -0.000 & -0.000 \\ -0.000 & -0.000 & -0.000 & 4.000 & 0.000 \\ 0 & 0 & -0.000 & 0.000 & 6.000 \end{bmatrix}$$

For $T_{(3)}$ we see that three iterations were required to reduce the off-diagonal elements in the third row to zero. The final matrix shown has the principal 2×2 matrix whose eigenvalues can be calculated directly. So, we have the five eigenvalues: 3.27, 4, 5, 6, 6.73 correct to three significant figures.

The theoretical justification of this algorithm is beyond our scope here. The interested reader is referred to Gourlay and Watson [18] or to the original paper of Wilkinson [50]. The important point in passing is that, with the shift strategy described here, the QR method has a *cubic* rate of convergence. Consequently, few iterations will be required to produce the eigenvalues of the symmetric tridiagonal matrix. We leave the reader to be convinced of this by attempting the exercises at the end of this section.

If the eigenvectors are required for a particular eigenvalue then application of the inverse iteration method described in section 2 will produce the required eigenvector.

Exercises

18. At any stage in the QR method rotations are applied to an $M \times M$ (say) tridiagonal matrix A to reduce the off-diagonal elements in the Mth row to zero. As the iterations progress the value of M, and hence the size of A, is reduced. In FORTRAN, to provide the ability of treating an array of differing size in a subroutine it is required to specify the dimension of the array as declared in the calling program. Let this be denoted MDIM (in ALGOL this facit is not required). Develop a subroutine (or procedure) QR (MDIM, M, A) which carries out a single iteration of the QR method on the tridiagonal $M \times M$ matrix A. The transformed matrix is to be returned in A. (Use the subroutine or procedure you developed in Exercise 4 of this chapter.)
19. Using QR from Exercise 18, write a program which will determine all the eigenvalues of a given tridiagonal matrix T. The main program should call QR successively so that the off-diagonal elements of the final row under consideration are reduced so that condition (viii) is satisfied with ε read as data.
20. Use your program of Exercise 19 to determine all the eigenvalues of the following matrices. Comment on the number of iterations.

$$(i) \quad \begin{bmatrix} 4 & 1 & & & \\ 1 & 4 & 1 & & \\ & 1 & 4 & 1 & \\ & & 1 & 4 & 1 \\ & & & 1 & 4 \end{bmatrix} \qquad (ii) \quad \begin{bmatrix} 5 & 1 & & & \\ 1 & 5 & 2 & & \\ & 2 & 5 & 3 & \\ & & 3 & 5 & 4 \\ & & & 4 & 5 \end{bmatrix}$$

21. For the matrices in Exercise 20, using the eigenvalues found determine the corresponding eigenvectors using inverse iteration (subroutine INVIT, Exercise 14).

5 QR for Hessenberg Matrices—the Implicitly Shifted Method

In the previous section we described the QR method for the symmetric tridiagonal matrix. If the matrix is unsymmetric but has real eigenvalues, no modification to our discussion is necessary. Should the matrix under consideration be upper Hessenberg but with real eigenvalues, then the algorithm described needs little modification. In the present section we require to address the question of complex eigenvalues—whether the given matrix be unsymmetric tridiagonal or Hessenberg. Since our original matrix A (before similarity transformations were applied to reduce the matrix to its current form) had real elements, we require to continue using real arithmetic even where the eigenvalues may be complex. Consequently we do not allow for the implementation of complex shifts $p_{(s)}$. This may sound a contradiction in terms where we know in the previous section that the shifts converged to the real eigenvalues. The fact that the QR algorithm can be so modified to allow for the calculation of complex eigenvalues without resorting to complex arithmetic is due to an ingenious observation by Francis [14].

Consider *formulating* two steps of the shifted QR method with shifts $p_{(s)}^{(1)}$ and $p_{(s)}^{(2)}$, where we assume $p_{(s)}^{(1)}$ and $p_{(s)}^{(2)}$ are complex conjugates. Suppose the initial Hessenberg matrix is denoted $H_{(1)}$. At "stage" s, the matrix $H_{2(s+1)}$ is proposed by the equations

$$H_{(2s)} - p_{(s)}^{(1)} = Q_{(2s)}R_{(2s)}$$

$$H_{(2s+1)} - p_{(s)}^{(1)} = R_{(2s)}Q_{(2s)}$$

$$H_{(2s+1)} - p_{(s)}^{(2)} = Q_{(2s+1)}R_{(2s+1)}$$

$$H_{(2s+2)} - p_{(s)}^{(2)} = R_{(2s+1)}Q_{(2s+1)}$$

(29)

where $Q_{(j)}$ are orthogonal and $R_{(j)}$ upper triangular. The shifts $p_{(s)}^{(1)}$ and $p_{(s)}^{(2)}$ are the eigenvalues of the bottom right-hand corner of $H_{(2s)}$.

Through direct substitution it is a simple matter to show that

$$Q_{(2s)}Q_{(2s+1)}R_{(2s+1)}R_{(2s)} = (H_{(2s)} - p_{(s)}^{(1)}I)(H_{(2s)} - p_{(s)}^{(2)}I)$$

(30)

and

$$H_{(2s+2)} = (Q_{(2s)}Q_{(2s+1)})^{\mathrm{T}}H_{(2s)}(Q_{(2s)}Q_{(2s+1)})$$

(31)

Equation (30) states that

$$QR = Q_{(2s)}Q_{(2s+1)}R_{(2s+1)}R_{(2s)}$$

(32)

is a real orthogonal decomposition of the right-hand side of (30) ($p_{(s)}^{(1)} + p_{(s)}^{(1)}$ and $p_{(s)}^{(1)}p_{(s)}^{(2)}$ are both real) and Eq. (31) states that $H_{(2s+2)}$ is orthogonally similar to $H_{(2s)}$.

Consequently, if we can determine Q and R in (32) without resorting to complex arithmetic, the stated objective will then be obtained. The algorithm to determine Q and R is dependent on the following theorem.

Theorem 7 *Let B, A, and Q be real $N \times N$ matrices. Let B be an upper Hessenberg matrix with all its subdiagonal elements nonzero. Let Q be the orthogonal matrix for which*

$$B = Q^{\mathrm{T}}AQ$$

Then, given the first column of Q, the remaining columns of Q and the elements of B are uniquely determined.

Proof. Suppose the first k columns of Q, denoted $\mathbf{q}_1, \mathbf{q}_2, \ldots, \mathbf{q}_k$, have been determined and the first $(k-1)$ columns denoted $\mathbf{b}_1, \mathbf{b}_2, \ldots, \mathbf{b}_{k-1}$ of B have been determined.

Since

$$B = Q^T A Q$$

and Q is orthogonal, we may note that the kth column $\mathbf{b}_k$ of B satisfies

$$Q\mathbf{b}_k = A\mathbf{q}_k$$

But B is upper Hessenberg so only the first $(k + 1)$ elements of $\mathbf{b}_k$ are nonzero. Hence

$$b_{1k}\mathbf{q}_1 + b_{2k}\mathbf{q}_2 + \cdots + b_{kk}\mathbf{q}_k + b_{k+1k}\mathbf{q}_{k+1} = A\mathbf{q}_k$$

where b_{jk} are the elements of $\mathbf{b}_k$. But the columns of Q are orthogonal as

$$b_{jk}\mathbf{q}_j^T\mathbf{q}_j = \mathbf{q}_j^T A\mathbf{q}_j$$

and since $\mathbf{q}_j^T\mathbf{q}_j = 1$, we have immediately

$$b_{jk} = \mathbf{q}_j^T A\mathbf{q}_j \qquad j = 1, 2, \ldots, k$$

Then

$$b_{k+1,k}\mathbf{q}_{k+1} = A\mathbf{q}_k - \sum_{j=1}^{k} b_{jk}\mathbf{q}_j$$

or

$$\mathbf{q}_{k+1} = \frac{1}{b_{k+1,k}}\left\{A\mathbf{q}_k - \sum_{j=1}^{k} b_{jk}\mathbf{q}_j\right\}$$

where $b_{k+1,k}$ is chosen so that $\mathbf{q}_{k+1}^T\mathbf{q}_{k+1} = 1$.

Consider Eq. (30) for $s = 0$, namely

$$QR = Q_{(0)}Q_{(1)}R_{(1)}R_{(0)} = (H_{(0)} - p_{(0)}^{(1)}I)(H_{(0)} - p_{(0)}^{(2)}I) \tag{33}$$

Since $H_{(0)}$ is upper Hessenberg, only the first three elements in the first column of $(H_{(0)} - p_{(0)}^{(1)}I)(H_{(0)} - p_{(0)}^{(2)}I)$ are nonzero. Furthermore, if we represent Q as $(\mathbf{q}_1, \mathbf{q}_2, \ldots, \mathbf{q}_N)$ and assume the elements of R are represented by r_{ij}, then

$$QR\mathbf{e}_1 = r_{11}\mathbf{q}_1 = (H_{(0)} - p_{(0)}^{(1)}I)(H_{(0)} - p_{(0)}^{(2)}I)\mathbf{e}_1$$

so that $\mathbf{q}_1$ is a multiple of the first column of

$$(H_{(0)} - p_{(0)}^{(1)}I)(H_{(0)} - p_{(0)}^{(2)}I)$$

Now, if we choose an orthogonal matrix U which annihilates all but the first element in the first column of $(H_{(0)} - p_{(0)}^{(1)}I)(H_{(0)} - p_{(0)}^{(2)}I)$ so that

$$U(H_{(0)} - p_{(0)}^{(1)}I)(H_{(0)} - p_{(0)}^{(2)}I)\mathbf{e}_1 = \alpha\mathbf{e}_1 \tag{34}$$

then direct computation reveals that

$$U\mathbf{e}_1 = \frac{1}{\alpha}(H_{(0)} - p_{(0)}^{(1)}I)(H_{(0)} - p_{(0)}^{(2)}I)\mathbf{e}_1$$

But since $\mathbf{q}_1$ is orthogonal $r_{11} = \pm\alpha$ so the first column of U^T, so defined by (34), is the same, up to the sign, as the first column of Q. To determine the matrix U we note that the nonzero elements in the first column of

$$(H_{(0)} - p_{(0)}^{(1)}I)(H_{(0)} - p_{(0)}^{(2)}I)$$

are given by

$$
\left.
\begin{array}{lll}
(1,1) & \text{element:} & h_{11}^2 - (p_{(0)}^{(1)} + p_{(0)}^{(2)})h_{11} + p_{(0)}^{(1)}p_{(0)}^{(2)} + h_{12}h_{21} \\[2mm]
(2,1) & \text{element:} & h_{21}[h_{11} + h_{22} - (p_{(0)}^{(1)} + p_{(0)}^{(2)})] \\[2mm]
(3,1) & \text{element:} & h_{21}h_{32}
\end{array}
\right\}
\tag{35}
$$

where we have denoted the elements of $H_{(0)}$ by h_{ij}. The shifts are chosen as the eigenvalues of

$$
\begin{bmatrix}
h_{N-1,N-1} & h_{N-1,N} \\
h_{N,N-1} & h_{N,N}
\end{bmatrix}
$$

Evaluating the characteristic polynomial of this matrix, we obtain

$$p_{(0)}^{(1)} + p_{(0)}^{(2)} = h_{N-1,N-1} + h_{N,N}$$

$$p_{(0)}^{(1)}p_{(0)}^{(2)} = h_{N-1,N-1}h_{N,N} - h_{N,N-1}h_{N-1,N}$$

which can be evaluated in terms of the (real) elements of $H_{(0)}$ and substituted into Eq. (35). Hence the orthogonal matrix U can be formed to annihilate the (2, 1) and (3, 1) elements in Eq. (35) and hence the similarity transformation $UH_{(0)}U^T$ formed.

The matrix UAU^T must then be reduced to Hessenberg form by the application of elementary reflectors—this to preserve, in the final transformation, the same first column $\mathbf{q}_1$. If we denote the reflectors U_j, $j = 1, 2, \ldots, N-2$, we have

$$H_{(2)} = U_{N-2}U_{N-3} \cdots U_2U_1UH_{(0)}U^TU_1^TU_2^T \cdots U_{N-3}^TU_{N-2}^T$$

since by Theorem 7

$$Q_0 = Q_{(0)}Q_{(1)} = (U_{N-2}U_{N-1} \cdots U_2U_1U)^T$$

Example 19 Apply the double implicit shifted QR method to the matrix

$$
H_{(0)} =
\begin{bmatrix}
2 & -1 & 0 & 0 \\
1 & 2 & -1 & 0 \\
0 & 1 & 2 & -1 \\
0 & 0 & 1 & 2
\end{bmatrix}
$$

From Eqs. (35), the first column of $(H_{(0)} - p_{(0)}^{(1)}I)(H_{(0)} - p_{(0)}^{(2)}I)$ is (0, 0, 1, 0).

Consequently the elementary reflector which reduces this column to a multiple of $\mathbf{e}_1$, using definitions in section 1, is found to be

$$U = \begin{bmatrix} 0 & 0 & -1.000 & 0 \\ 0 & 1.000 & 0 & 0 \\ -1.000 & 0 & 0 & 0 \\ 0 & 0 & 0 & 1.000 \end{bmatrix}$$

Then

$$UAU = \begin{bmatrix} 2.000 & -1.000 & 0 & 1.000 \\ 1.000 & 2.000 & -1.000 & 0 \\ 0 & 1.000 & 2.000 & 0 \\ -1.000 & 0 & 0 & 2.000 \end{bmatrix}$$

The process of reducing this matrix to upper Hessenberg form is now applied as described in section 1. We find, on continuing, that

$$U_1 = \begin{bmatrix} 1.000 & 0 & 0 & 0 \\ 0 & -0.707 & 0 & 0.707 \\ 0 & 0 & 1.000 & 0 \\ 0 & 0.707 & 0 & 0.707 \end{bmatrix}$$

is the reflector which annihilates the $(3, 1)$ and $(4, 1)$ elements in UAU. From this,

$$U_1 UAUU_1 = \begin{bmatrix} 2.000 & 1.414 & 0 & 0 \\ -1.414 & 2.000 & 0.707 & -0.000 \\ 0 & -0.707 & 2.000 & 0.707 \\ 0 & -0.000 & -0.707 & 2.000 \end{bmatrix}$$

which is already upper Hessenberg (this example is somewhat contrived; for a 4×4 matrix we would in general have to proceed to calculate U_2 and form $U_2 U_1 UAUU_1 U_2$ for the upper Hessenberg matrix $H_{(2)}$). This represents the completion of the *first* iteration. Using $H_{(2)}$ we then repeat the process outlined above to calculate $H_{(4)}$. It is found that

$$H_{(4)} = U_1 UH_{(2)} UU_1 = \begin{bmatrix} 2.000 & 1.591 & -0.000 & 0.000 \\ -1.591 & 2.000 & -0.273 & 0.000 \\ 0.000 & 0.273 & 2.000 & 0.629 \\ -0.000 & 0.000 & -0.629 & 2.000 \end{bmatrix}$$

where once again it is found unnecessary to construct U_2. (The reflectors are, of course, represented generically as U_j; they are different for each iteration.) Continuing, it is found that the sequence $H_{(2s)}$ very soon settles down to the matrix†

$$\begin{bmatrix} 2.000 & 1.618 & 0.000 & -0.000 \\ -1.618 & 2.000 & 0.001 & 0.000 \\ 0.000 & -0.001 & 2.000 & 0.618 \\ 0.000 & -0.000 & -0.618 & 2.000 \end{bmatrix}$$

The (3, 2) element is converging quickly to zero, at which point the matrix is split into the upper leftmost (2×2) matrix and the lower rightmost (2×2) matrix. The eigenvalues of the original matrix $H_{(0)}$ are then determined immediately as the roots of these two (2×2) matrices. From the matrix

$$\begin{bmatrix} 2.000 & 0.618 \\ -0.618 & 2.000 \end{bmatrix}$$

the eigenvalues are found to be $2 \pm 2i$ (0.309) and the eigenvalue of

$$\begin{bmatrix} 2.000 & 1.618 \\ -1.168 & 2.000 \end{bmatrix}$$

is found to be $2 \pm 2i$ (0.809).

The eigenvalues of $H_{(0)}$ are given in closed form as $2 \times 2i \cos(s\pi/5)$, $s = 1, 2, 3, 4$, which are precisely those determined.

As the example shows, the QR method with the double shift produces a sequence of matrices which converges to a matrix in which the bottom rightmost (2×2) matrix is left hanging, i.e., the elements in both corresponding rows become zero. At this stage the pair of complex conjugate eigenvalues can be evaluated, the matrix deflated, and the QR method with double shift applied in like fashion to the deflated matrix. The final matrix will contain either a 2×2 matrix in the upper leftmost corner (if N is even) or a single element (N odd). In either case the complete spectrum of eigenvalues will have been obtained.

Our treatment here is necessarily brief. The interested reader is directed to Wilkinson [49] for a full treatment of the practicalities and theoretical considerations.

Exercises

22. Using the subroutines you have created in this chapter construct a program which will determine the number of iterations necessary to produce the matrix given in Example 19.
23. Determine all of the eigenvalues of the following matrices using the double shifted QR method. Continue iterating until the $(k-2, k-1)$ element in the deflated $(k \times k)$ matrix being considered is less than 10^{-3} in magnitude.

---

† For display purposes there are only, at most, three decimal places shown; the actual calculations were performed in single precision on a Honeywell 66/60 which produces about 8 decimal places of accuracy.

(i)
$$\begin{bmatrix} 1 & -1 & & \\ 1 & 1 & -1 & \\ & 1 & 1 & -1 \\ & & 1 & 1 \end{bmatrix}$$

(ii)
$$\begin{bmatrix} 3 & -1 & & \\ 1 & 3 & -1 & \\ & 1 & 3 & -1 \\ & & 1 & 3 \end{bmatrix}$$

(iii)
$$\begin{bmatrix} 3 & -1 & 1 & 1 \\ 1 & 3 & -1 & 0 \\ & 1 & 3 & -1 \\ & & 1 & 3 \end{bmatrix}$$

(iv)
$$\begin{bmatrix} 1 & 2 & 0 & 0 & 3 \\ -1 & -1 & 0 & 0 & 0 \\ & 1 & -1 & 0 & 0 \\ & & -1 & 1 & 0 \\ & & & 1 & 1 \end{bmatrix}$$

where the elements not indicated are zero.

CHAPTER 11

The Numerical Solution of Ordinary Differential Equations

1 Introduction

In this chapter we are mainly concerned with solving the differential equation

$$y^{(1)}(x) = f(x, y) \tag{1}$$

where y is the unknown and f is a given function of the independent variable x and unknown y. Equation (1) may possess many solutions or none at all. To specify a particular one, the unknown y is required to satisfy a prescribed relationship, in addition to Eq. (1). For example, if we require to solve Eq. (1) on the interval $(0, 1)$ with $f = y$, we find that $y = A \exp(x)$ satisfies this equation for any constant A. To prescribe the relationship $y(0) = 1$ then yields a unique solution with $A = 1$, namely $y = \exp(x)$. This extra relationship which the solution of the differential equation is required to satisfy is called an *initial condition*. The whole problem

$$y^{(1)}(x) = f(x, y) \qquad x \in (0, 1); \, y(0) = 1 \tag{2}$$

is termed an *initial value problem*.

Given such a problem as Eq. (2) it is of interest to know in advance under what conditions a solution will exist, and in such a case whether the solution is unique. The following theorem, which we quote from Lambert [30], gives us conditions which will ensure existence and uniqueness.

Theorem 1 *Let $f(x, y)$ be defined and continuous for all points (x, y) in the region D defined by $a \leq x \leq b$, $-\infty < y < \infty$, a and b finite, and let there exist a constant L such that for every x, y, y^* such that (x, y) and (x, y^*) are both in D,*

$$|f(x, y) - f(x, y^*)| < L |y - y^*|$$

*Then if η is any given number, there exists a unique solution y(x) of the initial
value problem*

$$y^{(1)}(x) = f(x, y) \qquad y(a) = \eta \tag{3}$$

where y(x) is continuous and differentiable for all (x, y) in **D**.

In contrast to the examples of differential equations we meet in courses in
calculus, differential equations which arise in physical problems can mostly
never be solved analytically. So we have recourse to use approximate or
numerical methods. In the present treatment we will once again rely on the
idea of interpolation which we discussed in Chapter 4 and whose theory we
used in deriving methods for numerical integration and differentiation in
Chapters 7 and 8 respectively. For example, the solution to Eq. (3) at $x = X$
is given exactly by integrating formally the differential equation. Hence

$$y(X) - y(a) = \int_a^X f(x, y) \, dx$$

i.e.

$$y(X) = \eta + \int_a^X f(x, y) \, dx$$

where, recall, $y = y(x)$ in the argument of f. We are therefore faced with the
problem of evaluating the integral of a function (in general nonlinear) of y
when the behavior of y itself is unknown. Numerical integration naturally
suggests itself. However, instead of performing the integration over the
interval $[a, b]$ in a single step, we will propose evaluating the integration in
a stepwise procedure. Hence performing the integration in Eq. (3) stepwise
gives

$$\int_{x_n}^{x_{n+1}} y^{(1)} \, dx = \int_{x_n}^{x_{n+1}} f(x, y) \, dx \qquad n = 0, 1, \dots, N \tag{4}$$

where $x_0 = a$, $x_{N+1} = b$. We are therefore performing numerical integration
over subintervals of step lengths $x_{n+1} - x_n$, $n = 0, 1, \dots, N$. Let us assume
these step lengths are equal, and denote them by h. We may perform exact
integration on the left-hand side of Eq. (4), whence

$$y_{n+1} - y_n = \int_{x_n}^{x_{n+1}} f(x, y) \, dx \qquad n = 0, 1, \dots, N \tag{5}$$

where $y_n = y(x_n)$; or we may use an approximation, involving several point
values of y, to the integral on the left-hand side of Eq. (4). For Eq. (5) we
may propose a one-point numerical integration scheme (see Chapter 7) for
the integral. Hence

$$y_{n+1} - y_n = hf_I \qquad n = 0, 1, \dots, N$$

where $f_I = f(x_I, y_I)$ and $x_I = x_n + \alpha h$, $0 \le \alpha \le 1$, is some intermediate point in $[x_n, x_{n+1}]$. If $I = n$ ($\alpha = 0$) we have the rectangular integration rule and the resulting numerical scheme becomes

$$y_{n+1} - y_n = hf_n \tag{6}$$

which is *Euler's* rule. In contrast, if we use $I = n + \frac{1}{2}$ ($\alpha = \frac{1}{2}$) we have

$$y_{n+1} - y_n = hf_{n+1/2} \tag{7}$$

which is the *midpoint* rule. Choosing the $I = n + 1$ ($\alpha = 1$), we have

$$y_{n+1} - y_n = hf_{n+1} \tag{8}$$

Equations (6), (7), and (8) serve as examples of:
   an *explicit* method (since y_{n+1} can be obtained explicitly from Eq. (6));
   a method requiring a value of y at an *intermediate* point; and
   an *implicit* method (because of the occurrence of y_{n+1} in $f_{n+1} = f(x_{n+1}, y_{n+1})$).
   Method (6) can progress from $n = 0$, where $y_0 = \eta$, without further ado. Equation (7) will require some additional process to provide values of y at $x = (n + \frac{1}{2})h$, $n = 0, 1, \ldots$. For the scheme (8), a nonlinear equation must be solved for each value of n; the methods of Chapter 9 naturally suggest themselves here.
   Returning to Eq. (5) we may propose the trapezoidal rule for the integral occurring on the right-hand side. We therefore suggest

$$y_{n+1} - y_n = (h/2)[f_{n+1} + f_n] \qquad n = 0, 1, \ldots, N \tag{9}$$

Again this is an implicit method. If however f is a linear function of y, i.e., $f = g(x) \cdot y$, then the resulting numerical procedure can be used without the need for solving nonlinear equations. For example, Eq. (9) becomes

$$[1 - h/2g(x_{n+1})]y_{n+1} = [1 + h/2g(x_n)]y_n \qquad n = 0, 1, \ldots, N \tag{10}$$

A suggestion of using Simpson's 1/3-rule for the integral of Eq. (5), namely

$$y_{n+1} - y_n = (h/6)[f_n + 4f_{n+1/2} + f_{n+1}] \qquad n = 0, 1, \ldots, N \tag{11}$$

leaves us with both the problems of introducing an intermediate point $x_{n+1/2}$ and solving a nonlinear equation for each n. Alternatively, we can suggest using Eq. (11) over double intervals, namely

$$y_{n+2} - y_n = (h/3)[f_n + 4f_{n+1} + f_{n+2}]† \qquad n = 0, 1, \ldots, N-1 \tag{12}$$

where we retain the definition of $h = (x_{n+1} - x_n)$. If we apply Eq. (12) for $n = 1, 2, \ldots, N-1$, then what were intermediate points before will now be generated by the algorithm itself. We are left to specify y_1, as an extra starting value for this method. For example, we could suggest using the Euler or trapezoidal rule for the first interval. We will deal with this problem of starting values in more detail in section 5.

---

† Note this application is different to that advocated for numerical integration in Chapter 7.

Example 1 Calculate the solution of the ordinary differential equation

$$y^{(1)} = -y \qquad y(0) = 1$$

at the point $x = 1.0$ using in turn the values of $h = 0.05, 0.025, 0.001, 0.0001$. We will compare the methods (6), (7), (8), (9), and (12), where for the latter method we will use the trapezoidal rule to estimate y_1. The numerical method in each case will use $y_0 = 1.0$ as the initial condition. The solution at $x = 1.0$ obtained by the five methods is given in Table 1 together with the difference between the solution $y = \exp(-1)$ and the computed result. In method (7), the function value f has been replaced by $f(x_{n+1/2}, (y_n + y_{n+1})/2)$, which for the given function $f = -y$ produces just the trapezoidal rule, Eq. (9). Hence the results in columns 3 and 5 are identical.

In the preceding discussion, methods have emerged with the form

$$y_{n+k} - y_n = h \sum_{j=0}^{k} \beta_j f_{n+j} \tag{13}$$

where the coefficients β_j have been derived from replacing $\int_{x_n}^{x_{n+k}} f(x, y)\, dx$ by an appropriate numerical integration scheme. On the right-hand side of Eq. (13) we have a linear combination of the $k + 1$ values f_{n+j}, $j = 0, 1, \ldots, k$, whereas on the left-hand side there appear only the two values y_{n+k}, y_n. Without resorting to any connection with numerical quadrature, we can generalize the form of Eq. (13) to one which includes both the linear combination of the function values f_{n+j} *and* a linear combination of the solution values y_{n+j}, $j = 0, 1, \ldots, k$. That is, we propose to consider methods of the form

$$\sum_{j=0}^{k} \alpha_j y_{n+j} = h \sum_{j=0}^{k} \beta_j f_{n+j} \qquad n = 0, 1, \ldots, N-k+1 \tag{14}$$

where α_j and β_j are coefficients to be determined. We assume that $\alpha_k \neq 0$ and that not both α_0 and β_0 are zero. If $\beta_k = 0$ then the method is said to be *explicit*, whereas if $\beta_k \neq 0$ the method is called *implicit* (compare this with methods (6), (7), and (8)). In fact we may multiply through Eq. (14) by an arbitrary nonzero constant without changing the equations; it is clear that the coefficients α_j and β_j are arbitrary up to a constant multiplier. To remove this arbitrariness we will specify $\alpha_k = 1$. Numerical approximations of the form (14) are called *linear multistep methods with step number k*, or, in short, *linear k-step methods*. In the next three sections we will derive examples of the multistep methods, analyze accuracy, and consider conditions under which we can expect these methods to produce useful results. In these sections the reader will do well to bear Example 1 in mind.

Exercises

1. Write computer programs for the methods (6), (7), (8), (9), and (12) assuming $f(x, y) = g(x)y$ for some function g. For the ordinary differential equation

$$y^{(1)} = -x^2 y \qquad y(0) = 1$$

Table 1

h	Eq. (6)	Eq. (7)	Eq. (8)	Eq. (9)	Eq. (12)
0.05	3.584 86E−01 −9.393 55E−03	3.678 03E−01 −7.678 20E−05	3.768 89E−01 9.009 93E−03	3.678 03E−01 −7.678 20E−05	3.678 85E−01 5.085 02E−06
0.025	3.632 32E−01 −4.647 06E−03	3.678 60E−01 −1.936 41E−05	3.724 30E−01 4.551 02E−03	3.678 60E−01 −1.936 41E−05	3.678 80E−01 5.103 65E−07
0.001	3.676 94E−01 −1.859 14E−04	3.678 73E−01 −6.850 81E−06	3.680 61E−01 1.812 17E−04	3.678 73E−01 −6.850 81E−06	3.678 77E−01 −2.849 85E−06
0.0001	3.678 84E−01 −3.688 78E−05	3.678 12E−01 −6.785 24E−05	3.678 73E−01 −6.631 02E−06	3.678 12E−01 −6.785 24E−05	3.678 60E−01 −1.959 88E−05

For each value of h the upper number represents the computed solution, the lower figures represent the errors.

use your programs to conduct a similar set of experiments to those described in Example 1. $(y(x) = \exp(-x^3/3))$.)

2. The numerical methods (6), (7), (8), (9), and (12) were obtained in a manner so that the only error over a single application of the method arose from the approximation of the integral occurring on the right-hand side of Eq. (4). For these methods, using the error terms given in Chapter 7, write down the orders of accuracy of the methods (in the sense of Chapter 7). How do the errors obtained in the experiments of Exercise 1 compare with the theoretical estimates?

3. A means of avoiding solving nonlinear equations for the trapezoidal rule is to suggest the alternative method

$$y_{n+1} = y_n + h[(1 - 1/\alpha)f_n + 1/\alpha f_{n+1}^*]$$

where $y_{n+1}^* = y_n + \alpha/2hf_n$; α a parameter ≥ 1 and $f_{n+1}^* = f(x_n + \alpha/2h, y_{n+1}^*)$. Such a method is called a *predictor–corrector* method (see section 4). Write a program for this method and use it to solve the equation $y^{(1)} = \sqrt{1 - y^2}$; $y(0) = 0$ for $x = 1.0$ using values of $h = 0.1, 0.05$, and 0.025 for a series of values of α. Compare your results with the theoretical solution $y(1) = \sin(1)$.

4. Instead of the predictor–corrector rule advocated in Exercise 3, we may set up an iterative process to calculate the unknown y for each value of n. Such an iterative method is denoted by

$$y_{n+1}^j = y_n + \frac{h}{2}(f_n + f_{n+1}^{j-1}) \qquad j = 1, 2, \ldots; \qquad n = 0, 1, \ldots, N$$

where

$$f_{n+1}^{j-1} = f(x_{n+1}, y_{n+1}^{j-1}) \quad \text{and} \quad y_{n+1}^0 = y_n$$

Write a computer program for this algorithm for the differential equation of Exercise 3. Arrange to terminate the iteration at each step when

$$|y_{n+1}^j - y_{n+1}^{j-1}| \leq 10^{-4}$$

Compare your solution with those of Exercise 3.

2 Linear k-Step Methods

In this section we investigate theoretical properties of the k-step methods introduced in the previous section and denoted generally by Eq. (14). First, we require to define the concept of *order of accuracy* as applied to k-step methods. This definition will be similar to the other definitions of order of accuracy introduced in Chapters 4 and 7.

Define the linear difference operator $\mathscr{L}$ by

$$\mathscr{L}[Y(x); h] \equiv \sum_{j=0}^{k} [\alpha_j Y(x + jh) - h\beta_j Y^{(1)}(x + jh)] \tag{15}$$

where $Y(x)$ is an arbitrary smooth† function on $[a, b]$.

Equation (15) denotes the application of the k-step method to any smooth Y. That is, Eq. (14) is equivalent to

$$\mathscr{L}[y_n; h] = 0$$

---

† Smooth in the sense that Y has as many continuous derivatives as we need.

(but, for arbitrary smooth Y,

$$\mathcal{L}[Y(x); h] \neq 0)$$

Thus a Taylor series expansion of Eq. (15) will, for arbitrary Y, leave an error term. We therefore have the definition of order of accuracy:

Definition   Denote the Taylor series expansion of Eq. (15) as

$$\mathcal{L}[Y(x); h] = C_0 Y(x) + C_1 h Y^{(1)}(x) + \cdots + C_q h^q Y^{(q)}(x) + \cdots \tag{16}$$

where $C_0, C_1, \ldots$ are constants. The k-step method (14) is said to have *order* p if in Eq. (16) $C_0 = C_1 = \cdots = C_p = 0$, $C_{p+1} \neq 0$.

The definition of order here is similar to our previous definitions in the sense that if the linear operator $\mathcal{L}$ of order p is applied to any polynomial Y of degree p or less, then the difference equation

$$\mathcal{L}[Y(x); h] = 0 \tag{17}$$

will be exactly satisfied for this Y since $Y^{(q)}(x)$, $q = p+1, p+2, \ldots$, will be identically zero. Consequently any polynomial $Y(x)$ of degree p or less will be a solution of the difference equation (17). Thus if the solution of the differential equation (1) is a polynomial of degree p (or less), it will also be a solution of the difference equation;† i.e. the multistep method will exactly integrate polynomial solutions of degree p or less, which was how we defined order previously.

Example 2   What order does the trapezoidal method (9) possess?

Let us apply the difference operator $\mathcal{L}$ defined by Eq. (9) to an arbitrary $Y(x)$. Then

$$\mathcal{L}[Y(x); h] \equiv Y(x_{n+1}) - Y(x_n) - \frac{h}{2}[Y^{(1)}(x_{n+1}) + Y^{(1)}(x_n)]$$

$$= Y(x_n) + h Y^{(1)}(x_n) + \frac{h^2}{2} Y^{(2)}(x_n) + \frac{h^3}{6} Y^{(3)}(x_n) + \cdots - Y(x_n)$$

$$- \frac{h}{2}[Y^{(1)}(x_n) + h Y^{(2)}(x_n) + \frac{h^2}{2} Y^{(3)}(x_n) + \cdots + Y^{(1)}(x_n)]$$

$$= 0 \cdot Y(x_n) + 0 \cdot Y^{(1)}(x_n) + 0 \cdot Y^{(2)}(x_n) - \frac{h^3}{2 \cdot 3!} Y^{(3)}(x_n) + O(h^4)$$

Hence $C_0 = C_1 = C_2 = 0$; $C_3 \neq 0$, and hence the method is order two.

Exercises

5. Confirm that the Trapezoidal method is order two by showing that if $y(x) = x^2$, the method exactly integrates Eq. (1).

---

† Assuming that the initial conditions and starting conditions, of course, are chosen appropriately.

6. Determine the order of the methods (6), (7), (8), and (12) in an analogous manner to that described in Example 2. Compare your results in Exercise 2.

7. The two-step *Adams–Moulton* method

$$y_{n+2} - y_{n+1} = \frac{h}{12}(5f_{n+2} + 8f_{n+1} - f_n)$$

is obtained by replacing the right-hand integral in the identity

$$\int_{x_{n+1}}^{x_{n+2}} y^{(1)}\,dx = \int_{x_{n+1}}^{x_{n+2}} f(x, y)\,dx$$

by a linear combination of function values at x_n, x_{n+1}, and x_{n+2}. Determine the order of this method.

8. Determine the order of *Quade's* method:

$$y_{n+4} - \frac{8}{19}(y_{n+3} - y_{n+1}) - y_n = \frac{6h}{19}(f_{n+4} + 4f_{n+3} + 4f_{n+1} + f_n)$$

by considering the solution $y(x)$ to be x^i, $i = 0, 1, \ldots$.

The fact that methods are of order p if polynomials of degree p and less are integrated exactly by a k-step method suggests a method of determining the coefficients α_j and β_j in the formulation (14). This is, of course, just analogous to the method of undetermined coefficients which we describe for the derivation of numerical integration schemes described in Chapter 7. Hence let us determine the multistep method for $k = 2$ and for $k = 3$: the latter method we will require to be explicit, by requiring that Eq. (14) integrate exactly polynomial solutions of specified degree. Consider $k = 2$ in Eq. (14). Then we have

$$\sum_{j=0}^{2} \alpha_j y_{n+j} - h \sum_{j=0}^{2} \beta_j f_{n+j} = 0$$

We have five coefficients to determine (recall $\alpha_2 = 1$ by assumption). Hence substituting in turn $y(x) = x^i/i!$,[†] $i = 0, 1, \ldots, 4$, we obtain

$$\alpha_0 + \alpha_1 + \alpha_2 = 0 \qquad (\alpha_2 = 1)$$

$$\alpha_1 + \alpha_2 - (\beta_0 + \beta_1 + \beta_2) = 0$$

$$\frac{1}{2!}(\alpha_1 + 4) - (\beta_1 + 2\beta_2) = 0$$

$$\frac{1}{3!}(\alpha_1 + 8) - \frac{1}{2!}(\beta_1 + 4\beta_2) = 0 \tag{18}$$

$$\frac{1}{4!}(\alpha_1 + 16) - \frac{1}{3!}(\beta_1 + 8\beta_2) = 0$$

[†] Note we use $y(x) = x^i/i!$ rather than x^i in order that the coefficients here correspond to the C_q in the definition of order. For values of $i = 0, 1, \ldots, p$ (p the order of the method), this is of no consequence. However, for $i = p+1$, $C_{p+1} \neq 0$ and the resulting normalization $i!$ in $y(x)$ gives the corresponding coefficient in the definition of order.

The substitution of $y(x) = x^5/5!$ we expect not to be satisfied. This then would correspond to C_5 in Eq. (16), namely

$$C_5 = \frac{1}{5!}(\alpha_1 + 32) - \frac{1}{4!}(\beta_1 + 16\beta_2) \tag{19}$$

We find on solving the system (18) that

$$\alpha_0 = -1 \qquad \alpha_1 = 0 \qquad (\alpha_2 = 1) \qquad \beta_0 = \tfrac{1}{3} \qquad \beta_1 = \tfrac{4}{3} \qquad \beta_2 = \tfrac{1}{3}$$

Hence the two-step method of *maximal* order is

$$y_{n+2} - y_n = \frac{h}{3}(f_n + 4f_{n+1} + f_{n+2})$$

which is Simpson's rule, Eq. (12). Substituting the constants into Eq. (19) we find $C_5 = -1/90$ so Simpson's method is of order 4.

For $k = 3$ with $\alpha_3 = 1$ and $\beta_3 = 0$, we require that

$$y_{n+3} + \alpha_2 y_{n+2} + \alpha_1 y_{n+1} + \alpha_0 y_n - h[\beta_2 f_{n+2} + \beta_1 f_{n+1} + \beta_0 f_n] = 0 \tag{20}$$

exactly integrate polynomial $y(x) = x^i/i!$, $i = 0, 1, \ldots, 5$. Substituting in turn the polynomial y, we obtain the system of equations

$$\alpha_0 + \alpha_1 + \alpha_2 + 1 = 0$$

$$\alpha_1 + 2\alpha_2 + 3 - (\beta_0 + \beta_1 + \beta_2) = 0$$

$$\frac{1}{2!}(\alpha_1 + 4\alpha_2 + 9) - (\beta_1 + 2\beta_2) = 0$$

$$\frac{1}{3!}(\alpha_1 + 8\alpha_2 + 27) - \frac{1}{2!}(\beta_1 + 4\beta_2) = 0$$

$$\frac{1}{4!}(\alpha_1 + 16\alpha_2 + 81) - \frac{1}{3!}(\beta_1 + 8\beta_2) = 0$$

$$\frac{1}{5!}(\alpha_1 + 32\alpha_2 + 243) - \frac{1}{4!}(\beta_1 + 16\beta_2) = 0$$

whose solution we find is

$$\alpha_0 = -10 \qquad \alpha_1 = -9 \qquad \alpha_2 = 18 \qquad \beta_0 = 3 \qquad \beta_1 = 18 \qquad \beta_2 = 9$$

Thus our 3-step explicit method is

$$y_{n+3} + 18y_{n+2} - 9y_{n+1} - 10y_n - h[9f_{n+2} + 18f_{n+1} + 3f_n] = 0$$

Substitution of the coefficients into Eq. (20) with $y(x) = x^6/6!$ gives the error constant $C_6 = 1/20$. The method is of order 5.

Exercises

9. Determine the coefficients in the 3-step method of maximal accuracy with $\alpha_3 = 1$; $\alpha_1 = \alpha_0 = 0$; this is the 3-step *Adams–Moulton* method.

10. What is the error term for the method of Exercise 9?
11. Determine the coefficients in the explicit 4-step method of maximal accuracy with $\alpha_4 = 1$; $\alpha_2 = \alpha_1 = \alpha_0 = 0$; this is the 4-step *Adams–Bashforth* method.
12. What is the error term for the method of Exercise 11?
13. The methods of Exercises 9 and 11 are seen to be methods which fall into the category of Eq. (4) with the left-hand integral expressed as an exact expression and the right-hand integral expressed as a linear combination of point values without the interval. The methods therefore may be derived by passing the appropriate interpolating polynomial through these points.

   Rederive the methods of Exercises 9 and 11 by replacing the function f by the Newton interpolating polynomial of appropriate degree (see Chapter 4).
14. The Adams–Moulton method of Exercise 9 is an implicit method whereas the Adams–Bashforth method of Exercise 11 is explicit. To remove the implicitness use the Adams–Bashforth method as a predictor and the Adams–Moulton method as a corrector so that the combined algorithm is

$$y_{n+4}^{*} = y_{n+3} + \frac{h}{24}(55f_{n+3} - 59f_{n+2} + 37f_{n+1} - 9f_n)$$

$$y_{n+4} = y_{n+3} + \frac{h}{24}(9f_{n+4}^{*} + 19f_{n+3} - 5f_{n+2} + f_{n+1})$$

where starting values are required for $n = 0, 1, 2$, and 3. Write a computer program for this predictor–corrector method and use it to solve the differential equation of Exercise 3 for $x = 1.0$ using $h = 0.1, 0.05$, and 0.025. You may use the values of the theoretical solution $y = \sin(x)$ as starting values for the method. Compare your results with those obtained in Exercises 3 and 4.

3 Analysis of k-Step Methods

In the present section we require to investigate some of the theoretical properties of numerical methods for solving Eq. (1). In the linear multistep method we have essentially two parameters: n the step number; h the mesh length. Associated with these two parameters the following questions arise. Firstly, what happens if, for a fixed h, we let the number of steps n increase without bound? Does our difference solution y remain well behaved, in some sense, as a function of n? Secondly, what happens if we let the mesh size h tend to zero in some preassigned fashion so that the discrete problem approaches the continuous one? Does the difference solution approach the solution of the differential equation?

When we considered the convergence of the solutions of iterative methods in Chapter 9 it was clear how we would interpret such convergence; we had only the parameter n which we considered to tend to infinity. Here we have two parameters n and h so that there are several ways we might interpret convergence, depending upon the limiting processes associated with h and n. We could demand the difference between the solutions of the difference equation and the differential equation tend to zero as $h \to 0$ for a fixed number of applications of the multistep method. Alternatively, we could demand that the difference between the solution of the difference equation and the differential equation solution tend to zero as $n \to \infty$ for a fixed h. Both of these interpretations of convergence may be considered unsatisfactory in the sense that the first tells us only that the difference solution sufficiently close to a tends to y whereas the second requires us to go "far enough" along the x-axis. What we really require is to investigate the solution given by the difference equation at any prescribed point $x = X$ (say) and ask that it tend to the solution of the differential equation at $x = X$ as we progressively refine the mesh so that we continue to calculate the solution at the point $x = X$, i.e. the number of applications of the difference method will increase. Hence we have the following definition of convergence ([30]).

Definition The linear multistep method (14) is said to be *convergent* if, for all initial-value problems (1) subject to the hypothesis of Theorem 1, we have

$$\lim_{\substack{h \to 0 \\ nh = x - a}}^{\dagger} y_n = y(x)$$

holds for all $X \in [a, b]$ and all solutions y_n of the difference equation (14) satisfying the conditions $y_\mu = \eta_\mu(h)$ for which $\lim_{h \to 0} \eta_\mu(h) = \eta$, $\mu = 0, 1, \ldots, k - 1$.

The conditions $y_\mu = \eta_\mu(h)$ are the k starting values that are necessary to start the k-step method; we have in the past specified these either through another difference method or the theoretical solution evaluated at the appropriate abscissa points. All we require in this definition is that the starting values tend to η as $h \to 0$.

Example 3 Solve the differential equation $y^{(1)} = y$; $y(0) = 1$ using the multistep methods

$$y_{n+2} + 4y_{n+1} - 5y_n - 2h[2f_{n+1} + f_n] = 0 \tag{21}$$

and

$$y_{n+1} - y_n - hf_n = 0 \tag{22}$$

It is easily shown that Eq. (21) has a local error term $(h^4/6)y^{(4)}(\zeta)$ and Euler's method, Eq. (22), has a local error term $(h^2/2)y^{(2)}(\zeta)$. If both methods are convergent we would expect the error of Eq. (21) to tend to zero much more quickly as $h \to 0$ than the error of Eq. (22) (why?) However, using the theoretical solution for the starting values we see that the errors for Eq. (21) quickly grow in size, and for $h = 0.1$ exceed 100 in magnitude for x beyond 1.2 (see Table 2). In contrast, although the error for Eq. (22) is growing slowly as x increases, as h is reduced the size of the error at the same point is reduced; compare the errors at $x = 0.5$, 1.0, and 2.0 for $h = 0.1$ and 0.05. Thus we see from these results that the difference solution of Eq. (22) is converging whereas that of Eq. (21) is not.

† The limit here is to be taken as $h \to 0$ such that $nh = x - a$ is constant.

Table 2

Theo	Euler	Error	Eq. (21)	Error	x	h
						0.100
1.221 40	1.100 00	1.121 40	01.2214	00.0000	0.20	
1.349 86	1.210 00	0.139 86	01.3499	−00.0000	0.30	
1.491 82	1.331 00	1.160 82	01.4915	00.0003	0.40	
1.648 72	1.464 10	0.184 62	01.6499	−00.0012	0.50	
1.822 12	1.610 51	0.211 61	01.8163	00.0058	0.60	
2.013 75	1.771 56	0.242 19	02.0409	−00.0272	0.70	
2.225 54	1.948 72	0.276 82	02.0974	00.1281	0.80	
2.459 60	2.143 59	0.316 01	03.0620	−00.6024	0.90	
2.718 28	2.357 95	0.360 33	−00.1166	02.8349	1.00	
3.004 17	2.593 74	0.410 42	16.3425	−13.3384	1.10	
3.320 12	2.853 12	0.467 00	−59.4397	62.7598	1.20	
3.669 30	3.138 43	0.530 87			1.30	
4.055 20	3.452 27	0.602 93			1.40	
4.481 69	3.797 50	0.684 19			1.50	
4.953 03	4.177 25	0.775 78			1.60	
5.473 95	4.594 97	0.878 97			1.70	
6.049 65	5.054 47	0.995 18			1.80	
6.685 89	5.559 92	1.125 98			1.90	
7.389 06	6.115 91	1.273 15			2.00	
						0.050
1.105 17	1.050 00	0.055 17	01.1052	00.0000	0.10	
1.161 83	1.102 50	0.059 33	01.1618	−00.0000	0.15	
1.221 40	1.157 62	0.063 78	01.2214	00.0000	0.20	
1.284 03	1.215 51	0.068 52	01.2841	−00.0001	0.25	
1.349 86	1.276 28	0.073 58	01.3495	00.0004	0.30	
1.419 07	1.340 10	0.078 97	01.4210	−00.0020	0.35	
1.491 82	1.407 10	0.084 72	01.4823	00.0095	0.40	
1.568 31	1.477 46	0.090 86	01.6146	−00.0463	0.45	
1.648 72	1.551 33	0.097 39	01.4241	00.2247	0.50	
1.733 25	1.628 89	1.104 36	02.8232	−01.0899	0.55	
1.822 12	1.710 34	0.111 78	−03.4654	05.2875	0.60	
1.915 54	1.795 86	0.119 68	27.5666	−25.6510	0.65	
2.013 75	1.885 65	0.128 10			0.70	
2.117 00	1.979 93	0.137 07			0.75	
2.718 28	2.526 95	0.191 33			1.00	
4.481 69	4.116 14	0.365 55			1.50	
7.389 06	6.704 75	0.684 31			2.00	

The column headed Theo indicates the theoretical solution.
The blank entries indicate that the entries had exceeded 100 in magnitude.

The analysis of convergence of a multistep method, using the definition of convergence given, is in general very difficult. However, the condition for convergence of a k-step method is closely linked to certain algebraic properties of the coefficients through the concepts of *consistency* and *stability*. We will spend the remainder of this section discussing these two properties.

Definition   A *k*-step method is said to be *consistent* to a given differential equation if it has order $p \geq 1$.

From our definition of order of accuracy, we see that the *k*-step method is consistent if the constants $C_0 = C_1 = 0$. From Eq. (16) it is clear that for consistency we require the coefficients of the *k*-step method to satisfy

$$\sum_{j=0}^{k} \alpha_j = 0 \qquad \sum_{j=1}^{k} j\alpha_j - \sum_{j=0}^{k} \beta_j = 0 \qquad (23)$$

These conditions are therefore often referred to as the *consistency conditions*.

Example 4   Show that the difference method

$$y_{n+2} - 2y_{n+1} + y_n = \frac{3h}{2}(f_{n+1} - f_n).$$

is a consistent approximation to Eq. (1).
   Using Eq. (23) we see that

$$\sum_{j=0}^{k} \alpha_j = 1 - 2 + 1 = 0 \quad \text{and} \quad \sum_{j=1}^{k} j\alpha_j - \sum_{j=0}^{k} \beta_j = -2 + 2 - (-\tfrac{3}{2}) - \tfrac{3}{2} = 0$$

so that the consistency conditions are satisfied.

Thus for any particular method it is a simple matter to determine whether or not it is consistent.

Exercises

15. Show that the 2-step method

$$y_{n+2} - 2y_{n+1} + y_n = \frac{h^2}{2}(f_{n+1} + f_n)$$

   is not a consistent approximation to Eq. (1). Is it a consistent approximation to some other differential equation?
16. Show that the 2-step method (21) is a consistent approximation to Eq. (1).

To discuss the concept of stability we must digress and consider the general solution of a linear difference equation of order *k* with constant coefficients, namely the general solution of equations of the form

$$\sum_{j=0}^{k} a_j y_{n+j} = 0$$

In the calculus, we are familiar with the idea of looking for solutions of a given differential equation $y^{(1)} = Ay$ (*A* constant) of the form $y = \exp(\lambda x)$ where λ is a constant to be determined and may be either real or complex.

For example, for the differential equation

$$y^{(1)} = 6y \tag{24}$$

we find on substituting in Eq. (24) a trial function $Y = \exp(\lambda x)$ that Y is a solution if λ satisfies $\lambda = 6$. In this case a solution is $y = \exp(6x)$. The general solution would be $y = C \exp(6x)$ where C is a constant specified by some initial condition. If we consider a second-order differential equation

$$y^{(2)} = 6y \tag{25}$$

then again we propose a trial function $Y = \exp(\lambda x)$ and now find that λ satisfies $\lambda^2 = 6$, i.e. $\lambda = \pm\sqrt{6}$. Thus

$$Y = \exp(\sqrt{6}\, x) \quad \text{and} \quad Y = \exp(-\sqrt{6}\, x)$$

both satisfy the differential equation (25) and hence the general solution is $y(x) = C_1 \exp(\sqrt{6}\, x) + C_2 \exp(-\sqrt{6}\, x)$ where C_1 and C_2 are constants to be determined from the initial conditions. If we consider a linear differential equation of order † N, say,

$$\sum_{i=0}^{N} a_i y^{(i)} = 0 \tag{26}$$

we find that a trial function $Y = \exp(\lambda x)$ is a solution if λ satisfies the polynomial equation

$$\sum_{i=0}^{N} a_i \lambda^i = 0 \tag{27}$$

i.e. the functions $Y_j(x) = \exp(\lambda_j x)$, $j = 1, 2, \ldots, N$, will each be a solution of the differential equation (26) where λ_j are roots of Eq. (27). If the roots of (27) are distinct then the general solution of Eq. (26) will be

$$Y(x) = \sum_{j=1}^{N} C_j \exp(\lambda_j x)$$

where the constants C_j will be determined from N initial conditions.

In determining solutions to linear difference equations we again determine trial functions which satisfy the difference equation. For example, for the first-order difference equation

$$y_{n+1} = y_n \tag{28}$$

we try a function $Y_n = r^n$ where r is to be determined (similarly to λ in the differential equation trial function). We find that Y_n is a solution of Eq. (28) if r satisfies

$$r^{n+1} = r^n \quad \text{i.e. } r = 1$$

† The order of a differential equation is the order of the highest derivative in the equation.

Thus a solution is $Y_n = (1)^n$. The general solution is $Y_n = C(1)^n$, where C is determined from the starting conditions of the difference equation. Likewise, if we have a linear difference equation of second-order, for example

$$y_{n+2} = y_n$$

we suggest a trial function $Y_n = r^n$ and find that if $r^{n+2} = r^n$, i.e. if $r = \pm 1$, then $Y_n = (+1)^n$ and $Y_n = (-1)^n$ are both solutions of the difference scheme. The general solution will be

$$Y_n = C_1(1)^n + C_2(-1)^n$$

where C_1 and C_2 are determined from two starting conditions.

Example 5 Let us solve the difference equation

$$y_{n+2} = -y_n \qquad y_0 = y_1 = 1$$

Introduce $Y_n = r^n$ as the trial function. If r satisfies the quadratic equation

$$r^2 + 1 = 0$$

then Y_n is a solution. Clearly in this case $Y_n = (i)^n$ and $Y_n = (-i)^n$ $(i = \sqrt{-1})$ are solutions. The general solution is

$$Y_n = C_1(i)^n + C_2(-i)^n$$

Now for

$$n = 0, Y_0 = C_1 + C_2 = 1$$

and for

$$n = 1, Y_1 = i[C_1 - C_2] = 1$$

hence

$$C_1 = \tfrac{1}{2}(1 - i) \quad \text{and} \quad C_2 = \tfrac{1}{2}(1 + i)$$

The required difference solution is

$$Y_n = \tfrac{1}{2}(1 - i)(i)^n + \tfrac{1}{2}(1 + i)(-i)^n$$

which may be written in the neater form

$$Y_n = \cos n\frac{\pi}{2} + \sin n\frac{\pi}{2} \qquad n = 0, 1, 2, \ldots$$

Exercises

17. Solve the difference equation
$$y_{n+2} + y_{n+1} + y_n = 0 \qquad y_0 = 1 \qquad y_1 = \tfrac{1}{2}$$

18. Solve the difference equation
$$y_{n+2} - y_{n+1} - y_n = 0 \qquad y_0 = y_1 = 1$$

(This is the difference equation defining the sequence of Fibonacci numbers.)

19. Solve the difference equation

$$y_{n+2} + 4y_{n+1} - 5y_n = 0 \qquad y_0 = 0 \qquad y_1 = 0$$

20. Solve the difference equation of Exercise 19 subject to the starting conditions $y_0 = 0$, $y_1 = \varepsilon h$, ε a small constant.

21. Can the solution of the difference equation in Exercise 20 converge to the solution of the differential equation $y^{(1)} = 0$, $y(0) = 0$ in the sense of our definition of convergence given on page 375? Contrast this with the solution of the difference method and starting values of Exercise 19.

The difference method of Exercise 20 is subject to starting values which may be considered as perturbations of the starting values of Exercise 19. Exercise 19 is a consistent difference method (why?) for the differential equation $y^{(1)} = 0$ and $y(0) = 0$. Thus we may consider the difference method of Exercise 20 as the difference method of Exercise 19 but with a round-off error ε substituted for the second starting value. The differential equation $y^{(1)}(x) = 0$, $y(0) = 0$ has the trivial solution $y(x) = 0$. Hence we would expect the trivial solution from a convergent difference equation. The trial function in Exercise 20 produces two functions $(1/6)\varepsilon h(1)^n$ and $(-1/6)\varepsilon h(-5)^n$ as solutions of the difference equation so that the required solution is

$$y_n = \tfrac{1}{6}\varepsilon h(1)^n - \tfrac{1}{6}\varepsilon h(-5)^n$$

But $nh = x_n - x_0 = x_n$ so $n = x_n/h$. Hence

$$\lim_{\substack{h \to 0 \\ nh = x_n}} y_n = \lim_{\substack{h \to 0 \\ nh = x_n}} \tfrac{1}{6}\varepsilon h(1)^{x_n/h} - \lim_{\substack{h \to 0 \\ nh = x_n}} \tfrac{1}{6}\varepsilon h(-5)^{x_n/h}$$

Now

$$\lim_{\substack{h \to 0 \\ nh = x_n}} \tfrac{1}{6}\varepsilon h(1)^{x_n/h} = 0 \quad \text{but} \quad \lim_{\substack{h \to 0 \\ nh = x_n}} \tfrac{1}{6}\varepsilon h(-5)^{x_n/h} = \infty$$

Thus the difference method is not convergent. The cause of the problem is clearly the second root (-5). Through the consistency condition it is clear we will always obtain a root (1). We call this the *principal* root. However, with k-step difference schemes with $k > 1$ we will obtain additional roots which we call *spurious* roots. The spurious roots will introduce *extraneous* solutions which, as we have seen, may cause the difference scheme to be divergent. It is clear that the difference scheme will be convergent if we require the extraneous solutions to tend to zero as the number of applications of the method increases. For the k-step method

$$\sum_{j=0}^{k} \alpha_j y_{n+j} = 0 \tag{29}$$

$y_n = r^n$ is a solution if r satisfies the *first characteristic equation*

$$\rho(r) = \sum_{j=0}^{k} \alpha_j r^{n+j} = 0 \tag{30}$$

In terms of the roots of this polynomial equation we have the following definition.

Definition   The linear multistep method (14) is said to be *zero*† *stable* if no root of the first characteristic polynomial (30) has modulus greater than one and if every root of modulus one is simple.‡

Example 6   Show that any consistent one-step method is zero stable. Consider the one-step method written as Eq. (5),

$$y_{n+1} - y_n = \int_{x_n}^{x_{n+1}} f(x, y)\, dx \tag{5}$$

The coefficients of y_{n+1} and y_n must be as indicated since to be consistent by Eq. (23), $\alpha_1 + \alpha_0 = 0$ and by assumption $\alpha_1 = 1$. Thus no matter by what we replace the integral, the first characteristic polynomial is $\rho(r) = r - 1$ and clearly has just the zero $r = 1$. The family of one-step methods resulting from replacing the integral in Eq. (5) by a numerical quadrature is therefore zero stable.

Exercises

22. Show that Simpson's 2-step method (12) is zero stable.
23. Is Quade's method (Exercise 8) zero stable?
24. Is the method

$$y_{n+3} + 18y_{n+2} - 9y_{n+1} - 10y_n - h[9f_{n+2} + 18f_{n+1} + 3f_n] = 0$$

zero stable? Compute the solution (if you can) of the differential equation

$$y^{(1)} = \frac{3y}{2x} \qquad y(1) = 1$$

at $x = 2$ for a series of values of h.

We are now in a position to state the connection between the concepts of convergence, consistency, and stability. This is given by a fundamental theorem due to Dahlquist [7].

Theorem 2   *The necessary and sufficient conditions for a linear multistep method to be convergent are that it be consistent and zero stable.*

In other words, to prove that the solution of a difference equation converges to the solution of a given differential equation we may either prove the convergence directly or we may prove that the difference scheme is both consistent and zero stable. This of course implies that a consistent scheme which is not zero stable, or a zero stable scheme which is not consistent to a

---

† Zero stable because the definition considers the limit as $h \to 0$. In this limit the difference scheme (14) reduces to Eq. (29).

‡ I.e. if any roots have unit modulus they are not coincident in the complex plane.

given differential equation, cannot be convergent and hence such a scheme would have little practical value.

If we refer to the results of Table 2 we now see that the Euler method, being both zero stable and consistent, gives convergent results; whereas although the scheme given by Eq. (21) is consistent it is not zero stable; these latter results are divergent.

The theory we have described in this section constitutes a very small portion of the total theory of multistep methods. This complete theory is beyond the scope of this present book; the interested reader is referred to Henrici [21] for further details.

Exercises

25. Repeat the experiment depicted in Table 2 for the zero stable but inconsistent difference method

$$y_{n+2} + y_n - 2h[2f_{n+1} + f_n] = 0$$

Comment on your results.

26. Consider the two-step method

$$y_{n+2} - (1-\alpha)y_{n+1} + \alpha y_n - \frac{h}{2}[(3-\alpha)f_{n+1} - (1+\alpha)f_n] = 0$$

α is a constant. What order does the method possess for arbitrary α? What value of α achieves maximal order? Does this value of α give a method which is zero stable?

27. Compute the solution of the differential equation $y^{(1)} = y$; $y(0) = 1$, using the algorithm of Exercise 26 for a series of α and h to support the analysis carried out in Exercise 26.

28. (*Lambert* [30]) Find the range of α for which the linear multistep method

$$y_{n+3} + \alpha(y_{n+2} - y_{n+1}) - y_n - \frac{1}{2}(3+\alpha)h(f_{n+2} + f_{n+1}) = 0$$

is zero stable. Show that there exists a value of α for which the method has order 4 but that if the method is to be zero stable its order cannot exceed 2. Confirm your analysis by computing the solution of the differential equation in Exercise 27 for a series of α and h.

4 Predictor–Corrector Methods

In Exercises 3 and 4 we introduced the idea of a predictor–corrector method as a means of circumventing the need for the numerical solution of nonlinear equations which arise when we consider implicit k-step methods. Although we have not described the theory by which we can derive the result, it is known that a comparison of explicit and implicit methods of the same order yields the result that implicit methods have smaller error constants in absolute value and possess a considerable advantage in the size of the interval of stability.† In other words, implicit methods are not just of academic interest;

---

† The stability inferred here is, in fact, not zero stability which we described in the previous section, but a stability known as *weak stability*— see Lambert [30].

they have computational advantages. The main drawback of course is the necessity of solving a nonlinear equation to derive y_{n+k} and it is for this reason that predictor–corrector methods are important; they go some way to retaining the advantages of implicit methods but at the same time avoid the need to solve the nonlinear equations. Thus let us consider the implicit linear k-step method

$$y_{n+k} + \sum_{j=0}^{k-1} \alpha_j y_{n+j} - h\beta_k f(x_{n+k}, y_{n+k}) - h\sum_{j=0}^{k-1} \beta_j f_{n+j} = 0 \tag{31}$$

Following Exercise 4 we proposed the iterative method

$$y_{n+k}^{m} + \sum_{j=0}^{k-1} \alpha_j y_{n+j} - h\beta_k f(x_{n+k}, y_{n+k}^{m-1}) - h\sum_{j=0}^{k-1} \beta_j f_{n+j} = 0 \qquad m = 1, 2, \ldots \tag{32}$$

where y_{n+k}^{0} is to be determined by some additional algorithm. In Exercise 4 we proposed $y_{n+k}^{0} = y_{n+k-1}$ (for $k = 1$ in that case), whereas in Exercise 3 we proposed $y_{n+k}^{0} = y_{n+k}^{*}$ as given by a generalization of the Euler rule and applied the implicit formula just once. What we will propose here is, in some sense, a combination of the two ideas introduced separately in Exercises 3 and 4, i.e. we will predict y_{n+k}^{0} using an explicit method of the same order as the implicit method (32) which we call the corrector. Having decided upon the particular predictor method, the computational procedure is to evaluate y_{n+k}^{0} from this predictor, substitute this value into the argument of f and evaluate y_{n+k}^{1}. This in turn is substituted into f which is then re-evaluated and a new y_{n+k}^{2} produced. The corrector is applied until successive values of y_{n+k} agree to within a specified tolerance, ε, say. When this occurs n is incremented by one and the cycle is repeated. If we denote by P the predictor step, E the evaluation of f using either the predicted value or the most recent iterate from the corrector, and C as an application of the corrector, then at a particular value of n the predictor–corrector method can be denoted by

$$P(EC)(EC)\cdots$$

where the (EC) combination is repeated until

$$|y_{n+k}^{m} - y_{n+k}^{m-1}| \leq \varepsilon$$

is satisfied.

Example 7 Solve the differential equation $y^{(1)} = 2\sqrt{y - y^2}$, $y(0.5) = \sin^2(0.5)$ at $x = 1.5$ using the predictor–corrector pair

$$y_{n+1}^{0} = y_n + hf_n$$

$$y_{n+1}^{j} = y_n + hf_n^{j-1} \qquad j = 1, 2, \ldots \tag{33}$$

iterating the corrector until $|y_{n+1}^{j} - y_{n+1}^{j-1}| \leq \varepsilon$, where ε takes, in turn, the values 0.01, 0.001, and 0.0001, and $h = 0.1, 0.05$, and 0.025. Compare the numerical results, for the same values of h, with the midpoint rule

$$y_{n+1} = y_{n-1} + 2hf_n$$

The results of the predictor–corrector method for the sequence of values of the parameters ε and h are given in Table 3. The error denotes the

Table 3

h	ε	Error	IC
	0.0100		
0.100		$-1.704\,06\text{E}-02$	2
0.050		$-1.054\,17\text{E}-02$	1
0.025		$-4.169\,43\text{E}-03$	1
	0.0010		
0.100		$-1.280\,20\text{E}-02$	4
0.050		$-6.007\,60\text{E}-03$	2
0.025		$-3.015\,09\text{E}-03$	1
	0.0001		
0.100		$-1.226\,89\text{E}-02$	6
0.050		$-6.225\,22\text{E}-03$	3
0.025		$-3.127\,68\text{E}-03$	2

The column headed IC contains the average iteration count j in Eq. (33).

difference between the theoretical solution $\sin^2(1.5)$ and the computer solution. Since the space to print out the results for each step would be excessive, we have listed the average number of iterations of the corrector (33) to agree to the required tolerance ε. It is clear that in order to achieve the same accuracy with the larger values of h as that obtained by the smaller values of h, a greater number of iterations is required. However, for the same value of h it is seen for this example that the accuracy is not substantially increased by demanding a smaller tolerance. One concludes that to demand a tolerance which is exceedingly small for a given value of h can very well be unproductive. In comparison, the results obtained by the midpoint rule for the values of $h = 0.1, 0.05,$ and 0.025 are

$$2.221\,23\text{E}-03$$

$$4.204\,90\text{E}-04$$

$$8.363\,41\text{E}-05$$

respectively.

This example has some additional interesting features.

(i) If we had tried to solve the problem given the starting condition $y(0) = 0$ for $x \in [0, 1]$ we would find that the method would not converge. It is simple to see that for this initial condition the differential equation has (at least) the solutions 0 and $\sin^2(x)$, i.e. the solution of this differential equation is not unique. If we check the condition for uniqueness, given on page 365, for this problem, we find that for $x = 0$, $L = \infty$ i.e. the differential equation would not be expected to have a unique solution (as we have shown) and consequently the difference equation cannot converge. However, by choosing a different interval in which we require the solution (e.g. (0.5, 1.5)), the Lipschitz condition on page 365 is satisfied and we have a unique solution and a convergent difference scheme.

(ii) This example could give computational difficulties on the larger interval (0.5, 2.5), say, for then the theoretical solution $y(x)$ attains the value 1.0 and owing to the error of the difference solution y_{n+k} the function $2\sqrt{y_{n+k} - y_{n+k}^2}$ can become complex. To avoid this, an additional condition would have to be contained in the algorithm whereby $y_{n+k} > 1$ is immediately replaced by $y_{n+k} = 1.0$.

As we have described the predictor–corrector method, iteration is continued until $|y_{n+k}^j - y_{n+k}^{j-1}| \leq \varepsilon$. The questions naturally arise:

(i) When will the iteration converge for arbitrary y_{n+k}^0?

(ii) How small an ε should we choose? (We have seen in Example 7 that too small an ε can be unproductive.)

The answer to the first question is easily answered using the analysis we have described in Chapter 9. In the notation of that chapter, for any fixed mesh point $(n + k)h$,

$$y_{(j)} = \beta_k h f(y_{(j-1)}) + \text{constant}; \quad y_{(0)} \text{ arbitrary} \tag{34}$$

where

$$\text{constant} = \sum_{i=0}^{k-1} (\alpha_i y_{n+i} - h\beta_i f_{n+i})$$

plays no active role in the iteration, and where we have suppressed the dependence on x throughout. The conditions for the existence of a unique solution of the differential equation (1) given in Theorem 1 are clearly precisely those we demanded for a unique solution of the nonlinear equation $y - f(y) = 0$ given in Theorem 2 of Chapter 9. Thus using the Lipschitz constant L of Theorem 1, the condition for convergence of Eq. (34) to the solution y_{n+k} is simply

$$h|\beta_k|L < 1$$

This condition must be satisfied at each step of the k-step method and hence an h must be chosen so that

$$h < 1/|\beta_k|L$$

for all x in the interval in which the solution is being sought.

To answer the second question we proceed as follows. In the definition of order we required to determine the first nonzero C_q in the Taylor expansion of $\mathcal{L}[Y(x); h]$ (see Eq. (16)). Let this be C_{p+1} so that

$$\mathcal{L}[Y(x); h] = C_{p+1}h^{p+1}Y^{(p+1)}(x) + \cdots \tag{35}$$

The difference solution y_n is defined from the identity

$$\mathcal{L}[y_n; h] = 0 \tag{36}$$

whereas the difference operator $\mathcal{L}$ applied to the sufficiently smooth solution of the differential equation $y(x)$ at the point x_n gives, using Eq. (35),

$$\mathcal{L}[y(x_n); h] = C_{p+1}h^{p+1}y^{(p+1)}(x_n) + \cdots \tag{37}$$

If we use the Lagrange form of the remainder of the Taylor series (see Jones and Jordan [27]), Eq. (37) can be rewritten as

$$\mathcal{L}[y(x_n); h] = C_{p+1}h^{p+1}y^{(p+1)}(\zeta) \qquad \zeta \in (x_n, x_{n+k}) \tag{38}$$

Definition The *local truncation error* at x_{n+k} of the k-step method (14) is defined to be the expression $\mathcal{L}[y(x_n); h]$ given by (38).

Definition The *global truncation error* at x_{n+k} is defined to be $y(x_{n+k}) - y_{n+k}$.

The fact that these two quantities, the local and global truncation errors, are not the same is easily seen from the following.

By definition, y_n satisfies the equation

$$\mathcal{L}[y_n; h] = 0 \tag{39}$$

Hence subtract Eq. (39) from Eq. (38). Then

$$\mathcal{L}[y(x_n); h] - \mathcal{L}[y_n; h] = C_{p+1}h^{p+1}y^{(p+1)}(\zeta) \tag{40}$$

That is, the global error satisfies the inhomogeneous difference equation

$$y(x_{n+k}) - y_{n+k} + \sum_{j=0}^{k-1}(y(x_{n+j}) - y_{n+j}) - h\sum_{j=0}^{k}\beta_j\{f(x_{n+j}, y(x_{n+j})) - f_{n+j}\}$$
$$= C_{p+1}h^{p+1}y^{(p+1)}(\zeta)$$

Clearly we cannot equate $y(x_{n+k}) - y_{n+k}$ with $C_{p+1}h^{p+1}y^{(p+1)}(\zeta)$. If, however, we make the localizing assumption (see Lambert [30]) that no error has been made prior to the point x_{n+k}, then

$$y(x_{n+j}) = y_{n+j} \qquad j = 0, 1, \ldots, k-1$$

and hence Eq. (40) reduces to

$$y(x_{n+k}) - y_{n+k} - h\beta_k\{f(x_{n+k}, y(x_{n+k})) - f_{n+k}\} = C_{p+1}h^{p+1}y^{(p+1)}(\zeta) \tag{41}$$

Thus for an explicit method

$$y(x_{n+k}) - y_{n+k} = C_{p+1}h^{p+1}y^{(p+1)}(\zeta)$$

so that based on the localizing assumption the difference between the theoretical solution of the differential equation and the theoretical solution of the difference equation is just the local truncation error. For an implicit method then we may employ the mean value theorem in Eq. (41) to obtain

$$\left\{1 - h\beta_k \frac{\partial}{\partial y} f(x_{n+k}, \eta)\right\}(y(x_{n+k}) - y_{n+k}) = C_{p+1}h^{p+1}y^{(p+1)}(\zeta);$$

$\eta \in (y(x_{n+k}), y_{n+k})$ so that the difference between the two theoretical solutions $y(x_{n+k})$ and y_{n+k} this time is proportional to the local truncation error.† Thus based on the localizing assumption it is clear that we should choose ε to be of the same order as the local truncation error. Of course we do not know what this is precisely (what is the $(p+1)$th derivative; what is ζ?) but the following simple device can be used to obtain an approximation to it. Assume that both predictor and corrector formulas are of the same order (k will of course be different in general for the predictor and the corrector (why?)). Denote the value calculated by the predictor as y_{n+k}^0 and the value calculated by one application of the corrector as y_{n+k}^1. Then from the above analysis

$$y(x_{n+k}) - y_{n+k}^0 = C_{p+1}^0 h^{p+1}y^{(p+1)}(\zeta_0) \qquad \zeta_0 \in (x_n, x_{n+k}) \tag{42}$$

where C_{p+1}^0 denotes the error constant for the predictor formula. Similarly,

$$y(x_{n+k}) - y_{n+k}^1 = C_{p+1}^1 h^{p+1}y^{(p+1)}(\zeta_1) \qquad \zeta_1(\neq\zeta_0) \in (x_n, x_{n+k}) \tag{43}$$

where C_{p+1}^1 is the error constant for the corrector.

If we assume the function $y^{(p+1)}(\zeta)$ is constant, $\zeta \in (x_n, x_{n+k})$ then we have, by subtracting Eq. (43) from Eq. (42),

$$y_{n+k}^1 - y_{n+k}^0 = (C_{p+1}^0 - C_{p+1}^1)h^{p+1}y^{(p+1)}(\zeta)$$

i.e.

$$h^{p+1}y^{(p+1)}(\zeta) = \frac{y_{n+k}^1 - y_{n+k}^0}{C_{p+1}^0 - C_{p+1}^1} \tag{44}$$

Substituting from Eq. (44) into Eq. (43), we find

$$|y(x_{n+k}) - y_{n+k}^1| = \left|\frac{C_{p+1}^1}{C_{p+1}^0 - C_{p+1}^1}(y_{n+k}^1 - y_{n+k}^0)\right| \tag{45}$$

so that the local truncation error $y(x_{n+k}) - y_{n+k}^1$ is given approximately in terms of known (the C's) and calculated quantities. This means of estimating the local truncation error is known as *Milne's* device. Hence we would choose ε to be of the same order as the right-hand side of (45) and iterate the corrector until the iterates agree to within this tolerance.

Example 8   What is a suitable ε for the method of Example 7 in the light of the above analysis?

† It is the fact that the global truncation error $y(x_{n+k}) - y_{n+k}$ satisfies the imhomogeneous difference equation (40) which makes the direct proof of convergence often so difficult—we certainly cannot make the localizing assumption in the proof of convergence.

The Euler rule has a local truncation error $(h^2/2)y^{(2)}(\zeta_0)$ whereas the implicit equation $y_{n+1} = y_n + hf_{n+1}$ has a local truncation error $(-h^2/2)y^{(2)}(\zeta_1)$. Hence in this case a suitable ε would be

$$\left| \frac{-\frac{1}{2}}{\frac{1}{2}+\frac{1}{2}}(y_{n+1}^1 - y_{n+1}^0) \right| = \frac{1}{2}|y_{n+1}^1 - y_{n+1}^0|$$

The actual value of ε used for the example is given in Exercise 30.

We conclude this section with some general observations. As we described the predictor–corrector method, we have not specified an upper limit of the number of iterations. In the case where the function f is complicated and hence involves considerable computation for its evaluation, the strategy of allowing several (many?) iterations can produce considerable inefficiency. Thus we may specify for such cases that the maximum number of iterations be just two or three. In this case (unless convergence happens to have already been achieved), the local truncation characteristics and the stability conditions are complicated to ascertain (see Lambert [30]). If convergence occurs then the local truncation characteristics and stability conditions are just those of the corrector. In this case (as in the case of an upper limit on j) there is no advantage in specifying a predictor with a higher local truncation error than the corrector.

Exercises

29. Program the predictor–corrector pair of Exercise 14 where the corrector is to be iterated in the sense described in this section. Compute the solution of Example 7 using the theoretical solution to obtain starting values. Use values of ε and h given in Example 7. Compare your results. Is there a significant gain in accuracy for this method for decreasing ε?
30. Modify your program of Exercise 29 to include an estimate for the parameter ε using the analysis resulting in Eq. (45). Run the example of the previous exercise and comment on the resulting size of ε.
31. Suggest a suitable predictor for the trapezoidal rule. Include an automatic estimate for ε and compute the solution of Example 7. Compare your results, as well as the size of ε, with the previous exercise.
32. Suggest a suitable predictor for the method (12) and repeat the previous exercise. (Beware of the numerical results for this method—see Lambert [30] for an explanation.)

5 Runge–Kutta Methods

In our discussion of multistep methods in the previous two sections we have skirted the question of providing the extra starting values that a k-step method, $k > 1$ requires. In the present section we will describe a class of methods which will provide us with either such starting values or may be used as stand-alone numerical algorithms. In the strict sense these methods are 1-step methods but use intermediate values of the unknown y which are in turn used in the argument of f to provide higher accuracy than would be possible by conventional 1-step methods discussed in the previous two sections. These 1-step methods using intermediate values are called *Runge–Kutta* methods. Since the Runge–Kutta methods are to be 1-step methods, then from section 3 they will automatically be zero stable.

Consider the Euler method

$$y_{n+1} = y_n + hf_n$$

As we know, this method has order 1, its local truncation error is $(h^2/2)y^{(2)}(\zeta)$, $\zeta \in (x_n, x_{n+1})$. If we evaluate the argument of f at $x_{n+1/2}$ we obtain the midpoint rule

$$y_{n+1} = y_n + hf_{n+1/2} \tag{46}$$

where $f_{n+1/2} = f(x_{n+1/2}, y_{n+1/2})$. This method has order two with local truncation error $(h^3/24)y^{(3)}(\zeta)$. The basis of the increasing accuracy is that we are able to obtain a value of y at $x = (n+\frac{1}{2})h$. Clearly, provided we have a starting procedure to obtain $y_{1/2}$ (y_0 is given by the initial condition of the differential equation), we can use the midpoint rule itself on the intervals of size $h/2$. This would be a sensible thing to do computationally. However, it is not the only way we can get the intermediate values $y_{n+1/2}$. To obtain the increased order of the midpoint rule, just what do we need for $y_{n+1/2}$? Clearly whatever we substitute for $y_{n+1/2}$ must be such that when we expand the resulting expressions in Eq. (46), they must agree with the Taylor series expansion of $y(x_{n+1})$ up to and including terms of order h^2.

The Taylor expansion of $y(x_{n+1})$ is just

$$y(x_{n+1}) = y(x_n) + hy^{(1)}(x_n) + \frac{h^2}{2}y^{(2)}(x_n) + \frac{h^3}{6}y^{(3)}(x_n) + O(h^4)$$

where

$$y^{(1)}(x_n) = f(x_n, y(x_n))$$

$$y^{(2)}(x_n) = \left(\frac{\partial f}{\partial x} + \frac{\partial f}{\partial y}\frac{\partial y}{\partial x}\right)_{x_n, y(x_n)}$$

$$= (f_x + ff_y)_n$$

and

$$y^{(3)}(x_n) = \left(\frac{\partial}{\partial x} + f\frac{\partial}{\partial y}\right)(f_x + ff_y)_n$$

$$= (f_{xx} + f_x f_y + 2ff_{xy} + f(f_y)^2 + f^2 f_{yy})_n$$

where we have introduced the notation

$$f_x = \frac{\partial f}{\partial x} \qquad f_y = \frac{\partial f}{\partial y} \qquad f_{xx} = \frac{\partial^2 f}{\partial x^2} \qquad f_{xy} = \frac{\partial^2 f}{\partial x \partial y} \qquad f_{yy} = \frac{\partial^2 f}{\partial y^2}$$

and $(\ )_n$ means the enclosed expressions are evaluated at $(x_n, y(x_n))$. Thus our Taylor series expansion is

$$y(x_{n+1}) = \left(y + hf + \frac{h^2}{2}(f_x + ff_y) + \frac{h^3}{6}(f_{xx} + f_x f_y + 2ff_{xy} + f(f_y)^2 + f^2 f_{yy})\right)_n + O(h^4)$$
(47)

Let us as a first step assume $y_{n+1/2} = y(x_{n+1/2})$ in Eq. (46). Then

$$y_{n+1} = y_n + hf(x_{n+1/2}, y_{n+1/2}) = y_n + hf(x_{n+1/2}, y(x_{n+1/2}))$$
(48)

Using the Taylor series expansion of a function of two variables, then

$$f(x_{n+1/2}, y(x_{n+1/2})) = f(x_n + \tfrac{1}{2}h, y_n + \tfrac{1}{2}hy_n^{(1)} + O(h^2))$$

can be expanded as

$$f(x_n, y_n) + \tfrac{1}{2}h(f_x + ff_y) + O(h^2)$$

so that Eq. (48) can be written

$$y_{n+1} = y_n + hf(x_n, y_n) + \frac{h^2}{2}(f_x + ff_y)_n + O(h^3)$$

which agrees with the expansion (47) to terms of order h^2. The important point to note, however, is that in obtaining the required expansion we did not really need the theoretical $y(x_{n+1/2})$, we merely needed a y which was an $O(h)$ correct approximation to $y(x_{n+1/2})$. The most obvious approximation is just Euler's rule,

$$y_{n+1/2} = y_n + \frac{h}{2}f_n$$

Thus the algorithm

$$y_{n+1/2} = y_n + \frac{h}{2}f_n$$

$$y_{n+1} = y_n + hf_{n+1/2} \qquad f_{n+1/2} = f(x_{n+1/2}, y_{n+1/2})$$

is second order.

Let us continue this line of investigation a little further. Consider the trapezoidal rule,

$$y_{n+1} = y_n + \frac{h}{2}(f_{n+1} + f_n)$$

This method as we know is second order and has a local truncation error $(h^3/12)y^{(3)}(\zeta)$. The question again is, what y_{n+1} can be propose which will retain the order-two accuracy? Clearly the theoretical $y(x_{n+1})$ used in f will certainly do this but this is unrealistic since $y(x_{n+1})$ is what we are trying to find. If we carry through the steps described above, we find that all that is really needed is a first-order approximation to $y(x_{n+1})$. Again, Euler's rule will furnish this. The second-order method is hence

$$y^{[1]}_{n+1} = y_n + hf_n \qquad y_{n+1} = y_n + \frac{h}{2}(f^{[1]}_{n+1} + f_n) \tag{49}$$

where we have used [1] as a superfix on y_{n+1} produced by the Euler method to distinguish it as a first step in obtaining the second-order approximation from the trapezoidal rule.

The two derivations above are of course analogous to the predictor–corrector methods of the previous section. The point which has become clear, however, is that by using first approximations evaluated at some particular point (the particular point depending on the second step of the algorithm), we can increase the accuracy and retain an explicit 1-step method. For the two examples above we cannot obtain any higher order than two, in general. But the question naturally suggests itself: If we were willing to take a sequence of such steps, could we retain the one-step nature of the algorithm but increase the order by defining $y_{n+1} - y_n$ as a linear combination of the f's evaluated at these intermediate points?

The first step (and the only one we will derive for reasons which will become obvious as we proceed) of the generalization is to consider instead of Eq. (49) the scheme

$$y_{n+1} = y_n + h(c_0 f^{[0]}_{n+1} + c_1 f^{[1]}_{n+1} + c_2 f^{[2]}_{n+1}) \tag{50}$$

where

$$f^{[0]} = f_n$$
$$f^{[1]}_{n+1} = f(x_n + a_1 h, y_n + h b_{10} f^{[0]}_{n+1})$$

and

$$f^{[2]}_{n+1} = f(x_n + a_2 h, y_n + h b_{20} f^{[0]}_{n+1} + h b_{21} f^{[1]}_{n+1})$$

where c_0, c_1, c_2, a_1, a_2, b_{10}, b_{20}, b_{21} are constants to be chosen to achieve order[†] three from Eq. (50).

Substitute $f^{[0]}_{n+1}$ into $f^{[1]}_{n+1}$ and expand; then

$$f^{[1]}_{n+1} = (f + h(a_1 f_x + b_{10} f f_y)$$
$$+ \frac{h^2}{2}(a_1^2 f_{xx} + 2a_1 b_{10} f f_{xy} + b_{10}^2 f^2 f_{yy}) + O(h^3))_n$$

Now substitute this expression and $f^{[0]}_{n+1}$ into $f^{[2]}_{n+1}$ and expand once more. Then

$$f^{[2]}_{n+1} = (f + h[a_2 f_x + (b_{20} + b_{21}) f f_y] + h^2[b_{21}(a_1 f_x + b_{10} f f_y) f_y$$
$$+ \tfrac{1}{2}(a_2 f_{xx} + 2a_2(b_{20} + b_{21}) f f_{xy} + (b_{20} + b_{21})^2 f^2 f_{yy}) + O(h^3))_n$$

† We are using order here intuitively; we will define it presently for this type of method.

Now substitute these expressions into Eq. (50); then

$$y_{n+1} = \{y + h(c_0 + c_1 + c_2)f + h^2((c_1a_1 + c_2a_2)f_x + [c_1b_{10} + c_2(b_{20} + b_{21})]ff_y)$$
$$+ h^3(\tfrac{1}{2}(c_1a_1^2 + c_2a_2^2)f_{xx} + [c_1a_1b_{10} + c_2a_2(b_{20} + b_{21})]ff_{xy}$$
$$+ \tfrac{1}{2}[c_1b_{10}^2 + c_2(b_{20} + b_{21})^2]f^2f_{yy} + c_2b_{21}a_1f_xf_y + c_2b_{21}b_{10}ff_y^2) + O(h^4)\}_n$$

$$(51)$$

A comparison of expressions (51) and (45) reveals that the constants must satisfy the equations

$$c_0 + c_1 + c_2 = 1$$
$$c_1a_1 + c_2a_2 = \tfrac{1}{2}$$
$$c_1b_{10} + c_2(b_{20} + b_{21}) = \tfrac{1}{2}$$
$$c_1a_1^2 + c_2a_2^2 = \tfrac{1}{3}$$
$$c_1a_1b_{10} + c_2a_2(b_{20} + b_{21}) = \tfrac{1}{3}$$
$$b_{10}^2c_1 + (b_{20} + b_{21})^2c_2 = \tfrac{1}{3}$$
$$c_2b_{21}a_1 = \tfrac{1}{6}$$
$$c_2b_{21}b_{10} = \tfrac{1}{6}$$

This system gives rise to the reduced system of equations

$$c_0 + c_1 + c_2 = 1$$
$$c_1a_1 + c_2a_2 = \tfrac{1}{2}$$
$$c_1a_1^2 + c_2a_2^2 = \tfrac{1}{3}$$
$$c_2a_1b_{21} = \tfrac{1}{6}$$

$$(52)$$

where $b_{10} = a_1$, and $a_2 = (b_{20} + b_{21})$. Equation (52) involves four equations in six unknowns. This therefore indicates that we will not have a unique formula; rather we obtain a two-parameter family of such methods, each of which is order three.

Example 9 What is the explicit Runge–Kutta method of order three which arises from the constants $c_0 = \tfrac{1}{4}$, $c_1 = 0$, $c_2 = \tfrac{3}{4}$, $a_1 = \tfrac{1}{3}$, $a_2 = \tfrac{2}{3}$, $b_{21} = \tfrac{2}{3}$?
Substituting the constants we find

$$f_{n+1}^{[0]} = f(x_n, y_n)$$
$$f_{n+1}^{[1]} = f(x_n + \tfrac{1}{3}h, y_n + \tfrac{1}{3}hf_{n+1}^{[0]})$$
$$f_{n+1}^{[2]} = f(x_n + \tfrac{2}{3}h, y_n + \tfrac{2}{3}hf_{n+1}^{[1]})$$

$$(53)$$

and

$$y_{n+1} = y_n + \frac{h}{4}(3f_{n+1}^{[2]} + f_{n+1}^{[0]})$$

Example 10   What is the explicit Runge–Kutta method of order 3 which arises from the constants $c_0 = \frac{1}{6}$, $c_1 = \frac{2}{3}$, $c_2 = \frac{1}{6}$, $a_1 = \frac{1}{2}$, $a_2 = 1$, $b_{21} = 2$?

Substituting the constants we find

$$f_{n+1}^{[0]} = f(x_n, y_n)$$

$$f_{n+1}^{[1]} = f(x_n + \tfrac{1}{2}h, \ y_n + \tfrac{1}{2}hf_{n+1}^{[0]})$$

$$f_{n+1}^{[2]} = f(x_n + h, \ y_n - hf_{n+1}^{[0]} + 2hf_{n+1}^{[1]}) \tag{54}$$

with

$$y_{n+1} = y_n + \frac{h}{6}(f_{n+1}^{[0]} + 4f_{n+1}^{[1]} + f_{n+1}^{[2]})$$

We have been imprecise about our meaning of order in the context of Runge–Kutta methods, although it has become clear what we imply. In the above derivation, "intermediate" values of $f^{[0]}$, $f^{[1]}$, $f^{[2]}$ were used in a final combination to provide an approximation to y_{n+1}. Such a method is commonly called a three-stage algorithm. Let us define the R-stage algorithms by

$$f_{n+1}^{[0]} = f(x_n, y_n)$$

$$f_{n+1}^{[i]} = f\left(x_n + a_i h, \ y_n + h \sum_{s=0}^{i-1} b_{is} f_{n+1}^{[s]}\right) \qquad i = 1, 2, \ldots, R-1 \tag{55}$$

and

$$y_{n+1} = y_n + h \sum_{i=0}^{R-1} c_i f_{n+1}^{[i]},$$

the latter which we will write notationally as

$$y_{n+1} = y_n + h\phi(x_n, y_n, h)$$

For such an algorithm we thus quote the following theoretical results (see Lambert [30]) for an R-stage Runge–Kutta method.

Definition   The method (55) is said to have *order p* if p is the largest integer r for which $y(x+h) - y(x) - h\phi(x, y(x), h) = O(h^{p+1})$ holds where $y(x)$ is the theoretical solution of the initial value problem.

It is (tedious but) straightforward to show that the three-stage process (50) and (52) is third order in the context of our definition.

Definition   The method (55) is said to be *consistent* with the initial value problem if

$$\phi(x, y(x), h) = f(x, y)$$

This definition complies with our previous definition of consistency in its requirement of order greater than 0. Methods (53) and (54) are clearly consistent.

Since the one-step methods are automatically zero stable it would be reasonable to surmise that all we need (in addition) for convergence is that the Runge–Kutta method be consistent. This is indeed the case, as the result of the following theorem states.

Theorem 3 *Let the function $\phi(x, y, h)$ be continuous jointly as a function of its three arguments in the region D defined by $x \in [a, b]$, $y \in (-\infty, \infty)$, and $h \in (0, h_0)$, $h_0 > 0$.*

Let $\phi(x, y, h)$ satisfy a Lipschitz condition of the form

$$|\phi(x, y^*, h) - \phi(x, y, h)| \le M|y^* - y|$$

for all points (x, y^*, h), (x, y, h) in D.

Then the method (55) is convergent if and only if it is consistent.

Runge–Kutta methods are straightforward to program, requiring repetitive calls of a subprogram which calculates the function f for a given set of parameters. However, this function evaluation can often be time consuming and the Runge–Kutta methods attain their high degree of accuracy at the expense of many function calls per integration step. In contrast, a linear multistep method of the same order will comprise a considerably smaller number of function evaluations. For higher-order methods, however, these multistep methods require, as we have discussed previously, starting procedures, whereas the Runge–Kutta methods do not. This ability of achieving higher order without the need for starting procedures is perhaps the one main attribute of these methods—we can use them to produce the starting values for the linear multistep methods. The Runge–Kutta method would be used just a few times at the beginning of the computation and so the inefficiency from the function evaluations will not be a problem.[†]

The derivation of the Runge–Kutta methods is extremely tedious as the number of steps increases and as the required order increases the algebraic manipulation becomes increasingly heavy.[‡] Perhaps the most popular Runge–Kutta method is the four-stage, fourth-order method:

$$f_{n+1}^{[0]} = f(x_n, y_n)$$
$$f_{n+1}^{[1]} = f(x_n + \tfrac{1}{2}h, y_n + \tfrac{1}{2}hf_{n+1}^{[0]})$$
$$f_{n+1}^{[2]} = f(x_n + \tfrac{1}{2}h, y_n + \tfrac{1}{2}hf_{n+1}^{[1]})$$
$$f_{n+1}^{[3]} = f(x_n + h, y_n + hf_{n+1}^{[2]})$$

(56)

where

$$y_{n+1} = y_n + \frac{h}{6}(f_{n+1}^{[0]} + 2f_{n+1}^{[1]} + 2f_{n+1}^{[2]} + f_{n+1}^{[3]})$$

[†] A detailed comparison of predictor–corrector methods and Runge–Kutta methods can be found in section 9, Chapter 4, of Lambert [30].

[‡] A very pleasing method of short-cutting the work employing graph-theoretic methods has been described by Butcher [3].

The proof of the fourth-order accuracy is left to the more ambitious reader.

In general, Butcher has proved that the attainable orders of R-stage methods are given by

$$p(R) = R \qquad R = 1, 2, 3, 4$$
$$p(5) = 4$$
$$p(6) = 5$$
$$p(7) = 6$$
$$p(8) = 6$$
$$p(9) = 7$$
$$p(R) < R - 2 \qquad R = 10, 11, \ldots$$

where $p(R)$ is the attainable order of the Runge–Kutta methods with R stages. Thus the amount of work necessary to achieve a specified order greater than 4 becomes proportionately greater in terms of the number of function evaluations. It is for this reason that method (56) is possibly the most popular Runge–Kutta method.

Example 11 Compute the solution of $y^{(1)} = 2\sqrt{y} - y^2$, $y(0.5)$, at $x = 1.5$ using the Runge–Kutta methods (54) and (56). Compute the numerical results and the number of function evaluations necessary to achieve the solution for $h = 0.1, 0.05$, and 0.025. Compare the results obtained with the fourth-order Adams–Bashforth–Moulton predictor–corrector method

$$y_{n+4}^0 = y_{n+3} + \frac{h}{24}(55f_{n+3} - 59f_{n+2} + 37f_{n+1} - 9f_n)$$

$$\tag{57}$$

$$y_{n+4}^j = y_{n+3} + \frac{h}{24}(9f_{n+4}^{j-1} + 19f_{n+3} - 5f_{n+2} + f_n) \qquad j = 1, 2, \ldots$$

where the starting values y_1, y_2, and y_3 are given by the fourth-order Runge–Kutta method (56) and where the corrector is iterated until

$$|y_{n+4}^j - y_{n+4}^{j-1}| \le \varepsilon \tag{58}$$

where ε is estimated by the device described in Eq. (45). (The error constants in this case are $C_5^0 = 251/720$; $C_5^1 = -19/720$, and the condition for convergence of the corrector is

$$\frac{9h}{24}\left|\frac{\partial f}{\partial y}\right| < 1 \quad \text{where} \quad \frac{\partial f}{\partial y} = (1 - 2y)/\sqrt{y - y^2};$$

for y close to 0 or 1 this can clearly give problems.)

The results obtained using methods (54), (56), and (57) are given in Table 4. The columns headed ERROR54, ERROR56, and ERROR57 represent the numerical errors (the differences between the theoretical solution $\sin^2(1.5)$

and the computed results) using formulae (54), (56), and (57) respectively. The columns headed F54, F56, F57 represent the number of function evaluations needed to achieve the results by the methods (54), (56), and (57) respectively.

Table 4

h	ERROR54	ERROR56	ERROR57	F54	F56	F41
0.100						
	$0.752E-04$	$-5.461E-04$	$4.083E-06$	30	40	41
0.050						
	$1.518E-04$	$-2.863E-05$	$3.278E-07$	60	80	69
0.025						
	$1.920E-05$	$-1.587E-06$	$2.235E-08$	120	158	124

For this example the numerical accuracy is superior for the Adams–Bashforth–Moulton method. The reader should not draw any general conclusions from this as the particular accuracy obtained will depend upon the example considered. The greater number of function evaluations required for the predictor–corrector method than the Runge–Kutta method (54) (and (56) for the case of $h = 0.1$) is caused by the number of iterations required by the method in order to meet the specified criterion (58). Thus the earlier reference to the inefficiency of Runge–Kutta methods must be tempered by pointing out that the predictor–corrector methods may need several iterations to attain the specified tolerance. However, it is generally recognized that it is better to reduce the size of h to satisfy (58) rather than iterate more than once. Again, the function evaluation count of the predictor corrector method will, in practice, look less favorable than might appear at first sight. Finally, the device for choosing ε needs care in its implementation, for the following reason.

We require to continue iterating until

$$\left| y_{n+4}^j - y_{n+4}^{j-1} \right| \leq \varepsilon$$

This implies

$$-\varepsilon + y_{n+4}^{j-1} \leq y_{n+4}^j \leq y_{n+4}^{j-1} + \varepsilon$$

where ε is chosen according to (45). If in calculating, we find that ε is so small that

$$y_{n+4}^{j-1} - \varepsilon = y_{n+4}^j \ (\text{and} \ y_{n+4}^{j-1} + \varepsilon = y_{n+4}^{j-1}) \tag{59}$$

i.e. the relative magnitude of ε to y_{n+4}^{j-1} is so small that the addition of ε to y_{n+4}^{j-1} leaves y_{n+4}^{j-1} unchanged (this is clearly dependent on the word length of the machine and the precision with which the arithmetic is being performed), then we would require to iterate until

$$y_{n+4}^j = y_{n+4}^{j-1}$$

i.e. if ε is too small we will require exact (to the machine precision) agreement between successive iterates. This can clearly require substantial (infinite?) numbers of iterations. To ensure this does not occur, a sufficiently large ε should be chosen. In Table 4, ε was chosen according to (45) and a test was made to see if

$$y_{n+4}^0 + \varepsilon = y_{n+4}^0 \tag{60}$$

If the result is affirmative then the ε chosen by (45) is doubled until (60) is not true. For $h = 0.1$ and 0.05, an ε was always given by (45) so that (60) was not true. However, for $h = 0.025$, (58) occurred and ε had to be increased.

Exercises

33. Can you write down a one stage Runge–Kutta method of order 1? Is it unique?
34. Derive a two-stage Runge–Kutta method of order two.
35. Carry out the exercise described by Example 11 where the corrector is not iterated. Compare your results with Table 4. How many function evaluations are needed in this case by the predictor–corrector method to obtain a comparable accuracy with the Runge–Kutta method (56)?
36. Compute the solution of the differential equation of Example 11 using your method of Exercise 34. To obtain comparable results with those of Table 4, how small does h have to be chosen, and hence how many function evaluations are necessary?
37. In answering Exercise 33, the question of uniqueness will depend upon interpreting the form of a Runge–Kutta method as given by (55). Because at any stage the calculation is explicit, such methods are called *explicit Runge–Kutta methods*. It is a simple generalization of (55) to propose *implicit Runge–Kutta* methods by replacing $i-1$ by $R-1$ in the summation defining $f_{n+1}^{[i]}$. What is the one-stage implicit Runge–Kutta method? (See Lambert [30] for further discussion of implicit methods.)

6 Second-Order Equations

We conclude this chapter with a brief description of equations of the form

$$y^{(2)} = f(x, y) \tag{61}$$

Such equations are said to be of second order because the order of highest derivative occurring is 2. Equation (61) is a special case of the more general second-order equation

$$y^{(2)} + g(x)y^{(1)} = f(x, y)$$

where g is a function of x. However, we will keep things simple and assume $g(x) = 0$. Because the equation is second order, two extra conditions will be specified in order to guarantee a unique solution (contrast this with Eq. (2)). These extra conditions can now be of varying type:

(i) if in addition to (61) we require

$$y(0) = \alpha \qquad y^{(1)}(0) = \beta \tag{62}$$

then (61) and (62) constitute a second-order initial value problem and a solution will be sought for $x \in (0, \infty)$.

(ii) if in addition to (61) we require

$$y(0) = \alpha \qquad y(1) = \beta \tag{63}$$

then (59) and (63) constitute a second-order boundary value problem†

First, consider replacing the second derivative in Eq. (61) using the divided difference formula of Example 5, Chapter 8. Then we replace Eq. (61) by

$$y_{n+2} - 2y_{n+1} + y_n = h^2 f(x_I, y_I) \tag{64}$$

where I is one of the indices n, $n+1$, or $n+2$.

If we choose $I = n$, then applying (64) to the theoretical solution of the differential equation (61), assuming y is sufficiently smooth,

$$y(x_{n+2}) - 2y(x_{n+1}) + y(x_n) - h^2 f(x_n, y(x_n)) = h^3 y^{(3)}(\zeta) \qquad \zeta \in (x_n, x_{n+2})$$

and we have used Eq. (61). The local truncation error is thus $O(h^3)$. If we choose $I = n+2$, a similar result follows. However, if we choose $I = n+1$ then

$$y(x_{n+2}) - 2y(x_{n+1}) + y(x_n) - h^2 f(x_{n+1}, y(x_{n+1})) = \frac{h^4}{12} y^{(4)}(\zeta) \qquad \zeta \in (x_n, x_{n+2})$$
$$\tag{65}$$

and the local truncation error is $O(h^4)$.

Let us consider the implementation of (65), namely

$$y_{n+2} - 2y_{n+1} + y_n = h^2 f(x_{n+1}, y_{n+1}) \qquad n = 0, 1 \ldots \tag{66}$$

which is an explicit method. To start the computation we need y_0 and y_1. We have $y_0 = y(x_0) = y(0) = \alpha$; we also have $y^{(1)}(x_0) = \beta$. Thus we need a relationship between y_1 and $y^{(1)}(x_0)$. We know from Chapter 8, Eq. (13), that

$$y^{(1)}(0) = (y_1 - y_0)/h + O(h)$$

Hence if we replace $y^{(1)}(0)$ by $(y_1 - y_0)/h$ and use $y^{(1)}(0) = \beta$, then

$$y_1 = y_0 + h\beta + O(h^2)$$

or y_1 defined by $y_1 = y_0 + h\beta$ is a first-order approximation of $y(x_1)$. Hence we can use this equation to define y_1 for the second point in Eq. (66). The computational implementation is then

$$y_0 = \alpha$$

$$y_1 = \alpha + h\beta$$

$$y_{n+2} = 2y_{n+1} - y_n + hf(x_{n+1}, y_{n+1}) \qquad n = 0, 1, \ldots \tag{67}$$

† Because the boundary values are given at two points, such problems are more frequently known as two-point boundary value problems. Numerical methods for such problems are described in detail in Keller [28].

As we have seen, Eq. (66) has a local truncation error $O(h^4)$. But the starting formula (67) has a local truncation error $O(h^2)$. We would surmise therefore that this low-order starting procedure would affect the global truncation error. Hence a more accurate procedure would be desirable. From Eq. (14), Chapter 8,

$$y^{(1)}(x_0) = \tfrac{1}{2}h^{-1}[-3y_0 + 4y_1 - y_2] + O(h^2)$$

Thus we may define

$$-3y_0 + 4y_1 - y_2 = 2h\beta + O(h^3) \tag{68}$$

Hence Eq. (68) taken together with Eq. (66) produces

$$y_0 = \alpha$$

$$y_1 = \tfrac{1}{4}y_2 + \frac{h}{2}\beta + \tfrac{3}{4}\alpha$$

the latter of which is used to replace y_1 in (66) to give

$$y_2 = 2\beta h + \alpha + 2hf\left(x_1, \tfrac{1}{4}y_2 + \frac{h}{2}\beta + \tfrac{3}{4}\alpha\right)$$

which is nonlinear in y_2. Thus an iterative procedure would be needed to produce y_2 before progressing to y_3, whence the method is explicit. Once again, the global accuracy will be affected by the starting procedure for y_1. An alternative strategy is considered in Exercise 38.

More accurate replacements of Eq. (61) can be proposed. Without exception, such methods suffer from the need of complicated starting procedures. Thus let us consider the following alternative approach. We have

$$y^{(1)} = f(x, y) \qquad y(0) = \alpha, \ y^{(1)}(0) = \beta \tag{69}$$

Introduce the new dependent variables $v = y$ and $w = y^{(1)} \ (= v^{(1)})$. Hence Eq. (69) may be written

$$w^{(1)}(x) = f(x, v) \tag{70}$$

and

$$v^{(1)}(x) = w(x)$$

Introducing $\mathbf{y}(x) = (v(x), w(x))^T$ and $\mathbf{f}(x, v, w) = \mathbf{f}(x, \mathbf{y}) = \mathbf{f}(w, f(x, v))$, we can write Eq. (70) as

$$\mathbf{y}^{(1)} = \mathbf{f}(x, \mathbf{y}) \qquad v(0) = \alpha \qquad w(0) = \beta \tag{71}$$

which is a first-order system of ordinary differential equations. Equation (71) is clearly the vector form of Eq. (2) and we may propose linear multistep methods or Runge–Kutta methods for this equation following similar lines to those described in earlier sections of this chapter. For example, the Euler method becomes

$$\mathbf{y}_{n+1} = \mathbf{y}_n + h\mathbf{f}(x_n, \mathbf{y}_n)$$

(using the obvious notation), whereas the trapezoidal rule becomes

$$\mathbf{y}_{n+1} = \mathbf{y}_n + \frac{h}{2}[f(x_{n+1}, \mathbf{y}_{n+1}) + \mathbf{f}(x_n, \mathbf{y}_n)] \tag{72}$$

To solve for $\mathbf{y}_{n+1}$ directly from Eq. (72) would require the solution of nonlinear equations with the corresponding computing complexities described in Chapter 9. It is clearly much more efficient to consider the predictor–corrector version of Eq. (72) given by

$$\mathbf{y}_{n+1}^{0} = \mathbf{y}_n + h\mathbf{f}(x_n, \mathbf{y}_n)$$

$$\mathbf{y}_{n+1} = \mathbf{y}_n + \frac{h}{2}[\mathbf{f}(x_{n+1}\mathbf{y}_{n+1}^{0}) + \mathbf{f}(x_n, \mathbf{y}_n)]$$

which is explict.

Higher-order methods can be proposed in a straightforward manner for such equations and the additional starting values are provided by vector versions of the Runge–Kutta methods described in the previous section; the reader is referred to [30] for further details. Thus it would appear that treating the second-order equation as a system of first-order equations, when the initial conditions are given by (i) on page 397, leads to a promising computational algorithm. In fact this approach generalizes to an nth-order equation

$$y^{(n)}(x) = f(x, y)$$

or the more general equation

$$y^{(n)}(x) + \sum_{i=1}^{n-1} g_i(x)y^{(i)}(x) = f(x, y)$$

when the initial conditions are of the form

$$y^{(i)}(0) = \gamma_i \qquad i = 0, 1, \dots, n-1, \quad \gamma_i \text{ given numerical constants.}$$

An example may be found for $n = 3$ in Exercise 41.

The study of numerical methods for systems of differential equations plays a significant part in the solution of many physical problems. In particular, systems of differential equations which are stiff† are particularly difficult to solve and methods for the solution of such problems are currently the focus of much research and development (see Gear [15]; Lambert [30]).

---

† In a system of differential equations $\mathbf{y}^{(1)} = f(x, \mathbf{y})$, the Jacobian matrix of f with respect to the components of $\mathbf{y}$ has eigenvalues λ_i, $i = 1, 2, \dots, N$, say. The system of equations is said to be *stiff* if

$$\min_i |\mathrm{Re}\ \lambda_i| \ll \max_i \mathrm{Re}\ |\lambda_i|$$

Exercises

38. Consider the second-order equation $y^{(2)} = f(x, y)$ written as a first-order system. Propose an explicit method of order three for this linear system where the necessary starting values are provided by the vector version of the three-step, order-three method (54).

39. Use your algorithm of Exercise 38 to compute the solution of the equation $y^{(2)} = -y$; $y(0) = 0$; $y^{(1)}(0) = 1$ at the point $x = 1.0$ for a value of $h = 0.1$, 0.05, and 0.025.

40. For the linear second-order equation $y^{(2)} = f(x)y$, derive a third-order replacement of $y^{(1)}(0)$. Using this replacement and the third-order method (66) for this linear equation, compute the solution of the differential equation of the previous exercise. How does this algorithm compare in complexity with the approach based upon the first-order system? How do the numerical results compare?

41. Propose an explicit method for solving the third-order equation $y^{(3)} = -\sqrt{1 - y^2}$, $y(0.5) = \sin(0.5)$, $y^{(1)}(0.5) = \cos(0.5)$, and $y^{(2)}(0.5) = -\sin(0.5)$. Solve the equation for $x = 1.5$ for values of $h = 0.1$, 0.05 and 0.025.

Consider now the boundary value problem $y^{(2)} = f(x, y)$, $y(0) = \alpha$, $y(1) = \beta$. Assume the interval $[0, 1]$ is discretized with a mesh of size h so that $(N + 1)h = 1$. We again propose the difference method (64). This time, however, we cannot solve for the unknowns $y_1, y_2, \ldots, y_N$ until the difference method has been applied to each node x_i and the two boundary values are introduced. For $I = n + 1$ in (64), we obtain for $n = 1$,

$$y_0 - 2y_1 + y_2 = h^2 f(x_1, y_1)$$

which on using the boundary condition $y(0) = \alpha$, may be written as

$$-2y_1 - h^2 f(x_1, y_1) + y_2 = -\alpha$$

For $n = 2$,

$$y_1 - 2y_2 + y_3 - h^2 f(x_2, y_2) = 0$$
$$\vdots$$

For $n = i$,

$$y_{i-1} - 2y_i + y_{i+1} - h^2 f(x_i, y_i) = 0$$
$$\vdots$$

For $n = N$,

$$y_{N-1} - 2y_N - h^2 f(x_N, y_N) = -\beta$$

where again we have used the boundary condition.

Now even though the difference method was explicit, we must solve a system of nonlinear equations to derive the unknowns $y_1, y_2, \ldots, y_N$. This is clearly an area of application for the methods described in Chapter 9. If $f(x, y) = g(x)y$, then the system of equations becomes linear, the precise form being

$$
\begin{bmatrix}
-(2+h^2 g(x_1)) & 1 & & & \\
1 & -(2+h^2 g(x_2)) & 1 & & \\
& \ddots & & \ddots & \\
& & 1 & & -(2+h^2 g(x_N))
\end{bmatrix}
\begin{bmatrix}
y_1 \\ y_2 \\ \vdots \\ y_N
\end{bmatrix}
=
\begin{bmatrix}
-\alpha \\ 0 \\ \vdots \\ -\beta
\end{bmatrix}
$$

This is a tridiagonal system of equations and the special form of the Gaussian elimination method described in Chapter 3 (see page 69) is directly applicable.

Exericse

42. Consider the boundary value problem

$$y^{(2)} + y^{(1)} - 4y = \sin 2x - 2 \qquad y(0.5) = \sin^2(0.5) \qquad y(1.5) = \sin^2(1.5).$$

Derive an explicit method which is of order two. Write down the system of equations arising from applying the difference scheme at the points $(x_0 + ih)$, $i = 1, 2, \ldots, N$, where $x_0 = 0.5$ and $x_{N+1} = 1.5$, for $h = 0.1$, 0.5, and 0.025. Use the tridiagonal system solver of Chapter 3 to compute the solution of this system of equations.

References

1. E. T. Bell: *Men of Mathematics*, The Scientific Book Club, 1937.
2. E. K. Blum: *Numerical Analysis and Computation: Theory and Practice*, Addison Wesley, 1972.
3. J. C. Butcher: An algebraic theory of integration methods, *Mathematics of Computation*, **26**, 79–106, 1972.
4. B. Carnahan, H. A. Luther, and J. O. Wilkes: *Applied Numerical Methods*, John Wiley, 1969.
5. H. S. Carslaw and J. C. Jaeger: *Conduction of Heat in Solids*, Oxford, 1947.
6. S. D. Conte and C. de Boor: *Elementary Numerical Analysis: An Algorithmic Approach*, McGraw-Hill, 1981.
7. G. Dahlquist: Convergence and stability in the numerical solution of ordinary differential equations, *Math. Scand.*, **4**, 33–53, 1956.
8. P. J. Davis and P. Rabinowitz: *Methods of Numerical Integration*, Academic Press, 1975.
9. C. Dixon: *Applied Mathematics of Science and Engineering*, John Wiley, 1971.
10. T. Ekman and C. E. Fröberg: *Introduction to Algol Programming*, Oxford University Press, 1967.
11. G. E. Forsythe: *Pitfalls in Numerical Computations*, Stanford Computer Science Research Report, 1972.
12. G. E. Forsythe, M. A. Malcolm, and C. B. Moler: *Computer Methods for Mathematical Computations*, Prentice-Hall, 1977.
13. G. E. Forsythe and C. B. Moler: *Computer Solution of Linear Algebraic Systems*, Prentice-Hall, 1967.
14. J. G. F. Francis: the QR transformation I, II, *Computer Journal* **4**, 265–271, 332–345, 1961, 1962.
15. G. W. Gear: *Numerical Initial Value Problems in Ordinary Differential Equations*, Prentice-Hall, 1971.
16. J. A. George and J. H. Liu: *Sparse Matrix Methods*, Prentice-Hall, 1981.
17. F. Gerrish: *Pure Mathematics*, Vol. II, Cambridge University Press, 1960.
18. A. R. Gourlay and G. A. Watson: *Computational Methods for Matrix Eigenproblems*, John Wiley, 1973.
19. R. W. Hamming: *Numerical Methods for Scientists and Engineers*, McGraw-Hill, 1973.
20. P. Henrici: *Elements of Numerical Analysis*, John Wiley, 1964.
21. P. Henrici: *Discrete Variable Methods in Ordinary Differential Equations*, John Wiley, 1962.

404

22. F. B. Hildebrand: *Introduction to Numerical Analysis*, McGraw-Hill, 1974.
23. A. Householder: *Principles of Numerical Analysis*, McGraw-Hill, 1953.
24. E. Isaacson and H. B. Keller: *Analysis of Numerical Methods*, John Wiley, 1966.
25. L. W. Johnson and R. D. Riess: *Numerical Analysis*, Addison Wesley, 1977.
26. D. S. Jones: *The Theory of Electromagnetism*, Pergamon Press, 1964.
27. D. S. Jones and D. W. Jordan: *Introductory Analysis*, John Wiley, 1969.
28. H. B. Keller: *Numerical Methods for Two-Point Boundary Value Problems*, Blaisdell, 1968.
29. V. I. Krylov: *Approximate Calculation of Integrals*, Macmillan Company, 1962.
30. J. D. Lambert: *Computational Methods in Ordinary Differential Equations*, John Wiley, 1973.
31. D. H. Lehmer: A machine method for solving polynomial equations, *J. Assoc. Comput. Mach*, **2**, 151–162, 1961.
32. A. R. Mitchell and D. F. Griffiths: *The Finite Difference Method in Partial Differential Equations*, John Wiley, 1980.
33. A. R. Mitchell and R. Wait: *The Finite Element Method in Partial Differential Equations*, John Wiley, 1977.
34. A. Ralston: *A First Course in Numerical Analysis*, McGraw-Hill, 1965.
35. E. YA. Remes: Sur un procédé convergent d'approximations successives pour déterminer les polynômes d'approximation, *Comptes Rendues*, **198**, 2063–2065, 1934.
36. J. R. Rice: Experiments on Gram–Schmidt orthogonalization, *Mathematics of Computation*, **20**, 325–328, 1966.
37. H. Rutishauser: Der Quotienten-Differenzen-Algorithmus. *Mitteilungen aus dem Institüt für angew. Math.*, No. 7, Birkhauser, 1956.
38. M. R. Spiegel: *Theory and Problems of Advanced Mathematics for Engineers and Scientists*, McGraw-Hill (Schaum's Outline Series), 1971.
39. G. W. Stewart: *Introduction to Matrix Computations*, Academic Press, 1973.
40. G. Strang: *Linear Algebra and its Applications*, Academic Press, 1976.
41. A. H. Stroud and D. Secrest: *Gaussian Quadrature Formulas*, Prentice-Hall, 1966.
42. R. P. Tewarson: *Sparse Matrices*, Academic Press, 1973.
43. J. F. Traub: *Iterative Methods for the Solution of Equations*, Prentice-Hall, 1964.
44. R. Wait: *The Numerical Solution of Algebraic Equations*, John Wiley, 1979.
45. G. A. Watson: *Approximation Theory and Numerical Methods*, John Wiley, 1980.
46. G. N. Watson: *A Treatise on the Theory of Bessel Functions*, Cambridge University Press, 1922.
47. B. Wendroff: *First Principles of Numerical Analysis*, Addison Wesley, 1969.
48. J. H. Wilkinson: *Rounding Errors in Algebraic Processes*, Prentice-Hall, 1964.
49. J. H. Wilkinson: *The Algebraic Eigenvalue Problem*, Oxford University Press, 1965.
50. J. H. Wilkinson: Global convergence of tridiagonal QR algorithm with origin shifts, *Linear Algebra and its Applications*, **1**, 409–420, 1968.
51. R. A. Willoughby: *Sparse Matrix Proceedings*, IBM Research Report RAI, Yorktown Heights, New York, 1968.

Index